A Phenomenological Approach to Quantum Mechanics

A Phenomenological Approach to Quantum Mechanics

Cutting the Chain of Correlations

STEVEN FRENCH

OXFORD
UNIVERSITY PRESS

Great Clarendon Street, Oxford, OX2 6DP,
United Kingdom

Oxford University Press is a department of the University of Oxford.
It furthers the University's objective of excellence in research, scholarship,
and education by publishing worldwide. Oxford is a registered trade mark of
Oxford University Press in the UK and in certain other countries

Published in the United States of America by Oxford University Press
198 Madison Avenue, New York, NY 10016, United States of America

British Library Cataloguing in Publication Data
Data available

Library of Congress Control Number: 2023940916

ISBN 978–0–19–889795–8

DOI: 10.1093/oso/9780198897958.001.0001

Printed and bound by
CPI Group (UK) Ltd, Croydon, CR0 4YY

Contents

Preface

The initial prompt for this work was a review of Kostas Gavroglu's wonderful biography of Fritz London (Gavroglu 1995). I was very happy to take it on, not least because I had chosen superconductivity as my final year research topic in physics at the University of Newcastle-upon-Tyne, for which London's two-volume book *Superfluids* (London 1950 and 1954) was a major reference. At that time I wasn't aware of London's philosophical background in phenomenology, so Gavroglu's biography was a real eye-opener in that regard. I also wasn't previously aware that he had co-authored a piece on the so-called 'measurement problem' in quantum mechanics and so felt duty bound to take a look. Fortunately, an English translation of what Wigner called London and Bauer's 'nice little book' (London and Bauer 1939) is included in Wheeler and Zurek's magisterial collection (1983)[1] but reading it through the lens of London's phenomenological stance that Gavroglu exposed, I realized immediately that the standard appraisal of it as merely a summary of von Neumann's solution was wide of the mark[2]—a realization that was further reinforced when I read the original French version, a copy of which was tucked away in the Physics 'Stacks' of the University of Leeds Library.[3]

I soon discovered that this misappraisal ran throughout the literature of the measurement problem and in particular featured prominently in the debate between Margenau and Wigner on the one hand, and Putnam and Shimony on the other over the role of consciousness in the so-called 'collapse' of the wave-function. The general consensus seems to be that the latter pair came

[1] In his review of the collection, Greenberger mentioned the 'short, famous monograph by London and Bauer' and noted that '[o]btaining the rights to reprint this book from a recalcitrant publisher was a real coup, and the advantage the editors had of being able to combine three separate translations came about only because the first two could not get published' (1985, p. 193). We'll return to the issue of these translations later. Greenberger was friends with Shimony who was involved in the translation of the monograph and who will feature prominently in our narrative (see https://www.aip.org/history-programs/niels-bohr-library/oral-histories/25643 and also https://www.aip.org/history-programs/niels-bohr-library/oral-histories/34331).

[2] Cartwright, in her review of the Wheeler and Zurek collection, also noted the inclusion of the London and Bauer piece and wrote that it 'lays out the reduction of the wave packet in the way most readers of *Philosophy of Science* understand it today' (Cartwright 1985, p. 480). As we'll see, it really doesn't.

[3] It is now safely stored in the library's 'Special Collections' archive and when I went back to it in December 2021 to check some translations, I discovered that the last person to take it out was myself, in 1998!

away the victors and as a result the 'von Neumann' solution has been consigned to the scrapheap. However, it seemed to me that in the light of a phenomenological reading of London and Bauer's manuscript this consensus was undermined and that London and Bauer's work needed to be re-evaluated (French 2002). It would be hyperbolic to report that my paper 'fell dead-born from the press' but let's just say that the number of citations that it garnered in the first few years amounted to 'a bare handful'.

I did have a plan to set that re-evaluation in a broader context by way of a book with my former PhD student, Liz Hill but that came to nothing, for reasons I won't go into here. Some years later I decided to return to the idea, supported by a semester's leave granted by the School of Philosophy, Religion, and the History of Science at the University of Leeds. Things came even more into focus with the invitation by Harald Wiltsche and Philipp Berghofer to present something at their conference on 'Phenomenological Approaches to Physics: Historical and Philosophical Reflections' at the University of Graz in 2018. This was one of the most fruitful and congenial conferences that I've ever attended—a tribute not only to the organizers but also to the other participants, who were incredibly supportive. I left Graz thinking that with just a few more months work I'd have a draft of the book ready for feedback, but then other projects clamoured for my attention and after I'd put them to bed the Covid-19 pandemic intervened. So it wasn't until the summer of 2020 that I was able to slowly work my way back into the manuscript via various historical rabbit-holes which also delayed its conclusion. Now it's done and dusted in some sense of 'done', and I'm left with the feeling that philosophers of physics will look askance at the phenomenology, and philosophers of the latter inclination will recoil from all the physics! As always, there's a lot more to say about both, but it will have to be said by someone else, because I'm done too.

Acknowledgements

I am grateful to a number of colleagues for comments, advice, and encouragement including in particular Philipp Berghofer, Michel Bitbol, Otávio Bueno, Richard Francks, Liz Hill, James Ladyman, Michela Massimi, Tom Ryckman, Simon Saunders, Juha Saatsi, Blake Stacey, Laura de la Tremblaye, Bas van Fraassen, and Harald Wiltsche. I am also indebted to Matt Taylor for letting me have advance copies of chapters from his PhD thesis that helped inform my initial paper on this topic. Earlier versions of that paper were presented at the Informal Research Workshop of what was then the Division of History and Philosophy of Science, University of Leeds, the Graduate Philosophy of Physics Seminar of the Free University of Brussels, and the Philosophy of Physics Seminar of the University of Oxford. I would like to thank the participants at all of these seminars, the three referees of *Studies in History and Philosophy of Modern Physics*, which eventually published it, and the then-editor Jeremy Butterfield for their further suggestions and comments. Parts of the book are also derived from papers based on talks I gave at the Leeds Centre for Mind and Metaphysics in 2018, at the conferences *Phenomenological Approaches to Physics*, first in Graz in 2018 and then in Linköping in 2022, and at the conference *Metaphysics from a Human Point of View* in Edinburgh 2021. I am hugely grateful to Harald Wiltsche, Philipp Berghofer, and Michela Massimi respectively, for inviting me to these conferences and, together with all the participants, for their excellent and helpful comments. The same applies to all those who joined in the discussions of the QBism and phenomenology reading group, especially Blake Stacey who helped correct some misunderstandings of mine.

Harald and Philipp also commented on an earlier draft of this book and have been immensely encouraging and supportive throughout, as has Tom Ryckman who has been a human 'touchstone' for me on phenomenological matters for many years now. Summaries of the core idea have also been inflicted on members of the philosophy societies at the Universities of Glasgow and York who, I suspect, were not expecting me to do that when I was invited. As I have indicated, the bare bones of this book were laid down while I was on a semester's sabbatical; I am deeply grateful to my colleagues in the School for all their support during this period.

I would also like to thank the readers of Oxford University Press for their extensive and helpful comments, Christine Ranft for her excellent copyediting, Nadine Kolz for her help and patience, Saranya Ravi for shepherding the book through the production process and, of course, Peter Momtchiloff, also of Oxford University Press, for his unfaltering support and encouragement.

However, as always, I reserve my final and deepest acknowledgement of gratitude for Dena, Morgan, and a certain 'Ruffian', for keeping me on some sort of straight and narrow these past many years.

Having said all that, as always, the responsibility for any lack of appropriate historical or philosophical subtlety remains with me.

Some of the material presented here had its origins in the following papers:

'A Phenomenological Approach to the Measurement Problem: Husserl and the Foundations of Quantum Mechanics', *Studies in History and Philosophy of Modern Physics* 22, pp. 467–91, 2002.

'From a Lost History to a New Future: Is a Phenomenological Approach to Quantum Mechanics Viable?', in H. Wiltsche and P. Berghofer (eds.), *Phenomenological Approaches to Physics*, Springer, pp. 205–26, 2020.

'Putting Some Flesh on the Participant in Participatory Realism', in H. Wiltsche and P. Berghofer (eds.), *Phenomenology and QBism: New Approaches to Quantum Mechanics*, Routledge, 2023.

1
The Measurement Problem (Featuring the Usual Suspects)

1.1 Introduction

This book is going to be an exercise in intellectual exploration. It is one that I feel is worth undertaking, not least because the position arrived at offers a novel 'take' on quantum mechanics, one whose origins have been effaced in the long-running and highly contentious debate over the status of this theory. To have another understanding of it 'on the table', as it were, and in particular, one originally proposed by a leading 'post-revolutionary' quantum physicist, is something to take note of in itself. But in addition, this understanding emerges from a philosophical school of thought quite different from the one in which most current approaches are situated.

We've all learned now that the idea of a sharp division between 'Anglo-American', or 'analytic', and so-called 'Continental' forms of philosophizing is highly problematic. Nevertheless, the most widely considered interpretations of quantum mechanics currently in play are underpinned by the former, whereas the position explored here springs from the latter. Granted that it originates from a time when the division was, perhaps, less sharply pronounced than it is now, still, this feature is also noteworthy. Thus, I hope this book will also contribute to a reappraisal of the division and perhaps an appreciation of the 'Continental' tradition as also offering a set of potential resources for understanding modern physics. Finally, insofar as this tradition has effectively been 'airbrushed' out of the standard histories of (broadly) philosophical reflection on physics, I hope that this work will both challenge and supplement those histories.

Those histories of 'meta-level' thought and reflection are themselves typically tied to a particular history of the development of quantum mechanics itself, one that has now become canonical: that development began with Planck's work in 1900, the results of which were then applied to the structure of the atom by Bohr, who, together with Sommerfeld, developed what is often called the 'old' quantum theory. The cracks in this edifice began to show in the

A Phenomenological Approach to Quantum Mechanics: Cutting the Chain of Correlations. Steven French, Oxford University Press. © Steven French 2023. DOI: 10.1093/oso/9780198897958.003.0001

early 1920s and it was eventually replaced by Heisenberg's matrix mechanics, on the one hand, and Schrödinger's wave mechanics on the other. These were then shown to be equivalent by von Neumann who bequeathed us the framework of Hilbert space and operators and eigenvectors and so on that generations of physics students have grown to love (or not).

Sitting above this canonical history is an equally standard history of (broadly) philosophical reflections on what the theory tells us about the world: with the cracks widening in the Bohr–Sommerfeld 'picture' of the atom as consisting of electrons moving in elliptical orbits around the nucleus, Heisenberg, misunderstanding a quote from Einstein, insisted that all such pictures should be thrown away, whereas Schrödinger tried to sketch his own, based on (more or less) classical waves, until Einstein pointed out that once systems of more than one particle were taken into account, such waves would have to exist in some kind of multi-dimensional space. Bohr then took control of things again, imposing his own view according to which the theory should be understood in terms of 'complementary' accounts of how the world is: spatio-temporal on the one hand and causal on the other, tying these to Heisenberg's Uncertainty Principle (see Jähnert and Lehner 2022). And as long as you keep the twain distinct, some sort of sense can be made of what the theory tells us. This is taken to form the core of the so-called 'Copenhagen Interpretation' which then became the dominant hegemony (see Cushing 1994), or so the story goes.

Of course, some people have taken potshots at this edifice: Einstein famously debated Bohr over its central elements and was deemed to have lost the argument (until Bell's work in the 1960s prompted a reappraisal of that debate); de Broglie presented an alternative, wave-based, understanding, but was brutally shot down by Pauli (on faulty grounds, as it turned out); Bohm then effectively revived a form of de Broglie's view, introducing position as a 'hidden variable' into the formalism; Everett in turn eschewed all such 'add-ons', insisting that the theory should just be taken 'as is', with the wave-function encoding multiple possible 'branches' and so it goes, with an array of other accounts all thrown into the mix... Indeed, there are now so many interpretations of quantum mechanics, exemplifying so many different virtues, along so many different dimensions, that it can be difficult to pick one's way through the plethora (for an overview, see French and Saatsi 2020).

What I've just given are thumbnail sketches of the two intertwined 'standard histories' of the physics and philosophical reflections respectively, but they are just that—rude sketches. Indeed, the very origins of the theory have been contested, with Ehrenfest and Einstein, publishing five years after Planck's classic paper, given the accolade of 'founding fathers' (Kuhn 1978).[1] The developments in the mid-1920s were also not quite the major 'paradigm shift' that they are often presented as (see Seth 2010). Additional complexities enter the picture when we consider the elaboration of the underlying formalism, with not just Heisenberg and Schrödinger facing off, as it were, but also the likes of Dirac and Weyl throwing their respective hats into the ring.[2]

The Copenhagen Interpretation itself arose from a 'dialogical' process in which different principles and theoretical features were woven together under the pressure of the disparate forces powering the debates at the time (Beller 1999). Consequently, it has been claimed, we should abandon 'the very possibility of presenting the Copenhagen Interpretation as a coherent philosophical framework' (ibid., p. 173; see also Jacobsen 2011).[3] Indeed, the interpretation has been dismissed as a 'myth' (Chevalley 1999; Howard 2004) and it has been argued that the label itself only came to be established in the 1950s and early 1960s (Camilleri 2009).[4] Finally, the very distinction between 'classical' and quantum mechanics itself only emerged over a period of time that extended into the 1930s, as a way of legitimizing the new foundations for physics (Gooday and Mitchell 2013).

Woven throughout these contested, layered histories we find the 'measurement problem'. As far as the physics is concerned it is not even *that*—a 'problem'.[5] Partly this is due to the dormative virtue of the Copenhagen Interpretation;[6] partly, it is because of the way von Neumann formally described the issue, thereby effectively smoothing it away. At the philosophical level it became, and has remained, a convenient 'hook' on which to hang the

[1] By virtue of taking quanta to have an independent existence.

[2] Not to mention the considerable differences in textbook style and content, between not only the books of Dirac and Weyl, but those more commonly used by physics students, from Heitler's *Elementary Wave Mechanics* to Landau and Lifshitz's *Quantum Mechanics*. As Simon argues, the claim that such texts helped shape the relevant 'normal science', to use Kuhn's phrase, offers too crude a picture and the diversity and 'epistemological agency' of these textbooks should be acknowledged (Simon 2022, p. 724).

[3] Jacobsen adds a further dimension to this history by suggesting that the opposition to the Copenhagen Interpretation from the likes of Einstein and Schrödinger should not be allowed to obscure the fact that the 'upcoming' generation of physicists mostly decided to refrain from getting involved in any of the 'philosophical' debates, preferring instead to adopt a form of 'pragmatic instrumentalism' (2011, p. 376). As we shall see, not all such rising stars took this stance.

[4] None of the main protagonists used the label before that time, '[n]or did other major contributors to quantum mechanics such as H. Weyl or F. London and E. Bauer refer to a "Copenhagen Interpretation"' (Chevalley 1999, p. 62).

[5] A colleague once expressed the view, at a conference attended by both philosophers and physicists, including some Nobel Prize winners, that the measurement problem was like the fart at a party that everyone could smell but no one wanted to talk about. Needless to say, this comparison did not go down well.

[6] As Einstein put it, 'it provides a gentle pillow for the true believer from which he cannot very easily be aroused. So let him lie there' (Letter to Schrodinger, 31 May 1928; Przibram 1967, p. 31).

various interpretations briefly mentioned above.[7] However, reflecting these contentious histories, albeit in a focused or 'local' form, here too we find a 'standardized' story in which certain elements have been repurposed in the service of various aims and hence have lost their original philosophical grounding.

1.2 The Usual Story of 'The Measurement Problem'

Let us begin with an outline of the 'problem': consider an electron and one of its fundamental (quantum) properties, spin. This has the value of ½ and can take two orientations: 'up' and 'down'. According to the formalism of quantum mechanics (QM), prior to any measurement the spin state of the electron should be characterized as a superposition of 'spin-up' and 'spin-down'. Furthermore, if the electron interacts with another particle, or a bigger system, such as an atom, then the state of the joint system that results should also be characterized as a superposition of the relevant states and so on with each further interaction. *If* the formalism is then taken to apply to *all* physical systems, including measurement apparatuses (and as we shall see, this is contentious), then when the electron interacts with such an apparatus, the joint state should also be a superposition. However, and here's the nub of the problem, *we never observe such superpositions*. If we were to measure the spin of our friendly neighbourhood electron, we would always find it to be either 'up' or 'down'.

The standard way of dealing with the problem is to add a postulate to the formalism that, bluntly, states that when a measurement is undertaken the superposition 'collapses' into one or other definite state. However, this leaves the question open: what accounts for this 'collapse'? It can't be the dynamics as represented by Schrödinger's Equation because that can only generate further superpositions. As Albert has put it:

> [t]he dynamics and the postulate of collapse are flatly in contradiction with one another… the postulate of collapse seems to be right about what happens when we make measurements, and the dynamics seems to be bizarrely wrong about what happens when we make measurements, and yet the dynamics seems to be right about what happens whenever we aren't making measurements. (Albert 1992, p. 79)

[7] Bächtold identifies five different ways of describing the measurement problem in the context of what he calls the 'standard' interpretation of QM (Bächtold 2008).

The usual history has it that this tension was initially managed by situating the dynamics and collapse in different domains: for Bohr and his followers, these were the microscopic, where QM held sway, and the macroscopic, where classical physics prevailed; whereas for the likes of von Neumann and Wigner, it was, respectively, the physical and the mental. When it comes to the former view, the 'collapse' should be conceived of as a shift in the descriptive resources that we have available in those respective domains. In the case of the latter, however, the collapse was understood to be brought about by the insertion into the measurement process of the non-physical consciousness of the observer. Both alternatives effectively 'black boxed' the problem as far as the majority of physicists was concerned—they could get on with deploying the theory to explain and predict everything from conduction in metals to the formation of chemical bonds, leaving the issue of how this supposed 'collapse' was effected to those of a more philosophical inclination. Under sustained pressure, however, both options began to crumble by the late 1950s and early 1960s. That then opened up conceptual space for alternative ways of resolving the tension.

My aim in this book is to suggest that this story misses out a crucial component, as manifested in the 'little book' by London and Bauer (London and Bauer 1939; English trans. London and Bauer 1983). On the one hand this played a crucial role in shaping the above story, deployed extensively as it was by the main protagonists but, on the other, its grounding in a very distinctive philosophical tradition, namely Husserlian phenomenology, went unnoticed for many years. Excavating and further exploring this aspect then not only leads us to a deeper understanding of the history of philosophical engagement with twentieth-century physics but also expands the relevant conceptual space to accommodate an approach to the measurement problem that is quite different from all the others 'on the table'.

In the following two chapters the above 'consciousness causes collapse' solution will be presented in its historical context, and in particular, von Neumann's and Wigner's views will be covered in some detail. As will become apparent, it was Wigner who, in effect, appropriated London and Bauer's 'little book' for his own ends, thereby obscuring its central message. We will then consider Putnam's and Shimony's criticisms of this view, together with Margenau and Wigner's responses, again highlighting the importance of London and Bauer's work in 'shaping' this debate and its aftermath. This also included the attempt to relate the 'consciousness causes collapse' solution to parapsychological phenomena, in the context of which it was, briefly, acknowledged by Shimony that London and Bauer's approach had, in fact, been misunderstood.

In Chapter 4 I will lay the basis for the 'alternative' history, beginning with London's philosophical roots in Husserlian phenomenology.[8] This will involve an extensive discussion of the phenomenological understanding not only of consciousness, and of the ego in particular, but of science and 'reality' in general, all of which is presented in Chapter 5. I should say upfront that there is more to phenomenology than an emphasis on 'experience', however understood, and hence that following a phenomenological approach to QM will not lead to a position that might be described as 'Copenhagen-adjacent'.[9] As I hope you'll see, it is a lot more interesting than that.[10]

This will provide the philosophical background necessary for understanding London and Bauer's analysis of measurement, which we shall revisit in Chapter 6, together with Putnam's and Shimony's criticisms, which will now be seen to have missed the mark completely. The phenomenological understanding of objectivity is central to this analysis and in Chapter 7 we shall consider this in more detail. I shall suggest there that London and Bauer's work can be conceived as a kind of completion of Husserl's final work, *The Crisis of the European Sciences* (Husserl 1970b).

This issue of establishing objectivity will lead us nicely into a discussion of the so-called 'QBist' interpretation of QM, which, it has recently been argued, should be augmented by phenomenological considerations, due to its subjectivist underpinning. The work of another well-known phenomenological philosopher, Merleau-Ponty, has been cited extensively in this regard and as I shall indicate, he too drew heavily on London and Bauer's analysis. As a contrast, I shall suggest that the 'correlational' aspect of phenomenology,

[8] That he never forgot or abandoned these 'roots' offers a nice contrast to his contemporaries whose 'philosophical pronouncements . . . , no matter how strongly expressed, should not be taken as general and long-term commitments, but as context-dependent and flexible' (Kojevnikov 2020, p. 83).

[9] Crease and Sares, for example, write that 'Phenomenology is critical of the claims and pretentions of some of the realist interpretations of quantum mechanics—many worlds, hidden variables—because at least from a phenomenological perspective, they make some kind of leap to something that cannot be given and thus *known* to be the case' (2020, pp. 558–9; it is precisely for this kind of attitude that phenomenology has often been dismissed from the perspective of the philosophy of science; see Rouse 1987). As we'll see, that represents only one particular, and perhaps rather narrow, understanding of what phenomenology is all about. In response, Fuchs rightly laments that this is resonant of the reluctance of certain philosophers to learn from physics (ibid., p. 559). For an alternative view of the relationship between phenomenology and quantum physics, see Berghofer, Goyal, and Wiltsche (2021).

[10] In his critique of my account, Alves suggests that it is 'tantamount to explaining the less obscure by the more obscure and . . . compromise[s] phenomenology with a controversial interpretation—the so-called "Copenhagen interpretation"—where something like mysterious "collapses" appear as a postulate of the theory' (2021, p. 478). I shall respond to Alves' arguments later but let me just emphasise here that it is a mistake to situate London and Bauer's work within the 'Copenhagen Interpretation' (which is a retrospectively applied label anyway, as noted above); and relatedly, that the core feature of that work, at least so far as I understand it, effects a shift away from this notion of 'collapse'.

exemplified by the latter, invites comparisons with the Everettian and Relationalist interpretations, covered in Chapter 9. My tentative claim is that this approach offers the possibility of an alternative interpretation that accommodates both the emphasis on the relevant relations manifested by these interpretations and the significance of the role of the observer that is highlighted by QBism.

And then I shall wrap things up with a brief consideration of whether a phenomenological approach to QM should be regarded as an *interpretation* or *reconstruction* of the theory—as we'll see, it is really neither.

2

The Orthodox Solution, Its History and Multiplicity

2.1 Introduction

Even making good on the title of this chapter and outlining the 'orthodox solution(s)' to the measurement problem presents a bit of a historical problem.[1] First of all, precisely *when* this came to be seen as a 'problem' per se is unclear. Heisenberg certainly referred to the reduction of the wave-function during measurement as early as 1927 in his classic paper presenting the Uncertainty Principle (Heisenberg 1927; see Jähnert and Lehner 2022). There he wrote that in this process, measurement selects a definite value for an observable from the 'totality of possibilities' (Heisenberg 1927, p. 184). According to Beller, Heisenberg thereby 'inaugurated the notorious measurement problem of quantum mechanics' (Beller 1999, p. 67). Interestingly, the source of this idea, according to Heisenberg himself, lay in an analogy with Fichte's view of perception as involving the self-limitation of the ego: 'in every act of perception we select one of the infinite number of possibilities and thus we also limit the number of possibilities for the future' (Heisenberg 1952, p. 28);[2] a statement that is almost identical to the final words of Heisenberg's 1927 paper (Beller 1999, p. 67).[3] In a sense, then, the birth of the measurement problem is tied to considerations of the nature and role of the ego.[4]

[1] This is related to the aforementioned issue of delineating the Copenhagen Interpretation.

[2] The analogy was first pointed out by Heisenberg five years after the publication of the 1927 paper, in a talk given to the Academy of Science in Saxony (Beller 1999, p. 67).

[3] Fichte developed a radical form of Kantian transcendental idealism founded on a purely subjective basis according to which 'the I posits itself as self-positing' (Breazeale 2018). The notion of 'posits' here means 'to reflect upon', so the idea is that the essence of the ego lies in the assertion of its self-identity. Since such an assertion is both a 'doing' and a 'knowing' the ego here is not to be identified with any kind of substance, Cartesian or otherwise. As we shall see, this idea of the ego as crucially involving the act of reflection will feature prominently in our later discussions, as will Fichte's insistence that the 'I' actually exists only as embodied. Although he planned to develop a 'philosophy of nature' on this basis, Fichte didn't actually follow through, presenting only a 'very compressed' account of space, time, and matter (Breazeale 2018).

[4] For more on Heisenberg's reflections on the nature and role of the subject–object divide within QM, see Carson 2010.

A Phenomenological Approach to Quantum Mechanics: Cutting the Chain of Correlations. Steven French, Oxford University Press. © Steven French 2023. DOI: 10.1093/oso/9780198897958.003.0002

Notably, however, in his classic exposition of the foundations of the theory, von Neumann does not present the issue as a *problem* as such, referring to 'The Measuring Process' in chapter six (von Neumann 1932/1955); nor do London and Bauer, as we'll see. It also doesn't seem to feature in early presentations of the theory, such as Birtwhistle (1929), Darwin (1931),[5] or Dirac (1930); although he does argue for a different dynamics when it comes to observations; see Barrett (1999, pp. 27–30).[6] Bohr and Rosenfeld do refer to 'the usual quantum mechanical measurement problem' in 1933 but the context here is the measurability of electromagnetic field quantities (Bohr and Rosenfeld 1933; see also Jacobsen 2011). As Christian de Ronde has noted (in a post to the HOPOS email list), if you use Google's Ngram viewer, the phrase 'measurement problem' doesn't really take off until the late 1940s and 'quantum measurement problem' not until the late 1960s.[7] Freire Jr suggests that Wigner was one of the first to use the phrase (Freire Jr 2015, p. 142) and records that '[in] the second half of the 1950s there was a rise of studies on the measurement problem' (ibid., p. 86).[8]

Second, as Freire Jr has also noted, although '[t]he existence of an "orthodox view" of quantum mechanics was generally taken for granted since the 1930s, the meaning of such a label was far from being univocally determined' (Freire Jr 2015, p. 79). One avenue of approach is to delineate two rival claimants to the title, as sketched in the previous chapter: on the one hand, Bohr, and followers such as Rosenfeld, insisted on drawing a clear distinction between the microscopic domain, in which the system under observation is situated and which is appropriately described by QM, and the macroscopic measurement context, where we are constrained by our inescapable reliance on

[5] Recycling a quote from an earlier work by Darwin (1929), previously given in Bitbol (2000, p. 47), Alves has suggested that the former anticipated von Neumann's approach (Alves 2021, p. 459). However, as we'll see, this suggestion is problematic.

[6] Simon notes that such books played a major role in establishing the completeness and coherence of the theory (Simon 2022, p. 720).

[7] In response to another question posted to the HOPOS list about when the measurement problem was 'discovered', Kristian Camilleri wrote, 'My best answer to the question posed is that the idea that there was some "unsolved problem" only gradually crystallized in the 1950s, largely as a result of the new wave of challenges to the orthodoxy. But even here the "orthodoxy" was, in some sense an invention, characterized by different authors in various ways to suit their own agendas. Nevertheless, these post-war challenges did provoke a number of physicists to look at the matter in greater depth than it had been previously. It was during this time that we begin to see a sharp rise in the use of the term "measurement problem"' (private communication). As we'll note, Putnam played a role in pushing physicists to take the issue seriously.

[8] Having said that, one can find indications of concerns regarding what we would now call the 'collapse' of the wave-function in the discussions at the famous 1927 Solvay Conference, with contributions from Dirac, Einstein, and Heisenberg (see Barrett 1999, pp. 22–30; also Jähnert and Lehner 2022).

classical language;[9] on the other hand, von Neumann, accepting that the theory applied in both microscopic and macroscopic cases (a stance that underpinned his famous no-hidden-variables 'proof'[10] and which was further exemplified by London's work on superconductivity as a macroscopic quantum phenomenon, for example),[11] argued that where we draw the line is an arbitrary matter and, while not being explicit about the role of the extra-physical consciousness of the observer, opened the door to such considerations (fully thrown wide by Wigner, as we'll see).[12]

However, as Freire Jr has noted, Bohr's approach failed to gain much purchase in the textbooks that were crucial, of course, for educating the next generations of physicists (Freire Jr 2015, pp. 78–80). Indeed, in his report of 1957 on the possible translation of de Broglie's book, *La théorie de la mesure en mécanique ondulatoire*, Rosenfeld complained that '[t]here is not a single textbook of quantum mechanics in any language in which the principles of this fundamental discipline are adequately treated, with proper consideration of the role of measurements to define the use of classical concepts in the quantal description' (cited in Freire Jr 2015, pp. 78–9). He then went on to dismiss von Neumann's discussion of measurement (von Neumann 1932/1955) as creating 'unnecessary confusion and [raising] spurious problems' (cited in Freire Jr 2015, p. 80).[13] Nevertheless, it is von Neumann's analysis of the measurement situation that is perhaps most often cited, at least in philosophical discussions of the problem (see, for example, Jammer 1974, pp. 474–9; Barrett 1999, pp. 30–7).[14] This is founded on a division of all physical processes into two:

[9] For more on Bohr's interpretation of QM see Faye 2019 and for further consideration of his view of the micro–macro distinction in particular, see Zinkernagel 2015.

[10] Mitsch has argued that it is crucial for the proof that QM is taken to apply to the macroscopic measurement apparatus (Mitsch 2022).

[11] Anderson has suggested that London paid for this 'unpopular choice of subject matter' by being excluded from the Manhattan Project, after emigrating to the USA (Anderson 2005, p. 29). We shall consider London's career in Chapter 4 but he himself apparently felt that he was excluded from government projects because he was not yet a naturalized citizen (Gavroglu 1995, p. 192; this stands in contrast with the situation of his brother Heinz who remained in the UK and was recruited to the British atomic bomb project). It may also have been a case of who knew whom, with many participants in the Manhattan team being either an employee, alumni, or a student at UCLA Berkeley (where Oppenheimer and Fermi were employed) or the University of Chicago and even then it was often a case of being in the right place at the right time (thanks to Ann Bart of the National Museum of Nuclear Science and History for suggesting this; see also Oppenheimer 1965). Anderson also referred to the London and Bauer work as an 'obscure paper' that 'took on the notorious Bohr–Einstein debates' (ibid.) and wrote, '[t]his is the earliest paper I know of that expresses the most commonsense approach to the uncertainty principle and the philosophy of quantum measurement' (ibid.).

[12] The von Neumann variant, as elaborated by Wigner in particular, is sometimes called 'the Princeton Interpretation'.

[13] As Jacobsen has noted, 'Bohr never recognised von Neumann's axiomatic approach' (2011, p. 392).

[14] The extent to which this can be viewed as incompatible with Bohr's approach depends on which features of the two views are taken to be central, which obviously relates to what is understood by 'the

Processes of the First Kind: these are the processes involved in measurement (von Neumann referred to them as 'arbitrary changes') and are discontinuous, non-causal, and irreversible;

Processes of the Second Kind: these are the processes (or 'automatic changes') described by the equations of motion and are continuous, causal, and reversible.

According to von Neumann, processes of the first kind cannot be reduced to processes of the second kind and so the relationship between the two has come to characterize the heart of the measurement problem. Obvious questions now arise such as: *where* and *how* do processes of the first kind take place?

As to the 'where?', the answer will depend on which of the above two approaches—Bohr's (broadly understood) or von Neumann's—one adopts. According to the former, the transition between processes of the second kind and those of the first kind takes place in the shift from microscopic to macroscopic phenomena, whereas according to the latter, a 'chain' argument supports the answer that it takes place at the boundary between the subjective experience of the observer and the 'objective' world (see Jammer 1974, pp. 479–81).[15] Let us consider these two answers in a little more detail—as we'll see they do exhibit a certain commonality.

2.2 Heisenberg's 'Cut' and von Neumann's 'Chain'

An obvious concern about the 'Bohr approach' (again, broadly understood) has to do with how we should draw the distinction between 'microscopic' and 'macroscopic' systems in such a way that QM can be taken to apply to the former and classical mechanics to the latter. Heisenberg offered a response through the device of a 'cut' ('Schnitt'), articulated in most detail in his (unpublished) 1935 response to the famous 'Einstein–Podolsky–Rosen'

Copenhagen Interpretation' (see, again, Freire Jr 2015, pp. 80–1). Feyerabend (who we shall return to later) described the latter as a 'mixed bag' and hence, 'putting your hand into this bag you may come up with almost anything you want' (cited in Freire Jr ibid., p. 80 fn. 19). Howard, for example, has noted that the collapse of the wave-function is central to the popular image of the Copenhagen Interpretation but that Bohr never mentioned this 'or any of the other silliness that follows therefrom, such as a privileged role for the subjective consciousness of the observer' (Howard 2004, p. 669). Hence there exists only a 'tenuous relationship' between this interpretation and Bohr's notion of complementarity (ibid., pp. 670–1). Peres has suggested that '[t]here seems to be at least as many Copenhagen interpretations as people who use that term, perhaps even more' (Peres 2002, p. 29).

[15] This issue of the subjective–objective distinction, and in particular, how it should be characterized, will crop up again and again.

(EPR) argument, although he mentioned it earlier (in his famous Chicago lectures of 1929 for example; see Bacciagaluppi and Crull forthcoming, pp. 109–33).[16] The aim of this device was to separate those systems to which the quantum formalism should be applied from those—typically macroscopic—which fell under a classical description. Heisenberg argued, first, that the statistical or probabilistic aspect associated with QM must be situated at the cut but, second, that where that cut is drawn is arbitrary (and in that respect we see the commonality with von Neumann's argument).

The argument for the first conclusion is really straightforward: since deterministic equations hold on both sides of the cut, the probabilistic element has to be introduced at the cut itself (Bacciagaluppi and Crull forthcoming, p. 117). Furthermore, Heisenberg claimed that this element comes about due to the observation producing 'a fundamentally uncontrollable disturbance of the system' (ibid., p. 121).

The argument for the second, however, is a little more complex: Heisenberg began by imagining what was to become a commonplace set-up consisting of the system to be measured, a chain of measuring devices, and the observer. He then considered three possible scenarios: (i) the cut is placed between the system and the first measuring device; (ii) it is placed beyond that device but before the next; (iii) it is placed after the next device in the chain but before the observer. In the first case, the system is treated quantum mechanically, of course, and the probability of a given outcome is obtained via the Born Rule, which relates it to the modulus squared of the relevant wave-function (we shall return to consider the basis for this rule in Chapters 8 and 9). In case (ii), the quantum formalism is applied to the composite of the system and the first measurement device with the second device treated classically. Heisenberg then showed that the probability of the given outcome is the same as that in the first case; and likewise for the third. Since these cover all possible ways the cut can be made, Heisenberg concluded that the predictions of QM with regard to a given outcome are the same no matter where the cut is placed (Bacciagaluppi and Crull forthcoming, pp. 115–16).[17]

According to Bacciagaluppi and Crull, this device plays a different role in each of the above two arguments: as a means of demarcating the system to be

[16] In a letter to Heisenberg on this notion of a cut, Pauli wrote 'it seems to me that in a systematic construction of quantum mechanics one should *start from* the composition and separation of systems more so than has been done to date' (quoted in Bacciagaluppi and Crull forthcoming, p. 337).

[17] As Bacciagaluppi and Crull also point out, the placement of the cut is not completely arbitrary since if the first device is itself microscopic, it would have to be situated on the quantum mechanical side. Thus, the cut cannot be shifted arbitrarily 'towards' the system (Bacciagaluppi and Crull forthcoming, p. 116).

observed from that effecting the observation, 'the cut is an imaginary divide necessarily imposed by the experimentalist on any system to be investigated, classical or otherwise' (ibid., p. 117). However, as a mark of where probability enters the picture, in the move from quantum to classical mechanics, it has to be understood as a novel element introduced by virtue of the characteristics of the former. In neither case, however, should it be understood as a *physical* divide, since this would imply a discontinuity 'at some place between the macroscopic and the microscopic events [...] for which not the slightest indications are present either in experience or in the quantum mechanical formalism' (ibid., p. 120).

A rigorous proof of the movability of the 'cut' was subsequently given by von Neumann but although he acknowledged the 'similar considerations of Heisenberg' (von Neumann 1932, p. 262, endnote 208; 1955, p. 421, footnote 208), and referred to the latter's Chicago letters, von Neumann insisted that 'essential elements' of the discussion were derived from conversations with his colleague and compatriot Szilard (Baccigaluppi and Crull forthcoming, p. 117). Whatever the provenance, this idea of demarcating the quantum domain from the non-quantum was clearly 'in the air' at the time and provides a signiifcant commonality between the above two solutions to the measurement problem.[18] The difference, of course, is that von Neumann allowed for the cut to be made between the observer and the measurement apparatus-plus-system, thus opening the door for *consciousness* to play a role.

2.3 The von Neumann–London–Bauer Theory

When it comes to the question of *how* a process of the first kind occurs, the formal side of the response is represented by the so-called Projection Postulate, that essentially captures the idea of wave-function collapse:[19] in terms of the formalism, a 'projection operator' is introduced that, as the name suggests, projects the wave-function onto a sub-space of the relevant Hilbert

[18] According to Zinkernagel Bohr also thought the cut was movable, in effect, since he argued that any system could be treated quantum mechanically in principle, but that not all could be treated so simultaneously, since in any given experimental context, some part of the total system must be regarded in classical terms (Zinkernagel 2015, p. 8).

[19] The term 'projection postulate' was introduced by Margenau, who argued—in the 1930s—that the process it represented was in fact unnecessary and dispensable (Jammer 1974, p. 481, fn. 17). Even if this 'absurdity', as he called it, were justified by the introduction of consciousness or the ego, he insisted that QM would have to show rather more competence in the psychological realm before the proposal could be taken seriously (Margenau 1937). We'll come back to Margenau's views.

space associated with the result of the (strictly ideal) measurement, where that transition occurs with a certain probability, corresponding to that of obtaining the given result (see Goldstein 2009). With the mechanism of projection understood as having something to do with consciousness, this was understood as reducing the solution to the measurement problem in QM to the solution of the mind–body problem in general and, as we'll see, concerns arising from the latter were then carried over to the former.[20]

Now, this is one of the points where our narrative of 'the usual story' splits into multiple iterations. Many popular accounts ascribe this introduction of consciousness to von Neumann himself (see, for example, Brooks 2012). However, more nuanced analyses acknowledge that von Neumann actually said very little about the nature and role of consciousness and suggest that it was actually London and Bauer and, subsequently, Wigner who emphasized its significance.[21]

Thus, Herbert, for example, has stated that:

> von Neumann himself merely hinted at consciousness-created reality in dark parables. His followers, notably London, Bauer and Wigner, boldly carried von Neumann's argument to its logical conclusion: If we wholeheartedly accept von Neumann's picture of quantum theory, they say, a consciousness-created reality is the inevitable outcome. (Herbert 1994, p. 249)[22]

Likewise, according to Gavroglu:

> von Neumann did not include the consciousness of the observer to [*sic*] the measuring chain. The novelty of the London–Bauer treatment was the explicit claim that the reduction of the wave function was the result of the conscious activity of the human mind.
>
> (Gavroglu 1995, p. 171; cf. Shimony 1963, p. 758)[23]

[20] Thus, von Neumann's approach has been described as dualistic (see Jammer 1974, p. 482).

[21] Jammer notes that von Neumann was 'rather reticent' when it came to the details of processes of the first kind (1974, p. 481).

[22] Herbert can be taken to be representative of a certain view of the role of consciousness in QM, one that has been associated with books such as *The Tao of Physics* and *The Dancing Wu Li Masters* (see Marin 2009). He was a member of the 'Fundamental Fysiks Group' which also explored the relationship between QM and telepathy which we shall touch on later. Herbert proposed a method for sending signals faster than the speed of light using quantum entanglement, the refutation of which led to the famous 'no-cloning theorem', proved by Wooters, Zurek, and Dieks (see Kaiser 2011).

[23] See also the Wikipedia article on the 'Von Neumann–Wigner Interpretation'; https://en.wikipedia.org/wiki/Von_Neumann–Wigner_interpretation

Although such presentations highlight the significance of London and Bauer's approach, which, as understood (incorrectly as it turns out) by Wigner, did bring consciousness into philosophical prominence,[24] they have not only failed to grasp the overall philosophical 'shape' of that approach, and as a result have misunderstood its radical nature, but have perhaps also given von Neumann *too little* credit in this regard.[25]

Thus, Herbert again went on to claim that London and Bauer's work is a mere 'elaboration' of von Neumann's[26] and characterized the argument in the following way: according to QM, prior to an observation being made the world is nothing but 'pure possibility' (Herbert 1994, p. 249); but then, 'out of what solid stuff do we construct the device that will make our first observation?' (ibid., p. 249); either there must be certain physical systems that do not fall within the remit of QM or there are non-physical systems that possess 'single-valued actuality'; the former is ruled out by experiment, whereas the existence of one example of the latter is incontestable: consciousness. Hence, he insisted, London and Bauer concluded that consciousness is required to 'create reality' in the sense of bringing 'an actual world into existence, out of the all-pervasive background world of mere possibilities' (ibid., p. 250).

Similarly, de Broglie, in his treatise on the treatment of measurement in QM (de Broglie 1957), referred repeatedly to the 'von Neumann–London–Bauer' theory, taking London and Bauer's work to be no more than a presentation of the core concepts of von Neumann's approach.[27] De Broglie placed this in the section entitled, 'Less-admissable consequences of the theory of measurement in the present interpretation of wave mechanics', and maintained that some of these consequences are 'truly difficult to accept' (ibid., p. 30). Thus, he wrote that, '[i]n the von Neumann–London–Bauer theory, one must even say that it

[24] Hooker, for example, has noted, '[e]xamples of even well-informed scholars who nonetheless write as if there is more or less a single school of "orthodox" thought [include] Wigner... who lumps von Neumann, Heisenberg and London and Bauer with Bohr' (Hooker 1972, p. 262, fn. 51). Hooker himself understood London and Bauer as re-presenting von Neumann's 'subjectivist' approach (ibid., p. 75).

[25] Likewise, in his generally excellent historical study, Freire Jr refers to London and Bauer's 'little book' as 'intended to clarify the puzzling aspects' of von Neumann's work (2015, p. 86).

[26] Gavroglu has also portrayed London and Bauer as undertaking to 'analyze further the role of the observer which von Neumann had not fully elaborated' (Gavroglu 1995, p. 171). Likewise, Becker contends that von Neumann was not very clear on this but that '[s]ome took him to be saying that consciousness itself causes the collapse of the wave function; this was a view promoted by physicists Fritz London and Edmund Bauer in a book they wrote several years later, heavily influenced by von Neumann's work' (Becker 2018, p. 68). Finally, similar sentiments can be found scattered among the essays in a recent collection (Gao 2022), with the exception of Bitbol (2022), which I shall come to in Chapter 6.

[27] This work was published in the series 'The Great Problems of Science', edited by Paulette Fevrier-Destouches who we shall encounter again in Chapter 8. It is significant that a work on measurement in QM was published in a series on *problems* in science.

is the awareness of the macroscopic phenomenon by the observer that localizes the corpuscle... However, that seems truly unacceptable!' (ibid., p. 66).[28] And the reason is that if the observer were to close her eyes, a definite macroscopic outcome would still result (see also p. 77, where he stated that the approach is inadmissible because '[s]omething that happens in the perception of an observer cannot provoke a physical effect at a distance').[29]

Of course, as we'll see, these are not accurate characterizations of London and Bauer's view by any means: it is not the case that they held that the pre-observed world is one of mere possibility, with consciousness as the only actuality through the action of which reality is created; nor did they understand the role of consciousness in their account to be that of producing some sort of collapse, effecting the 'localization' of the corpuscle or whatever, or more generally, a reduction from one kind of process to another, as von Neumann framed it. Before we consider the details of their view, however, we should look at von Neumann's work a little more closely.[30]

2.4 Psychophysical Parallelism

Von Neumann is, of course, renowned as an outstanding mathematician, having made significant contributions across a wide range of fields, including physics (see Bhattacharya 2021 and https://en.wikipedia.org/wiki/John_von_Neumann). He studied chemical engineering at the Swiss Federal Institute of Technology at Zurich, where Einstein had been both a student and a lecturer and at the same time obtained a PhD in mathematics at the Pázmány Peter University in Budapest. After working with Hilbert in Göttingen[31] he became the youngest ever Privatdozent at Berlin and briefly worked in Hamburg before being invited to Princeton, where, after a few years, he was offered a lifetime Professorship at the Institute for Advanced Study.

[28] De Broglie's preferred solution was to adopt his version of the 'causal' interpretation of QM which is typically regarded as an early form of Bohmian mechanics.

[29] De Broglie did at least note London and Bauer's insistence that it is not some 'mysterious interaction' that produces a new wave-function for the system but rather (spoiler alert!) there is a separation of the 'I' from the correlation (de Broglie 1957, p. 30). However, he went on to say that he finds this 'separation' to be much more mysterious than any such interaction. Of course, it is my intention here to dissipate this air of mystery!

[30] See also Atmanspacher: 'By contrast to von Neumann's fairly cautious stance, London and Bauer (1939) went much further and proposed that it is indeed human *consciousness* which completes quantum measurement... In this way, they attributed a crucial role to consciousness in understanding quantum measurement—a truly radical position' (Atmanspacher 2015).

[31] According to Nordheim, it was here that he became interested in QM (Nordheim 1962).

When it comes to quantum physics, von Neumann is most well known for his introduction of what is now regarded as the 'standard' Hilbert space formalism for the theory[32] in his *Mathematical Foundations of Quantum Mechanics*, published in 1932 (in German; and in English with Beyer as translator in 1955; it was then further revised by Wheeler in 2018).[33] Within this framework, observables are represented by linear operators acting on the vectors representing states of the system and both the matrix mechanics of Born, Heisenberg, and Jordan and Schrödinger's wave mechanics 'drop out' as particular representations of the formalism (see Muller 1997a and b).[34] It is in this work that von Neumann presented his (in)famous 'no-hidden-variables' proof, the flaws in which were set out at the time by the mathematician and neo-Kantian philosopher Grete Hermann[35] and subsequently rediscovered by Bell (see Crull 2022 and Bacciagaluppi and Crull forthcoming, pp. 121–5).[36]

More significantly for us, of course, it is within this framework that von Neumann identified the afore-mentioned processes of the first and second kind and presented the 'chain argument' in the context of his 'psychophysical parallelism'. Let's look at this argument in a little more detail, after which we'll

[32] He also played a crucial role in the application of the mathematics of group theory, guiding and supporting Wigner in this respect (see Chayut 2001).

[33] Shortly after publication of his book, von Neumann expressed his dissatisfaction with this formalism due to concerns with the interpretation of quantum probabilities (Rédei 1996; see also Bueno 2016). Nevertheless, he decided to retain it when he prepared the book for translation, even though that required the text to be 'extensively rewritten' since the 'peculiar scope' of the work, tying together as it does mathematical-physical considerations with those of a 'philosophical-epistemological' nature, 'requires a very specific and sensitive use of the language' (von Neumann to the publisher, 1949, in Rédei 2005, p. 91). He went on to note that he 'practically had to rewrite Dr Beyer's translation' (ibid.), taking him six months, including preparation.

[34] Recalling our earlier discussion, it is in the context of this framework that the 'reduction' or 'collapse' of the wave function, previously introduced by Heisenberg, came to be understood as a manifestation of the measurement problem (see Jähnert and Lehner 2022).

[35] Hermann studied with the neo-Kantian philosopher Leonard Nelson who was a friend of Hilbert's and worked with Husserl.

[36] However, Mitsch (2022) has argued that Bell's dismissal of the proof as 'foolish' is unduly harsh and that criticisms of it fail to appreciate the context in which von Neumann was working: rather than attempting to achieve an axiomatic reconstruction of quantum mechanics, he was applying Hilbert's methodology in order to arrive at a form of 'axiomatic completion' of the theory, 'where "quantum mechanics" refers to a *specific* theory of quantum phenomena rather than, vaguely, to *any* theory of quantum phenomena' (ibid., p. 84). So, the idea was to take the qualitative core of the theory, in terms of the combination of probability, uncertainty, and measurable quantities and develop an appropriate mathematical framework that would embrace these essential ingredients. Within that framework there simply is no room for hidden variables. Thus, in section IV.1 of the book he offered a qualitative proof and then asked: does the Hilbert space formalism bear this out? The answer was given in section IV.2 and was no, of course, but *trivially so*; the proof demonstrates that the Hilbert space framework is the unique formalism for QM, *as so considered*. However, as von Neumann recognized, one could consider 'the' theory, insofar as one can talk about such, in other terms and that would yield a different formalism, as in the case of Bohmian mechanics say (this obviously bears on issues as to what we take the referent of 'quantum mechanics' to be; see French 2020). Stöltzner has noted that von Neumann attended a lecture by Bohm at Princeton and did not raise any objections (see Stöltzner 1999).

consider some further remarks von Neumann made about the role of the observer (see also Barrett 1999).

So, in chapter VI of *Mathematical Foundations* entitled 'The Measuring Process', von Neumann began by setting out his processes of the first and second kind, the latter, we recall, being governed by Schrödinger's Equation and the former occurring when a measurement is made. He then invited the reader to compare the formal presentation with the circumstances that 'actually exist in nature, or in its observation' (2018, p. 272). He noted, first, that:

> it is inherently correct that measurement or the related process of subjective perception is a new entity relative to the physical environment, and is not reducible to the latter. Indeed, subjective perception leads us into the intellectual inner life of the individual, which is extra-observational by its very nature, since it must be taken for granted by any conceivable observation or experiment. (ibid.)[37]

He then introduced his Principle of Psychophysical Parallelism:

> it must be possible so to describe the extra-physical process of subjective perception as if it were in the reality of the physical world; i.e., to assign to its parts equivalent physical processes in the objective environment, in ordinary space. (2018, p. 272)

In other words, the Principle requires that there be some form of *correlation* between this 'extra-physical process of subjective perception' and physical events such that our physical theory can describe, at least in coarse-grained terms, the former (see Barrett 1999, p. 47). Von Neumann went on to say that 'in this correlating procedure there arises the frequent necessity of localizing some of these processes at points which lie within the portion of space occupied by our own bodies. But this does not alter the fact of their belonging to the "world about us", the objective environment referred to above' (von Neumann 2018, p. 272).

[37] Here we find a particular difference between the German original and the later English translation, as von Neumann originally stated that the process of 'subjective apperception' leads into the 'uncontrollable' mental inner life of the individual; I am grateful to Michael Stöltzner for pointing this out to me and for suggesting that this can be related to the notion of the 'pre-reflexive cogito', as discussed by Fichte for example. The idea here is that the reflective form of self-awareness, in which the self takes itself as an object, presupposes a non-reflective form in which the self posits its own existence by merely existing, since 'it is necessary for the reflecting self to be aware that the reflected self is in fact *itself*' (Smith 2020).

The process of measuring temperature is a useful example here:[38] we can begin by looking at the mercury in a thermometer and declaring, 'This is the temperature as measured by the thermometer.' But we can go further and by taking the relevant properties of the mercury, together with the relevant laws, calculate the length of the mercury column, and say 'This length is seen by the observer.' Pressing on, we can consider the light source and track the path of the photons and say, 'This image is registered by the retina of the observer' and of course we can go even further, to consider the relevant chemical reactions and electro-chemical impulses in the brain. However far we go, at some point we have to stop and declare, 'And this is perceived by the observer':[39]

> That is, we are always obliged to divide the world into two parts, the one being the observed system, the other the observer[40] .... The boundary between the two is arbitrary to a very large extent.
>
> (von Neumann 2018, p. 272)

So, we can always push the boundary between the two as far as we like 'into the body of the observer', according to the Principle of Psychophysical Parallelism. However, if that Principle is not to be vacuous, the boundary has to be placed somewhere and, according to von Neumann, 'experience only makes statements of this type: "An observer has made a certain (subjective) observation", and never any like this: "A physical quantity has a certain value"' (2018, p. 273). Indeed, he continued, the Principle will be violated unless it is

[38] As Barrett has suggested, this 'everyday' example serves the rhetorical purpose of encouraging us not to worry about the two different kinds of processes involved in measurement in the quantum context (Barrett 1999, p. 47).

[39] As noted earlier (fn. 5), Alves has claimed that this approach was anticipated by Darwin, who considered a similar chain involving α-particle decay and the scintillations produced, to conclude that 'we can put the inexplicable feature of the quantum theory, the irreconcilability of wave and particle, in exactly the place where we have got in any case to have an inexplicability, in the transfer from objective to subjective' (Darwin 1929, p. 393). However, as the passage cited by Alves (2021, p. 459, fn. 16) and Bitbol (2000, p. 43) before him, makes clear, Darwin was primarily concerned with accounting for the appearance of particle-like behaviour in terms of a *wave-based* ontology. Thus, he argued that there is no need to invoke such behaviour at any point in the chain, *until* it reaches the consciousness of the observer, after which it becomes possible to 'infer back' and describe what happened using particle language. The 'transfer' from the objective to the subjective, then, has to do with accommodating wave-particle duality, rather than the measurement problem per se. Interestingly, Darwin developed an early form of 'wave-function realism' in terms of a multi-dimensional 'sub-world' in which the wave-function expresses everything that could possibly happen and there is no mention of observation at all (1929, p. 393). Our consciousness then, in effect, 'cuts sections' of this world of potentialities when it makes observations, which are then described in a language 'foreign' to it (ibid., p. 394; for more on Darwin's interpretation of QM and his insistence on retaining a visual representation, see Navarro 2009).

[40] This of course is von Neumann's version of the 'cut', discussed previously (see Stöltzner 2006, p. 505).

accepted that the boundary between system and observer can be arbitrarily shifted, given that the 'duality' represented by the above two processes is fundamental to the theory.[41]

To illustrate this further, von Neumann invited the reader to consider the world as divided into three parts (and here we can note the similarity with Heisenberg's approach):[42] I the system actually observed; II the measuring instrument; and III the actual observer. In the comparison of the first and second cases of the above example, I is the given system, II the thermometer, and III the light plus the observer; in the comparison of the second and third cases, I is the system plus the thermometer, II is the light plus the eye of the observer, and III is the observer, from the retina on. However, in the comparison of the third and fourth cases, I is everything up to the retina; II is the retina, optic nerve, and brain of the observer; and III is her 'abstract ego' (von Neumann 2018, p. 273).[43] Thus, in one case a process of the second kind, covered by Schrödinger's Equation, is to be applied to I and that of the first kind applies to the interaction between I and II + III; in the other case, the second kind applies to I + II and the first kind applies to the interaction between I + II and III. 'In both cases', von Neumann noted, 'III remains outside of the calculation' (ibid., p. 273).

He then showed that both cases yield the same result by carefully considering the composition of systems[44] and so 'quantum mechanics poses no problem for the Principle of PsychoPhysical Parallelism' (Barrett 1999, p. 51). Along the way von Neumann proved that the 'non-causal' nature of processes of the first kind cannot be attributed to incomplete knowledge of the state of the observer—in the sense that the information available to the observer regarding her own state might be limited in some way—and thus

[41] In a footnote, interestingly, von Neumann recognized Bohr as the first to note, in 1929, that 'the duality which is necessitated by quantum formalism, by the quantum mechanical description of nature, is fully justified by the physical nature of things, and that it may be connected to the Principle of PsychoPhysical Parallelism' (ibid., p. 273, fn 207).

[42] It is here that von Neumann acknowledged his conversations with Szilard in identifying the essential elements of the discussion to follow. Szilard had recently published his paper on thermodynamics in which he concluded that the Second Law could be violated by an intelligence with knowledge of the instantaneous state of a system, such as Maxwell's infamous 'demon'. According to Jammer, this created a space for consideration of the 'physical intervention' of consciousness upon physical systems (Jammer 1974, p. 480). However, Heisenberg had earlier—in 1928—stressed the role of the observer in the reduction (see the discussion in Barrett 1999, pp. 26–7). The role of consciousness was also discussed as early as 1927 in private conversations at the Solvay Congress (see Marin 2009). I'll return to this interaction with Szilard later.

[43] This is also the translation of the original German 'abstraktes "Ich"' (1932, p. 224; in the 1955 translation, this part of the text is placed in parentheses). Bitbol has also noticed this use of the phrase 'abstract ego' and takes it to be 'quasi-Husserlian' (Bitbol 2021, p. 569; 2022, p. 271).

[44] Here we may recall Pauli's letter to Heisenberg as quoted in fn. 16.

he assumed 'in all that follows that the state of the observer is completely known' (von Neumann 2018, p. 284). As we'll see, London and Bauer also made this assumption, albeit giving it a phenomenological gloss.

So, von Neumann did indeed say little about the role of consciousness in processes of the first kind (corresponding to his 'process I'). However, any expectation that he *should* say something more only arises because he has been interpreted, mistakenly, as advocating a form of physical collapse of the wave function, resulting from the action of the mind (see also Bueno 2019, pp. 130–1).[45] Nevertheless, as we've seen, he did emphasize that, 'measurement or the related process of subjective perception is a new entity relative to the physical environment' which is not reducible to the latter but leads us to the 'inner life of the individual', which by its very nature is 'extra-observational' and hence must be taken for granted in any observation. Furthermore, the Principle of Psychophysical Parallelism embodies the idea of a 'correlative procedure' between 'the extra-physical process of subjective perception' and 'the reality of the physical world'.[46] Now, as von Neumann also insisted, this idea is grounded in a kind of 'dualism' expressed by the difference between processes of type I and II. However, this should not, of course, be identified with the more familiar 'mind–body' dualism, since according to the above Principle, the 'cut' between the two types of processes can be made arbitrarily anywhere in the sequence from the system under investigation up to the 'abstract ego' of the observer.

As we'll see, London and Bauer also explicitly referred to a 'cut' in the 'chain of statistical correlations' which, as with other aspects of their 'little book', clearly encouraged casual readers to take it to be a mere summary of von Neumann's work. However, whereas von Neumann declined to articulate further how this 'cut' is effected, *London and Bauer present it in phenomenological terms*. Indeed, von Neumann's account has been criticized on the grounds that although he succeeded in showing that the theoretical predictions about the outcome of a measurement do not depend on where in the chain processes of type I take place, 'he did not show that when it is applied is

[45] According to Bueno, von Neumann 'does not advance an interpretation of the issues beyond what is strictly required, and for which there is evidence. (In this respect, von Neumann is a good empiricist...)' (2019, p. 131).

[46] The nature of the correlation here has been disputed. For Barrett (1999, pp. 47–8) it is a kind of lining up of sequences of events, mental on the one hand and physical on the other. In private correspondence (Becker 2004, p. 127), however, he has taken it to express the supervenience of the mental on the physical. However, as Becker has argued, in the temperature example there is only one mental event involved, namely the observation of the temperature of the liquid and that 'von Neumann's contention is that the mental event, which he treats as a single event, can be associated with any of a series of physical events arbitrarily' (ibid.).

empirically irrelevant in general' (Barrett 1999, p. 51). It is Wigner who is typically acknowledged as recognizing that *when* type I processes are taken to occur actually makes a difference and hence this must be stipulated in order for QM to be complete (ibid.).[47] In this regard—or at least so the usual story goes—Wigner deployed his famous 'Friend' argument, which we shall consider shortly, to conclude that such processes come into play by virtue of the observer being conscious. This is then couched in terms of mind–body dualism, sparking the debate with Putnam and Shimony. It is in this context that Wigner drew on London and Bauer's 'little book';[48] however, for London, at least, the placement of such processes of type I at the point where the conscious observer, or 'ego' as von Neumann himself would have it, enters the chain was a *phenomenological* requirement and so from that perspective, Wigner's argument appears superfluous.[49]

However, the above criticism of von Neumann's analysis has been rejected as misplaced, on the grounds that he did not regard such processes as involving some kind of *physical* collapse but instead maintained a form of 'relative-state' account, according to which such a process marks a shift in the observer's relation to the system being measured (Becker 2004). We recall von Neumann's insistence that the *content* of the Principle of Psychophysical Parallelism is embodied in the claim that the boundary between the observed system and the observer can be pushed arbitrarily into the body of the latter. However, this makes no sense if the shift from one type of process to the other is regarded as *physical*.[50] Thus, although the role of the mind is essential here,

[47] Koehler has reported that 'Abner Shimony tells me that this chapter [of von Neumann's book] was written by Eugene Wigner—later Shimony's teacher at Princeton—who was close to *both* von Neumann and Szilard' (Koehler 2013, p. 129, fn. 37). However, this seems implausible given both the style and the content of this part of the book. Stöltzner suggests that it might be 'a kind of "Stille Post" (Chinese Whispers) with exaggerations and that initially the point could have been that Wigner might have contributed some ideas about the physics, and such ideas that Abner would have found really important' (private email). There is nothing in Shimony's AIP Oral History interview (Shimony 2002) to support the claim. What Shimony does say, as we'll see later, is that Wigner liked very much London and Bauer's 'little book' and that London and Bauer were 'more explicit about the intervention of mentality in the measurement process than von Neumann is' (Shimony 2002). So it may be that Shimony said something like that to Koehler and that this is another (meta) example of the conflation of London and Bauer's position with von Neumann's. I have written repeatedly to Professor Koehler asking about this but he has not deigned to reply.

[48] In his outline of Wigner's argument, Barrett writes that 'Wigner believed that he clearly had "direct knowledge" of his own sensations' (1999, p. 52), the significance of which he obviously took from London and Bauer.

[49] At the end of his discussion of this approach to the measurement problem, Barrett essentially repeats the concerns of Putnam and Shimony, which we shall consider shortly (Barrett 1999, p. 55).

[50] Likewise, the temperature example makes no sense if the collapse is understood as physical, since here, of course, there is no such collapse; rather, Becker has argued, the example 'points out a dualism between a purely objective way of describing the world, and a subjective observer's point of view' (2004, p. 128).

it cannot be *causal*; rather it must be understood as merely descriptive (Becker 2004, pp. 128–9). On this interpretation, the dualism inherent in the two kinds of processes is just a manifestation of the difference between the physical description of an observer-independent world and the subjective language of what is observed. Furthermore, to insist that von Neumann has not demonstrated that when one applies the 'cut' is empirically irrelevant, is to miss the point:

> He does not say that he will prove that in general we cannot tell that collapse has not occurred in various situations. What he says is that if we observe the result of a measurement, it is arbitrary where in the physical process we apply the collapse. The kinds of counterexamples that would allow us to test for collapse would all get in the way of the observation in question being completed, and so they would not be cases in which we ever observe the result of a measurement. (Becker 2004, p. 129)

All that von Neumann required, on this view, is that the formalism itself does not commit us to any kind of physical collapse and given that, we can describe the situation in terms of superposition as far along the measurement 'chain' as we like—up to the point where the 'abstract ego' enters the picture. Of course, this may appear to make von Neumann's argument weaker than it is usually seen to be, but that's only because it has been taken to support 'physical collapse'; under the alternative interpretation, it is strong enough.

Given this, it would be not quite correct to say either that von Neumann declined to explain how the shift from a process of type II to that of type I comes about, or that he provided a mechanism in terms of some form of mind–body dualism (Barrett 1999, pp. 36–7; countered by Becker 2004, p. 132). Nevertheless, this still leaves a residual concern which is not fully addressed by suggesting that von Neumann advocated an early version of the Everettian relative-state formulation (Becker 2004, p. 134), albeit with instrumentalist overtones as when he wrote, 'the states are only a theoretical construction, only the results of measurements are actually available, and the problem of physics is to furnish relationships between the results of the past and future measurements' (von Neumann 1955, p. 337).[51] If we grant that he took the shift from processes of type II to those of type I to be a manifestation of the shift from an 'objective' language, appropriate for physical processes, to

[51] Cf. Stacey (2016) who has insisted that 'von Neumann treats quantum states as physical properties held by objects themselves'.

a subjective one, appropriate for the mental states of the observer, there is still the question of what it is that requires or motivates such a shift. Of course, we could always point to the apparently classical circumstances of the measurement situation, as Bohr did, and argue that these circumstances require us to use 'ordinary' language in which the relevant outcomes are definite. But then, leaving aside all the well-known issues with this stance, including that of the intrusion of quantum phenomena into such circumstances as in the cases of superconductivity and superfluidity, this would leave it unclear why Bohr's account and the usual understanding of von Neumann's were perceived as being at odds, with both competing for the title of 'the orthodox solution'.

It may be, of course, that the conflict was due to a misperception of von Neumann's work, or that it was appropriated by those keen to give it a gloss in terms of an explicit role for consciousness in yielding a physical collapse. Certainly, the book was not extensively cited[52] and those reviews that did appear in physics journals tended to focus on the formal aspects, for obvious reasons, with no mention made of psychophysical parallelism, much less of any role for consciousness in measurement.[53] Thus Dyson records that he was 'surprised to discover that nobody in the physics journals ever referred to Johnny's book' (Dyson 2013, p. 157).[54]

Interestingly, one review that did focus on von Neumann's account of measurement was written by the philosopher of science Feyerabend, who, as we shall see, was concerned with such matters at the time (Feyerabend 1958). After critically presenting the main features of the book,[55] Feyerabend argued that von Neumann already had all that he needed to complete the account by drawing on his earlier work on thermodynamics in the quantum context.

[52] It seems to have been regarded as highly technical and difficult. Jammer notes that, with the exception of reviews by Margenau and Bloch, both in 1933, it was not reviewed until 1957, two years after the publication of the English translation (Jammer 1974, p. 272).

[53] Mitsch has pointed out that von Neumann's book was published in Courant's series of textbooks for lay-mathematicians, *Basic Teachings of Mathematical Science* and that although this is clear in the frontmatter of the German (Springer) publication, it was obscured in subsequent printings (Mitsch 2022). Having said that, as noted earlier, von Neumann later wrote that the book wove together 'mathematical-physical' and 'philosophical-epistemological' considerations 'which gives it a content not covered in other treatises, written by physicists or by mathematicians, on quantum mechanics' (Rédei 2005, p. 92). The latter aspect is significant given that, as we shall see, London and Bauer insisted that QM should be regarded as a *theory of knowledge*.

[54] It is perhaps worth noting that there may be some Anglo-American bias underneath this comment; certainly, von Neumann's book had an impact on physicists in the Soviet Union where, in 1939 for example, Mandelstam delivered a series of lectures based on it (Kuzemsky 2008, p. 138; see also p. 157).

[55] Not the first two chapters, however, where von Neumann presented his extension of the Hilbert space formulation, as Feyerabend insisted that this part of the book 'has found little appreciation among physicists as it 'involves a technique at once too delicate and too cumbersome for the . . . average physicist' (E.C. Kemble)' (Feyerabend 1958, p. 343).

At the core of this is the idea that 'the measuring process is closely connected with the problem of the evolution of a large body towards its state of thermodynamic equilibrium' (Daneri, Loinger, and Prosperi 1962, p. 298). From this perspective, this process should be seen not as an 'inseparable chain' between object and subject but as one holding between the microscopic domain and the macroscopic—and in this regard it is claimed that the approach is in harmony with the ideas of Bohr (ibid., fn.†).[56]

In this context it is Szilard, who wrote his doctoral thesis on thermodynamic fluctuations, who is credited as exerting an influence on von Neumann, as we have already noted.[57] In considering the possible violation of the Second Law of Thermodynamics by an intelligence with knowledge of the state of a system, Szilard noted that in order to gain this information such an intelligence would have to expend energy, leading to an increase in entropy that would outweigh the decrease in entropy associated with the gas alone (Szilard 1929). This has subsequently generated considerable discussion (Landauer 1961; see also Ladyman 2018) but here I just want to note it has been claimed that Szilard's claim had an impact on not only von Neumann but also London and Bauer:

> Szilard's conjecture mentioned above has led many commentators [here a reference is made to London and Bauer] to believe that the measuring process in quantum mechanics is connected in an essential manner with the presence of a conscious observer who registers in his mind an effect and that this conscious awareness is responsible for the oft-discussed, paradoxical 'reduction of the wave packet'. (Jauch and Baron 1972, p. 221)[58]

Likewise, in a review article that examines the historical reasons for the 'identification' of information-theoretic and thermodynamic entropy,[59] Skagerstam wrote:

> In his famous book on quantum mechanics von Neumann transferred the Gedanken experiment of Szilard's [involving the demon] to the

[56] Here it is evident that the association of von Neumann's account with an explicit role for consciousness has been cemented into place, as the authors write that 'It is clear that von Neumann's theory is founded on a radically subjectivistic (solipsistic) philosophy' (1962, p. 303).

[57] Szilard and von Neumann became friends in Berlin (along with Wigner and Polanyi, who will appear in the narrative later) and taught seminars together (as well as with Schrödinger).

[58] They went on to argue that the presence of a conscious observer is in fact not necessary.

[59] An identification that Carnap, for example, criticized, only to face fierce resistance from von Neumann (and also Pauli; see Anta 2022). Carnap's critical analysis was eventually published posthumously, having been edited by none other than Shimony (ibid., p. 55), who, as we shall see, was also critical of what he saw as von Neumann's introduction of consciousness into QM.

> measurement problem... F. London and E. Bauer thus claimed in 1939 that the measurement process in quantum mechanics 'is connected in an essential manner with the presence of a conscious observer who registers in his mind an effect, and that this conscious awareness is responsible for the oft-discussed, paradoxical "reduction of the wave packet". We mention this because it is very interesting to see that "old problems" in the theory of the measurement process are connected, at least historically, to the problem of giving entropy some "deeper" meaning on a microscopical level.'
>
> (Skagerstam 1975, p. 457)

However, this claim has been resisted by Timm, who has suggested that:

> mentions of German physical chemist Fritz London and French physicist Edmond Bauer (1939), on collapse of the wave function and consciousness, do not seem to be directly relevant, Neumann's interest in this aside, to the historical etymology of the manner in which 'information' came to be allegedly associated with 'entropy', but rather seem to be later adumbrations made by theorizers seeking to discern the quantum mechanical nature of information and or information theory. (Timm 2012, p. 78)[60]

Still, the point about the influence of Szilard on von Neumann's thinking remains. In a 1949 paper, 'On the Process of Measurement in QM',[61] Jordan argued that von Neumann's 'subjective view', when deployed in the context of Szilard's 'phenomenological thermodynamics', has the unfortunate implication that entropy must be regarded as relative.[62] However, as a beam-splitting thought experiment shows, a real physical process must be involved and the entropy associated with that 'has an objective meaning, independently of the mental processes of any observer' (Jordan 1949, p. 276).[63]

[60] Perhaps because of this association with information theory, von Neumann has also come to be regarded as an early 'Quantum Bayesian'. However, as Stacey has noted, 'if *von Neumann had seen quantum states in anything like the QBist fashion,* it is difficult to find a rationale for why *MFQM*'s entire chapter on "the measuring process" assumes the shape it does' (Stacey 2016; he also argues that von Neumann's understanding of probability aligns closer to Keynes' 'logical' interpretation).

[61] This work was influenced by Margenau who organized the symposium in which it was presented.

[62] Jordan played a crucial role in the development of quantum theory, co-authoring a number of fundamental contributions with Born (under whom he studied in Göttingen) and Heisenberg on matrix mechanics before pioneering the development of quantum field theory. He also independently discovered Fermi–Dirac statistics and developed transformation theory (which deals with the transformations that a state vector undergoes in Hilbert space as the state changes with time), noting in his paper that a 'very clear and transparent treatment' had also been presented by London in a work that Jordan received after he had completed his own manuscript (see Schroer 2003, p. 2).

[63] Jordan's own approach involved the addition of a 'special axiom' to express the 'empirical fact' that 'each large accumulation of microphysical individuals always shows a well-defined state in space

2.5 Von Neumann and Consciousness

Granted all this, and also the fact that *Mathematical Foundations* is indeed less than forthcoming on the details, von Neumann did retain an interest in consciousness and the relationship between the mind and the brain throughout his life.[64] In February 1939 he wrote to his friend and fellow Hungarian Rudolf Ortvay, also a physicist, thanking him for an earlier letter and noting that the topic of 'the brain anatomical, in a phenomenological point of view', interested him very much (von Neumann to Ortvay in Redéi 2005, p. 198). In a subsequent letter he suggested that processes essentially connected to life cannot be described spatially and went on to note that:

> I have thought a great deal since last year about the nature of the 'observer' in quantum mechanics. This is a kind of quasi-psychological, auxiliary concept. I think I know how to describe it in an abstract manner divested from its pseudo-psychological complications, and this description gives a few quite worthwhile insights regarding how it might be possible to describe intellectual processes (therefore ones essentially connected to life) in a non-geometrical manner (without locating them spatially)'. (ibid., p. 201)

Such thoughts were pursued in his unfinished book *The Computer and The Brain* (von Neumann 2000; based on lectures from 1956), in which he concluded that the 'logics and mathematics in the central nervous system, when viewed as languages, must structurally be essentially different from those languages to which our common experience refers' (ibid., p. 82; see Leydesdorff 2016). Of course, such a difference might not necessarily be regarded as an obstacle to considering the role of consciousness in resolving the measurement problem (even assuming a broadly physicalist account of consciousness) since the mathematics of QM is also not that of common experience and is 'non-spatial' (at least as von Neumann conceived of it) just as he took that of intellectual processes to be.[65]

and time—that a stone never, unlike an electron, has indeterminate coordinates' (1949, p. 272); here he mentioned Schrödinger's 'cat' thought-experiment which is ruled out by this new axiom and which we shall consider later.

[64] Stöltzner suggests that 'von Neumann abandoned the Copenhagen thinking of the 1930s, not by giving a new interpretation, but by caring less about it since he would take a wider view in virtue of his progress in mathematics and his belief where the real problem of quantum physics was' (email). Further 'circumstantial evidence' for this is given by his shift away from the Hilbert space formalism to an alternative underpinned by quantum logic.

[65] As Stöltzner puts it, 'the ego somehow becomes mathematizable by its own mathematics' (private email).

2.6 Conclusion

Wrapping things up, von Neumann's view of the nature and role of the observer is certainly more nuanced that is typically appreciated. Nevertheless, in its lack of specificity of the manner in which that role is manifested, it remains distinct from the account given by London and Bauer.[66] However, in the debate with Putnam and Shimony that spelled the end of this approach as a serious contender for solving the measurement problem, it was understood as having been made more explicit through London and Bauer's work and, again, framed as the 'received' or orthodox view.

The principal agent behind this framing, I claim, was Wigner[67] who, as we have already noted, described the London and Bauer pamphlet (1939, 1983) as 'a very nice little book... which summarizes quite completely what I shall call the orthodox view' (Wigner 1963a, p. 7; in Wheeler and Zurek 1983, p. 325).[68] Wigner, of course, is a major figure in the history of modern physics, among many things widely credited (with Weyl), for the introduction of group theory into QM and, thereby, illuminating the role of symmetry principles in elementary particle physics more generally, for which

[66] Although acknowledging this lack of specificity, Alves has also maintained that '[t]he monograph [by London and Bauer] was a kind of digest of the hard, fundamental mathematical work done by von Neumann in his 1932's *Mathematische Grundlagen der Quantenmechanik*' (2021, p. 453).

[67] Thus, in a brief note, Moldauer wrote that, '[i]n a number of recent articles Wigner has reminded us of the logical consistency of the orthodox view of the role of measurement in quantum mechanics which has become associated principally with the name of von Neumann' (Moldauer 1964, p. 172). He then mentioned concerns about the loss of 'objective reality' raised against this view (suggesting it collapses into solipsism) and suggested that 'conclusions regarding the existence of an objective reality must be based on the properties of sense perception (or the results of physical measurements)' (ibid.). And Moldauer concluded that given that QM itself ensures that two observers will always agree on the results of physical measurements, objective reality in this sense will always be preserved: 'Accordingly, it appears to be unnecessary to sacrifice either objective reality, or a physical explanation of consciousness, or the orthodox interpretation of measurement in quantum mechanics as has been suggested by Margenau' (ibid.; and he thanks both Margenau and Wigner for their clarificatory communications).

[68] Later Wigner wrote that London and Bauer introduced the collapse postulate 'with even greater clarity than von Neumann' (Wigner 1971, p. 15) and in a letter from 1980 to Stapp he wrote, 'I liked the book of London and Bauer very much; in fact I wanted to have it translated and published in English but the editors did not give permission' (I can't give any further details about this letter as I have no idea how I obtained it!). Stapp also offered a 'theory of psychophysical phenomena' that he related to Wigner's view but which also draws on Whitehead's philosophy as well as developments in neuro-science (see Stapp 1982; also 2009). According to Stapp's theory, the selection of certain 'mutually exclusive self-sustaining patterns of neural excitations' (1982, p. 385) as the 'image' of the physical world, as represented by QM, is a 'creative act from the realm of human consciousness' (ibid.). We shall return to this element of creativity in the context of the London and Bauer account.

he received the Nobel Prize (see https://www.nobelprize.org/prizes/physics/1963/wigner/lecture/).[69] As we'll see in the next chapter, he also had a deep interest in the nature and role of consciousness but by leveraging London and Bauer's account into his own framework, he effectively distorted the former, generating the fundamental misapprehension as to its import and significance that continues to this day.

[69] Weyl's book (1928) set group theory at the foundations of QM but was widely regarded as too dense and difficult to understand; Wigner's (1931) on the other hand, offered more in the way of applications and was seen as more accessible. It was from the latter that the founders of what is now known as the Standard Model drew their inspiration in the 1950s and 1960s. Wigner recollects that he never interacted with Weyl who was somewhat dismissive of his work (https://www.aip.org/history-programs/niels-bohr-library/oral-histories/4965). Certainly, when considering who to invite to the Institute for Advanced Study in 1945, Weyl placed Wigner in the 'second rank', along with Bethe, Gamow, and Heitler. Further details on the Weyl and Wigner 'programmes' can be found in Bueno and French (2018).

3

The Debate about Consciousness

3.1 Introduction

Through the 1950s and 1960s, Wigner attempted to re-shape the conception of the 'orthodox view' of quantum theory by displacing Bohr's approach in favour of von Neumann's (Freire Jr 2015, pp. 149–61).[1] He did this in the context of historians' growing interest in the origins and development of the theory (ibid., p. 153), as indicated by the formation of the *Archives for the History of Quantum Physics* by Kuhn and others.[2] And at the heart of this re-shaping, of course, he set London and Bauer's 'very nice little book', which summarized 'quite completely' Wigner's vision (Freire Jr 2015, p. 152; see also Bueno 2019, pp. 132–3).

The principal opposition to this attempted re-orientation of the 'orthodox view' came in the form of Léon Rosenfeld, who was a staunch advocate of the Bohrian line (Freire Jr 2015, pp. 154–62).[3] The differences crystallize in Rosenfeld and Wigner's alternative takes on the measurement process (ibid., p. 155). Wigner followed von Neumann, not only with regard to the chain argument, but also with the adoption of the general framework of exposing the axiomatic foundations of the theory in order to better appreciate its implications (ibid.). Rosenfeld distrusted such approaches, insisting that

[1] Wigner's antipathy to Bohr's philosophy of complementarity is apparent in his argument that the duality inherent in complementarity is not reflected in the formalism where one can easily find three operators that do not commute, such as in the case of spin (Wigner 1963d). Given Bilban's argument about the relationship between Bohr's thought and Husserl's (Bilban 2013 and 2020), this displacement may be construed as a further effacement of the phenomenological approach.

[2] For a useful account of the establishment of this archive, see te Heesen 2020 and 2022. te Heesen has noted that Kuhn's participation in the project began just as he was finishing *The Structure of Scientific Revolutions* (which perhaps explain why the latter is so slight when it comes to the history of QM) and also that he became increasingly disappointed with the form of the interview for not revealing relevant details of moments of confusion, late night discussions, or the growing sense of crisis in general. For Kuhn the unhelpful responses of the interviewees revealed the limitations of the form. Seth, on the other hand, sees in these responses a reluctance to accept the Kuhnian framework of 'crisis' and 'revolution' which informed the questionnaire (Seth 2010, p. 269).

[3] Rosenfeld was initially 'mystified' by Bohr's ideas (Jacobsen 2011, p. 377) but after the two started collaborating on the measurability of the electromagnetic field, in the context of the emerging quantum field theory, he came to accept Bohr's 'great truth that we are not only spectators, but actors in the drama of existence' (quoted in Jacobsen 2011, p. 387).

A Phenomenological Approach to Quantum Mechanics: Cutting the Chain of Correlations. Steven French, Oxford University Press. © Steven French 2023. DOI: 10.1093/oso/9780198897958.003.0003

'*in the last resort* we must here appeal to *common experience* as a basis for common understanding' (from a letter about the Everett interpretation, quoted in Freire Jr 2015, p. 155).[4] And as we have seen, he even went so far as to dismiss von Neumann's *Mathematical Foundations* on the grounds that its 'unhappy presentation' of the measurement problem had created confusion and raised spurious problems (ibid.).[5] For Rosenfeld, it wasn't the role of consciousness, or, more generally perhaps, the relationship between subject and object[6] that was epistemologically significant about quantum theory but rather, complementarity.[7]

The clash between these views came to a head in the context of the so-called 'ergodic approach' to measurement, touched on in the previous chapter. According to this, the interaction between a system and the measurement device triggers a thermodynamic amplification of the signal, and as a result the state of the measurement device can be described in classical terms. Rosenfeld endorsed this approach because it was clearly compatible with Bohr's framework.[8] Wigner, on the other hand, took this to be in stark conflict with QM, given the application of the latter to macroscopic phenomena such as

[4] And in this regard, we may detect some resonance with the Husserlian notion of the 'lifeworld' that we shall consider in some detail later.

[5] Having said that, Rosenfeld studied von Neumann's papers in which the axiomatic approach was first presented, initially regarding it as helpful and even lecturing on it. However, he came to believe that 'the danger of formalizing is that you lose the physical content of it, or at least you are in danger of losing it so it's always a double-edged sword' (Rosenfeld 1963). He went on to say, 'I'm glad to have had that grounding in "Neumannistics" because when it is corrected by the influence of Bohr, then it is a sort of skeleton which helps to get a precise expression for the ideas' (ibid.). Von Neumann, on the other hand, thought that the scientific method was primarily 'opportunistic' and that axiomatization could be fertile even (and perhaps especially) in cases where the core concepts of the theory were unclear and the evidence still weak (see Stöltzner 2013 and also Mitsch 2022).

[6] Landé, for example, saw this as laying at the heart of the 'subjectivistic trend' which he subsequently identified with Wigner and dismissed on the grounds that he found it impossible to understand why the measurement of the velocity of an electron can be described only by reference to consciousness but that of the velocity of a falling stone need not be (Lande 1965/2015, pp. 134–5).

[7] Rosenfeld recalled that von Neumann gave a talk at Copenhagen on issues to do with measurement but that Bohr was less than impressed (ibid.). Interestingly, Rosenfeld made the acquaintance of Bauer during his stay in Paris in 1926–27, immediately following his graduation from the University of Liège (Jacobsen 2012, p. 17). While there he was supervised by Langevin (who encouraged him to go to Göttingen, where he became Born's assistant) and on his return visit in 1931, stayed in the Langevins' home (ibid.).

[8] Jacobsen has noted the role of thought experiments in this context, with Bohr acknowledging that his analyses of the measurement process were imaginary and 'could only be used for providing a logical justification of the theory's statements, not for shooting down or arguing for modifications, of the formalism' (2011, p. 388). As far as he was concerned, the issue of whether or not these situations could be realized in practice was irrelevant: 'The purpose of his analyses was to pair the *definitions* derived from the quantum formalism with the possibility of *measuring* these properties in an idealised experimental set-up to arrive at an interpretation of them. In this way his sole purpose when examining measurement problems in quantum theory was to provide meaning and limitations to the formalism' (ibid., pp. 380–1). This then became a source of contention since such idealizations allowed him to consider individual systems, contrary to what his opponents insisted should be considered in practice.

superconductivity and superfluidity, as demonstrated by London (Freire Jr 2015, p. 157).[9] At the core of this dispute (see Freire Jr ibid., pp. 156–61) we find, again, this issue of the limits of the applicability of QM, with Wigner insisting that the line should be drawn along the mind–body distinction, since 'the conscious content of the mind is not uniquely given by its state vector' (ibid., p. 160).[10]

This is all by way of setting out the context in which we can place the further debate between Wigner and Margenau on one side and Putnam and Shimony on the other, with the London and Bauer manuscript at its heart. Before we finally get stuck into the to-and-fro, however, let's consider in a little more detail how these two sides lined up.

3.2 Wigner

The trajectory of Wigner's career in physics is well known (Szanton 1992). After graduating from the same high school as von Neumann, Wigner entered the Technical University of Berlin to study chemical engineering. While there he attended the weekly colloquia of the German Physical Society, featuring the likes of Einstein, Heisenberg, and Pauli and he also met Szilard, who became one of his closest friends. After returning to Budapest to work in his father's tannery, in 1926 he accepted an offer of a job at the Kaiser Wilhelm Institute back in Berlin, studying X-ray crystallography (on the recommendation of Polanyi who he had met earlier and who we shall also return to).

It was while in Berlin that Wigner met London whom he regarded as 'a very thoughtful, very industrious, thorough, imaginative person' (Wigner 1963c).[11] Wigner subsequently moved to Göttingen and while there, became concerned with how one should represent measurements within QM. In particular, he was bothered by the question:

> Why is it that we always see positions macroscopically? Position operator is just an operator like every other operator. What is it that makes our minds principally think in terms of position operators? Why are there macroscopic

[9] Wigner and his co-authors also noted that there are counter-examples of measurements—such as so-called 'negative-result' measurements—which do not proceed according to this schema (Jauch, Wigner, and Yanase 1967, p. 186).

[10] It was after this dispute that the label 'Princeton School (or interpretation)' came to be used. According to Moreira dos Santos and Pessoa Jr, members of the latter included G. Süssmann, Josef Jauch, and Wigner's students, such as Huzihiro Araki and Mutsuo Yanase (Moreira dos Santos and Pessoa Jr 2011, p. 627).

[11] London himself went on to work on the applicability of group theory in QM and Wigner described his papers in this area as 'very nice'.

> bodies? Why do they have definite positions rather than having another, arbitrary, wave function, or another, arbitrary, operator measured? I may be completely wrong, but I do feel that there is some mystery here not completely cleared up. Several times I've had ideas on this but nothing really convincing. I've discussed that with Johnny [von Neumann] also.
>
> (Wigner 1963d)

Subsequently he became 'perturbed' by the fact that:

> we make an entirely—or largely—idealistic epistemology and it is not fair to do that without knowing much more about the mind. This did not bother me much at that time; what did bother me was the behavior of the macroscopic bodies, that it is always a concentrated wave packet with the position operator 'sharp'. (Wigner 1963d)

Now, the process of 'decoherence' offers a way forward here. When a system interacts with a measuring apparatus, say, because the latter has many more degrees of freedom than the former, there is suppression of interference between certain states. These can then be considered to be robust in the sense that information about them is stored redundantly in the environment so that an observer can then recover that information without further disturbing the system (Bacciagaluppi 2016). These preferred states are related to position because the relevant interaction potentials are functions of that property. Thus, the states effectively picked out by decoherence tend to be localized in position, or position and momentum, and so may be regarded as kinematically classical. It is important to note, however, that decoherence in and of itself does not 'solve' the measurement problem, because the combination of system + apparatus + environment remains in a superposition (Bacciagaluppi 2016).[12] Wigner was aware of this but according to Shimony was initially 'antipathetic' to such an approach (Shimony 2002), although his resistance softened towards the end of his career.[13]

[12] And so alternative outcomes remain possible, albeit with very low probabilities.

[13] According to Jha, 'Abner Shimony has explained that Wigner considered hypotheses other than the hypothesis I discuss, that the reduction of the superposition is the work of consciousness, but did not choose among them. One of Wigner's proposed tentative solutions ("Wigner's solution") to the various problems in the quantum theory of measurement was that consciousness may play a role in the reduction of the wave packet, but while evaluating H.D. Zeh's observation that the macroscopic measuring apparatus is not a closed system, he was skeptical that this observation could solve the reduction of the superposition' (2011, fn. p. 339).

Although he also disdained philosophy in its modern form (Szanton 1992, p. 308),[14] when it came to consciousness, Wigner recalled being impressed, as a young engineering student, by Freud's book *The Interpretation of Dreams.*[15] This led him to think deeply about the nature of consciousness, reporting later that, '[c]onsciousness is that thin layer of experience no greater in ourselves than is our small planet in the mighty universe. Yet when we speak of ourselves, we refer almost exclusively to this thin layer. I have never lost my fascination with human consciousness' (Szanton 1992, p. 67).

This fascination emerges in a number of Wigner's more reflective pieces, including certain crucial papers in which he directly drew on London and Bauer's work. This is explicitly so in the 'Wigner's Friend' argument, mentioned previously, which he took to offer perhaps the most powerful case for the role of consciousness in the measurement process (Wigner 1962; Wheeler and Zurek 1983, pp. 168–81).[16]

3.2.1 Wigner's Friend

Here Wigner began with the usual set-up in which a measurement device is set to measure the value of some observable with regard to a given system but the apparatus is replaced with a (conscious) 'friend'. The system under consideration is assumed to have only two states, with corresponding observable values—spin 'up' and 'down', for example. After the measurement, Wigner then asks his friend whether he saw spin 'up', say, and the Born Rule gives the well-known probabilities for a positive and negative answer. If Wigner then asks his friend what he saw *before* he was asked, the friend, Wigner insisted, will say 'I already told you. I saw spin "up"', since 'the question whether he did or did not see the [corresponding] flash was already decided in his mind before I asked him' (Wheeler and Zurek 1983, p. 176). It is at this point that Wigner cited a crucial phrase from London and Bauer: 'He [the friend] possesses a characteristic and quite familiar faculty which we can call the "faculty of

[14] In his interview for the AIP Oral History project, Shimony says 'Wigner was a man of incredible intelligence, acuity, and dependence of thought. He was not well read in philosophy, but he knew a lot of philosophy. Of course, lots of philosophical ideas are in the air, so he must have heard discussions of Kant, and discussions of logical positivism.... Then secondly, he just thought about things himself' (Shimony 2002).

[15] Indeed, he even thought 'vaguely' about becoming a psychiatrist (Szanton 1992, p. 68).

[16] Wigner read London and Bauer's 'little book' sometime in 1960 (Moreira dos Santos and Pessoa Jr p. 628; Moreira dos Santos and Pessoa Jr repeat the standard view—due to Wigner himself as I have claimed (see, for example, Wigner 1971, p. 15)—that this was nothing more than a development or clarification of von Neumann's position).

introspection". He can keep track from moment to moment of his own state.'[17] (As I shall argue, this invocation of introspection is not as straightforward, philosophically speaking, as Wigner and others took it to be.)

Since the issue as to what he saw was already decided in his friend's mind before the question was asked, Wigner concludes that the state immediately after the interaction between his friend and the system cannot be a superposition. He wrote: 'It follows that the being with a consciousness must have a different role in quantum mechanics than the inanimate measuring device' (ibid., p. 177); and he then went on to pursue the issue of the interaction between consciousness and physical systems (Wigner 1964).

This thought experiment went on to be widely cited (for a recent elaboration, see Frauchiger and Renner 2016) and the emphasis on the role of consciousness led to Wigner being generally regarded as a dualist. However, Esfeld has noted that unlike Descartes, say, Wigner did not take mind and body to be distinct entities; rather he insisted that the content of consciousness was the primary reality (Esfeld 1999a, no page number). This is emphasized in Wigner (1964) where he argued that the existence of physical objects is relative to consciousness by virtue of their construction in terms of that content.[18] It is the latter that is immediately accessible to us and hence even the existence of other people has to be regarded as on a par with that of physical objects. Again, as we'll see, there are obvious similarities with a phenomenological stance here, although Wigner remained concerned about the possibility of a slide into solipsism (see, for example, Wigner 1962).

It was partly as a result of these concerns, according to Esfeld, that Wigner eventually shifted his position, moving closer to the Bohrian form of

[17] Jammer acknowledged that Wigner incorporated the London and Bauer treatment into his account (1974, p. 499), as did Atmanspacher in his encyclopaedia article on consciousness and quantum physics (Atmanspacher 2015) but Barrett, for example, cited the passage where Wigner made reference to their work without noting it at all (1999, p. 53). Esfeld, however, did note the reference in his review of Wigner's collected papers, writing that 'Wigner thereby elaborates on a suggestion by London and Bauer (1939, §11): consciousness randomly selects one product state out of the superposition of product states and it thereby effects a state reduction' (Esfeld 1999b, pp. 147–8). He went on to acknowledge that Wigner referred explicitly to London and Bauer's work but wrote that Wigner conceded that we don't have a description of how that state reduction is actually effected by consciousness (in an earlier draft of this review, co-authored with Primas, the point about Wigner referring to the London and Bauer piece is omitted and instead it is stated that, 'In reviewing von Neumann's theory, Fritz London and Edmond Bauer unmistakably attribute the capacity to select a product state out of the superposition to the human consciousness' (Primas and Esfeld unpublished, p. 9) and later on the claim that 'consciousness reduces the state vector' is referred to as the 'London–Bauer–Wigner idea' (ibid., p. 14)).

[18] In a letter to Wigner, Shimony wrote that he found his work on the mind–body problem to be 'extremely stimulating' and one of the few treatments that considers it to be a 'legitimate subject for scientific investigation' (Freire 2015, p. 153).

orthodoxy (Esfeld 1999b, p. 149; see also Freire Jr 2015, pp. 166–170).[19] Relatedly, he became more sympathetic to the decoherence approach, through the early work of Zeh, and in particular the implication that it was not possible, practically, to consider macroscopic objects to be isolated systems (Freire Jr 2015, p. 168).[20] He took this to raise a significant problem for physics, particularly with regard to the role of the observer, noting that if the observer is macroscopic, as she must be, then given decoherence, she cannot be regarded as separate from the rest of the world (Wigner 1972). Thus, the dualistic imposition of a dividing line between the observer and the world must be abandoned (a conclusion that is compatible with London and Bauer's phenomenological analysis, as again we'll see).

Indeed, by 1984 he had come to the view that his earlier belief that the role of the physical apparatus can always be described by QM:

> implied that 'the collapse of the wave function' takes place only when the observation is made by a living being—a being clearly outside the scope of our quantum mechanics. The argument which convinced me that quantum mechanics' validity has narrower limitations, that it is not applicable to the description of the detailed behaviour of macroscopic bodies, is due to D. Zeh.
>
> (1984, p. 78)[21]

If such a body cannot be considered an isolated system, then it is not a system to which our current equations of physics apply, which means that 'a radical departure from the established principles and laws of physics is needed' (ibid.; see also Wigner 1972; this is an issue that came up in the debate with Putnam, as we'll see). Speculatively, he considered a possible equation for the change of state of a non-isolated system but concluded that insofar as it would not describe mental phenomena, it would, at best, only 'extend quantum

[19] Nevertheless, Shimony insisted that, 'He's not a Bohrian.... Wigner calls that the orthodox interpretation of quantum mechanics. You would think that when a man uses the term "orthodox" something or other, that's what he believes in. That's not true. It's orthodoxy, but it's not his orthodoxy. He really wasn't a Bohrian. What he was a man who thought we don't understand how events occur. We don't know the limits of the validity of quantum mechanics' (Shimony 2002).

[20] In (Wigner 1971), he suggested that decoherence might be more acceptable than collapse but acknowledged that whether it could form the basis of a solution to the measurement problem was not yet clear (ibid., p. 18). It has often been commented that Wigner needed to obtain a result himself before being convinced of it and so he set up his own thought experiment of a tungsten cube in 'intergalactic space' to demonstrate that even in this sparse environment, such a macroscopic body could not be considered as isolated.

[21] This issue of the limits of validity of the theory is a frequent motif in Wigner's work. Indeed, on the previous page he noted that despite the 'marvelous' successes of QM, it faced serious problems, not least that it does not describe 'the fact of *consciousness*', which motivates us to consider where those limits lie (Wigner 1984, p. 77).

mechanics to those limits which many of us thought it had already reached' (Wigner 1984, p. 81). This idea that new principles would be needed in order for consciousness to be covered by the theory became a recurrent theme in Wigner's later work.[22] As he recorded in his autobiography, his chief scientific interest over the previous twenty years was to 'somehow extend theoretical physics into the realm of consciousness....It has never been properly described, certainly not by physics or mathematics. It is shrouded in mysteries. And what I know of philosophy and psychology suggests that those disciplines have never properly defined consciousness either' (Szanton 1992, p. 309).

3.2.2 Polanyi and 'Tacit' Knowledge

Wigner shared these concerns with his colleague and compatriot Polanyi who, as mentioned earlier, he met in Berlin (Szanton 1992, pp. 76–81), where he studied crystal symmetries (which led to the later work on group theory; see Chayut 2001) and chemical reaction rates in Polanyi's laboratory (see also Wigner 1963b).[23] Polanyi's broad range of interests made a deep impression on Wigner and in subsequent years they talked and corresponded extensively on issues relating to the role of consciousness and the mind–body problem. Within the philosophy of science, at least, Polanyi is perhaps best known for his introduction of the notion of 'tacit' knowledge and his emphasis on its significance for scientific research. Insofar as this is associated with pre-scientific experience, his claim that it is the foundation of all knowledge has

[22] Again, Shimony, perhaps recalling his earlier engagement with Wigner, has stated that 'he is unequivocally against a physicalistic treatment of mentality. And he says it is possible that the locus of the breakdown of validity of the Schrödinger equation is when systems endowed with mentality are involved, like Wigner's friend. That is, the paradox of Wigner's friend still being suspended between having seen a red light or a green light, would be resolved if the Schrödinger equation doesn't govern the mentality of the friend. So stochastically, one or the other of these possible visions is picked out' (Shimony 2002). Shimony then noted two consequences: first, the integration of physics with psychology; and second, that with mentality as the locus of the breakdown of the Schrödinger equation, Wigner's view slides back towards Bohr's. However, there would still be 'subtle' differences, in that Bohrian orthodoxy 'makes the fixed points of physics to be sharp, clear observations made on experimental apparatus, like what number you read on a scale, or whether a bell rings or does not ring. Whereas in a real integration of physics and psychology, you must take into account the whole range of psychic phenomena, including sleep, the unconscious, peripheral vision, many things that are not sharp' (ibid.). These phenomena also feature in Shimony's contribution to the debate with Wigner and Margenau.

[23] Wigner recalls that he read 'a great deal of serious quantum theory inasmuch as there was serious quantum theory at that time' (Wigner 1963c), but that Polanyi was not well acquainted with the theory—indeed, no one in Berlin at that time was.

led to comparisons with Husserl's idea of the 'life-world', that we will come to later (Bennett 1978).[24]

Polanyi illustrated what he meant by 'tacit' knowledge using the example of planetary motion: in practice observations typically deviate from the corresponding theoretical predictions and the issue arises of determining whether these deviations are random or indicative of some trend. Even granted all the tools of statistics, it is down to the astronomer to decide whether there is a 'significant shape' or not to these deviations, and for that she must bring tacit or intuitive knowledge into play. As Polanyi put it, 'to perceive an object is to solve a problem' (Polanyi 1962, p. 3) and as part of that solution we need to distinguish the object from the background via this 'tacit power' (here Polanyi draws on Gestalt psychology;[25] ibid. pp. 5–6). This illustrates the general principle that:

> whenever we are focusing our attention on a particular object, we are relying for doing so on our awareness of many things to which we are not attending directly at the moment, but which are yet functioning as compelling clues for the way the object of our attention will appear to our senses. (ibid., p. 8)

Thus, 'perception is performed by straining our attention towards a problematic centre, while relying on hidden clues which are eventually embodied in the appearance of the object recognised by perception' (ibid., p. 12). These 'clues' are hidden within the body and cannot be experienced in themselves by the perceiver (ibid., p. 9), and indeed, if identified and held up as such would lose their suggestive power.[26]

According to Polanyi, this 'straining' of perception towards a problematic centre is also how science works; hence the structure of scientific intuition is the same as that of perception. Furthermore, the hidden meaning to which these clues point is an aspect of reality that may manifest via an 'indeterminate range of future discoveries' (ibid., p. 13). There are obvious comparisons to be made here with Husserl's idea that our perception is permeated with a horizontal intentionality that intends the hidden or absent profiles of an object: although I perceive directly only the screen and keyboard of my

[24] Comparisons have also been made with Merleau-Ponty's extension of Husserlian phenomenology, with its central claim of the primacy of lived perception (Takaki 2011). We shall also consider Merleau-Ponty's work later, in Chapter 8.

[25] Gestalt psychology will make further appearances, particularly as it informed the views of London's friend and fellow phenomenologist, Gurwitsch.

[26] Polanyi refers to this tacit act of making sense as an act of 'indwelling' and Jha sees this as analogous to the ideas of Merleau-Ponty on the body in action (2011, p. 342).

Macbook Air, as a *perceived object* it has a co-intended back. Furthermore, I may anticipate my perception of the back of the computer and the open manifold of such anticipations constitutes what Husserl calls the *intentional horizon*.

Polanyi applied his notion of tacit knowledge to the mind–body problem, arguing that there is a continuum of 'levels of existence' from objective reality to the mental, with each level reliant for its workings on the laws of the levels beneath it but with those workings not explicable in terms of those lower-level laws (see Jha 2011, p. 334). He also believed that this offered a solution to the measurement problem, whereby measurement is understood as a case of the observer 'making sense' of the hidden clues and particulars via which we come to perceive the object (Jha 2011, pp. 336–7). Wigner, for his part, confessed he was unclear about some of the details, but did acknowledge the importance of tacit knowledge in observation in a letter to Polanyi where he recalled von Neumann's 'chain' argument and concluded that to avoid 'an endless process' we must:

> admit that we have some knowledge which developed in our unconscious, as your tacit knowledge, without conscious observations. (ibid., p. 338)

Although their correspondence continued, clear divergences emerged in their attitudes: Wigner understood that what he was engaging in was epistemology, in the traditional sense, and regarded Polanyi's approach, with its emphasis on Gestalt theory, as falling under psychology; whereas Polanyi took epistemology to incorporate the latter, together with background knowledge, yielding the act of 'meaning-making' (Jha 2011, p. 342). From this perspective, he viewed natural science as an extension of lived perception and here perhaps Wigner missed the point in a way that parallels his approach to the London and Bauer material.[27] Eventually Wigner became frustrated by Polanyi's use of 'neologisms and multitiered analogies' (ibid., p. 347) and the exchange ended with the two talking past one another.

Having lost this opportunity to bring a more nuanced perspective to the subsequent debate with Putnam and Shimony, Wigner had another chance through his relationship with Margenau, who was also better equipped, philosophically, to grasp what was at stake. However, as we'll see, Margenau effectively suppressed his own views when it came to the exchange itself.

[27] Jha goes on to present Putnam, in the context of the debate with Margenau and Wigner, as criticizing the latter's 'Polanyian' epistemology (2011, pp. 344–5) but given Jha's own discussion of the above distinctions, I can't see any basis for such a claim.

3.3 Margenau

Well known both for his textbooks in physics and his work in the philosophy of science,[28] Margenau obtained a PhD in physics at Yale.[29] While there he was awarded a fellowship that allowed him to visit Sommerfeld in Munich, followed by Born in Göttingen, before meeting with Schrödinger in Berlin. There he also met London who was working on van der Waals forces and Margenau extended the latter's approach to dipole molecules.[30] After returning to Yale, Margenau gave a course on 'Foundations of Physics' covering QM and also took on students in the philosophy of physics, including, perhaps most famously, Adolf Grünbaum.[31] Prompted by Northrop,[32] and encouraged by Cassirer, with whom he and Northrop taught a course on Kant, Margenau also became interested in the nature of scientific constructs (Margenau 1964).[33] It

[28] According to Google Scholar, his most cited work is his co-authored textbook, *Mathematics of Physics and Chemistry* from 1943, followed by *The Nature of Physical Reality: A Philosophy of Modern Physics*, published in 1950 (reprinted without changes in 1977) and still cited by philosophers of physics today.

[29] Significantly, Margenau read a good deal of philosophy during his early years and became a 'devotee' of Kant (Margenau 1964).

[30] Margenau said that 'Fritz London himself was a strange person. He was always very diffident about meeting people. He was a very, very hard taskmaster, great perfectionist, who would not allow anything to slip by in the work of his students. He was not very happy about the paper, the manuscript which I showed him. However, he did not discourage me from publishing it. In fact, he said I should publish it but I had to do more than I have done. And that was that' (Margenau 1964). Margenau subsequently wrote letters on London's behalf that he thought might have helped with the latter's emigration to the USA (Gavroglu 1995, p. 191). However, there appears to have been no philosophical interaction between the two although Margenau did recall that he attended lectures by Reichenbach while in Berlin. In *The Nature of Physical Reality* London is only mentioned once, in the context of Heitler and London's demonstration that valence forces are reducible to so-called 'exchange forces' (Margenau 1950, p. 92; such 'forces' arise from the particles' non-classical indistinguishability).

[31] Grünbaum is perhaps most well known for his work in the philosophy of space and time. He subsequently criticized his former professor's approach to the measurement problem and concluded, '[i]nsofar as quantum mechanics does raise the question concerning interaction not only between a physical system and a measuring device, viewed as ontologically real, but also concerning the role of the sensed events in the observer's experience, Mr. Margenau would have to offer a theory of the observer as well so as to round out his epistemology of quantum mechanics' (Grünbaum 1950, p. 32). As we'll shortly see, Margenau believed he had such a theory in what he took to be London and Bauer's account.

[32] Although Northrop became well known for his work in comparative philosophy, his book *Science and First Principles* (Northrop 1931; first delivered as lectures in 1929) contains a potted history of matrix and wave mechanics, covering not only Heisenberg's and Schrödinger's works but also Dirac's. However, he argued, given the problems with the theory, physics needed to go back to 'first principles', as exemplified by the case of Special Relativity. As we'll see, a similar attitude has been adopted by the advocates of QBism who tend to have a better grasp of the physics than Northrop (about whom Lenzen wrote, '[t]he author's exposition of contemporary physics is not trustworthy' (Lenzen 1933, p. 321)).

[33] In his (1950) Margenau wrote that the methodological modifications introduced by quantum physics directly affect our idea of reality and that '[f]ew have seen this more clearly than the late Professor E. Cassirer [and here he gives a reference to Cassirer's *Determinism and Indeterminism*] with whom the author had the pleasure and the good fortune often to discuss his views' (p. 14). Further references to the aforementioned book as well as to some of Cassirer's other works are scattered across various chapters. Margenau later recorded that his return to his earlier philosophical interests was triggered by the arrival at Yale of Cassirer whom he described as his 'hero' (1978c, p. xxvi). He was also

was Cassirer who stimulated the development of Margenau's own epistemology of 'constructionalism'[34] which offered a broad understanding of 'experience' that included not just perception, but also rational and emotional features, the significance of which has to be determined by explicating the relevant procedures, as exemplified in science (Margenau 1950, p. xxvii). 'Only in this way', Margenau insisted, 'can we avoid the difficulty which Husserl raised for Kant by asking: why should not all clear contents of consciousness be allowed to compete with sense data in determining reality?' (1949, p. 288).

The latter are those aspects of experience that can be distinguished from others by their spontaneity, their relative independence, and their irreducibility. So, the sensory aspect of the experience of seeing a tree is the residuum that remains when all the rational aspects and 'mnemonic associations' are deleted. It is because such a residuum has withdrawn itself from rational manipulation, that it cannot figure as such in our theories but must be rationalized through being translated into wavelengths, geometrical figures, and the like. Having said that, Margenau insisted that 'sensation as part of the process of knowledge is not wholly *sui generis*, and that there may well be a gradation from those qualities which signify an act of clear perception into those that characterize pure thought' (1949, p. 290). It is by understanding the nature of this gradation, he argued, that we can grasp how classical and modern physics are related. So, returning to the example of the tree, we move from the 'perceptual tree' to the tree as a physical object by supplementing our visual, tactile, and kinaesthetic impressions with, first, those qualities that go beyond our immediate impressions, to those we recall from memory, of the back of the tree, for example (and here we might think of Husserl's 'horizon' again), or of its roots and so on, and second with qualities such as permanence or continuity of existence.[35]

This movement is not mere integration but *construction* where such constructs may include not only trees and electrons, say, but also ghosts, mirages and the luminiferous ether (ibid., p. 293). What is crucial, obviously, is to be able to identify those constructs that are deemed to be 'valid' which involves,

involved in the preparation of the revised, English edition of *Determinism and Indeterminism* whose bibliography, prepared in 1945, included the London and Bauer monograph. Margenau also supplied the preface after Cassirer's death.

[34] See for example (Margenau 1935) where in the context of the recently developed formalism of QM, he considered the replacement of the 'pseudo-sensible' construct of the electron with an abstract construct articulated in terms of operators and Hamiltonians. There is then a correspondence, if only indirect, between such constructs and the relevant data, in terms of which the physical universe can be divided (although Margenau gave two examples where such a correspondence was, at that time, absent: Dirac's postulate of the infamous negative energy sea and Fermi's of the neutrino).

[35] Thus, he noted (1950/1977 p. 59), Natorp and Cassirer both saw clearly that the tree, qua object, is not 'given' (*gegeben*) but 'posed' as a problem (*aufgegeben*).

in addition to the satisfaction of certain metaphysical requirements, justification via the 'circuit of empirical confirmation' (Margenau 1949, p. 295). Here we begin with sense data and then proceed into the 'field of constructs' via rules of correspondence that, in effect, unite experience, data, and constructs and in the scientific context, take us from the blueness of the sky to a particular wavelength. The only difference between quantum and classical mechanics, then, has to do with the span of these rules and their statistical character (ibid.). Within this field one can move from one set of constructs to another via logical or mathematical theorems and then back, via a similar rule of correspondence, to the 'field of sensation', yielding a prediction which, if true, provides validation to the set of constructs in the circuit. Constructs that have been validated a sufficient number of times become 'verifacts' and physical reality is then conceived of as the class of all such 'verifacts' (ibid., p. 295).[36]

Interestingly, Margenau also applied this 'constructionalist' approach to the self and concluded that:

> the reflecting (not experiencing) ego is initially a construct to be verified, a construct of remarkable universality, enabling a self-reference of every part of experience. That such self-reference is possible, and hence that the ego construct can be verified... may indeed be the most noteworthy fact of our experience; but it is not thereby exempt from rational and empirical examination. (1950, p. 455)[37]

In certain respects, this may not appear to be so far removed from a phenomenological conception, as we'll see, although Margenau argued that whereas scientists adopt a fallibilist attitude towards empirical data (and have developed theoretical criteria for the rejection of illusory data), the phenomenologist is guilty of the uncritical admission of introspective evidence which was regarded as stable and indubitable (and thus had no similar criteria for excluding 'abortive introspections'), something he regarded as 'wholly disastrous' (Margenau 1950, p. 463; this is based on his earlier 1944 essay 'Phenomenology and Physics', reprinted in Margenau 1978a, pp. 317–28).[38]

[36] Wave functions are also understood to be such a validated abstract and hence are part of physical reality (ibid., p. 299).

[37] For a critical analysis of 'constructionalism' see Werkmeister (1951) and for a reply, see Margenau (1952).

[38] Margenau distinguished his epistemology from both Berkeleian and Kantian idealism through the establishment of rules that 'certify' what is objective in things (1950/1977, p. 48). His core claim is that objectivity has to be 'discovered by procedures of which we all are vaguely cognizant and which reach highest precision in the methods of the physical sciences' (ibid.).

This characterization of the phenomenological attitude towards introspective evidence is of course too crude for the criticism to be taken seriously.

3.3.1 Interjectivity

Margenau's own philosophy cannot be said to have had a major impact, although it is worth noting that a 'constructivist' or 'constitutive-phenomenological' interpretation of QM was later pursued by Gauthier (1971), for which Margenau provided 'helpful comments' (ibid., p. 429, fn. 1).[39] This gives a central role to the notion of 'interjectivity', in the sense of 'the relation in which first a subject and an object can arise and upon which inter-subjective structures become possible' (ibid., p. 431). On this view, consciousness does not *interact* with a system but, rather, makes it possible for phenomena to occur in the first place. Both the observer, as a public entity, and the joint physical system of the object of the measurement and the apparatus are then to be understood as the twin poles of these inter-relational structures (Gauthier 1971, p. 432). As we'll see, this is strongly reminiscent of the London and Bauer account.[40]

According to Gauthier, then, the observer should not be conceived as a psychological subject but rather as 'a nexus of structures of experience' (ibid., p. 432) that define the constitutive conditions of a physical phenomenon.[41] 'The world', then, is not, naively, 'out there' but must be seen as a kind of approximation in the constructive endeavour as we approach, through deeper and larger structures, the ideal totality of all structures; that is the universe as a whole (ibid., p. 433).[42]

[39] Gauthier has insisted that, '[o]f course there is a Husserlian background in my 1970 paper' (private communication) and records that he wrote a longish paper (now lost) on Husserl in the 1960s, on the basis of which Gadamer agreed to be his 'Doktorvater' or PhD supervisor (but see Gauthier 2019).

[40] Gauthier also writes that 'In 1970, I didn't know about London and Bauer' but that he subsequently (in 1991) discussed the issue with Wigner (private communication).

[41] In a footnote he writes that his position could also be called 'structuralist' (Gauthier 1971, p. 432, fn. 5); see also Gauthier (1969) where he articulates the notion of 'structure' in terms of both linguistic and mathematical structuralism.

[42] I was not aware of Gauthier's work when I wrote (French 2002) and his discussion note appears to be little known, having been cited only seven times, six of them by himself. Subsequently, he seems to have developed an interactionist account, incorporating the notion of a 'local observer', that might be usefully compared to Rovelli's 'relationist' interpretation (which we shall consider in Chapter 9), although Gauthier himself insists that his approach has priority (Gauthier, private communication).

3.3.2 Margenau on Measurement

Returning to the issue of how to understand measurement in QM, Margenau joined Wigner in rejecting Bohrian orthodoxy,[43] with its emphasis on complementarity and the latter's account of measurement prompted him to return to and clarify his own approach, the core concepts of which he understood to be not so different from Wigner's (Freire Jr 2015, p. 163; Margenau 1963a).[44]

Thus, he sought an account that did not rely on classical models and defined a measurement as:

> any observational ingression into the 'state' of a physical system which reveals a number, a number guaranteed by experience with the apparatus employed and by theoretical consistency to have relevance for the state of the system before measurement, and guaranteeing nothing with respect to the state afterwards. (1963, p. 472)

In these terms, what are often called measurements—such as the passage of an electron beam through an inhomogeneous magnetic field—should rather be regarded as state preparations, since observation is not involved (see also 1937, p. 359).[45]

He then set out the 'most general mathematical features of every measurement' (1963, p. 474), following 'in essence', he noted, the treatment given in London and Bauer (and which he had previously followed in his earlier papers). Significantly, however, Margenau rejected von Neumann's 'Projection Postulate', taking it to offer a 'positive seductive risk for philosophic misinterpretations' (1963, p. 476; see also 1937, p. 356). One such is the supposed loss of objectivity due to the 'projection', or 'collapse' being understood in terms of a shift in our knowledge (see also Margenau

[43] Margenau first met Wigner in Berlin in 1932 and then again in 1939 at the Institute for Advanced Study at Princeton, after Wigner had fled to the USA. He records that he saw him 'off and on' in subsequent years; see: https://www.aip.org/history-programs/niels-bohr-library/oral-histories/4757.

[44] Feyerabend claimed that Margenau's view was actually indistinguishable from Bohr's (appropriately understood) 'except for some fancy terminology' (Feyerabend 1968, p. 311, fn. 8). And Putnam argued, as we'll see, that it was incompatible with Wigner's. Both are wrong but in interesting ways. Margenau and Wigner actually speculated about writing a book together, although it seems that Wigner may have been put off by Margenau's interest in extrasensory perception and his concern with reconciling science and religion (Freire Jr 2015, pp. 163–4, fn. 48). Nevertheless, Margenau invited Wigner to join the editorial board of *Foundations of Physics*, where he played an influential role (Freire Jr 2015, p. 164; see also Murgueitio Ramírez 2022, p. 761).

[45] In his (1937), Margenau addressed the issue of the impact of QM on the notion of 'state', arguing that the state function should still be regarded as referring to the system rather than our knowledge; see also (1950/1977, p. 350).

1937, pp. 367–8).[46] If, instead, we interpret it in terms of *selection*, then we regain objectivity and, moreover, can apply it to both measurements per se and state preparations. In a 'Philosophic Postscript' (1963, pp. 482–4), he reflected on the role of the observer in this process and reiterated that given this unwarranted interpretation of the Postulate, the speculation that is based upon it, namely that consciousness is involved, becomes suspect.[47] In particular, he insisted, the physical situation—in terms of pure states in the full Hilbert space—remains unaltered if there is no conscious observer included. What a measurement involves is the deliberate selection of 'specific items of knowledge' (ibid., p. 483) and consideration of that selection has no bearing on the mind–body problem.

In a later work Margenau considered the introduction of consciousness as a kind of hidden variable that would effectively render QM classically deterministic[48] and noted that Wigner had hinted at such a move 'when suggesting the need for implementation of quantum mechanics to render it applicable to physiological and psychological processes' (1978b, p. 373). He then proposed a philosophical argument to give this suggestion a 'measure of credibility' although as he admitted, it is 'highly metaphysical', 'surprising', and 'unconventional'.[49] The argument begins by noting that free will requires there to be elements of chance and choice. The former is guaranteed by QM and the latter has to do with the will. However, he continued, the human will does not select among the alternatives presented by 'the ket [wave function] of the universe' as this would run counter to the stochastic nature of the theory. Hence, any consciousness that could make such a choice cannot be human; either physical systems themselves have a 'sovereign will', which leads to panpsychism, or there must be some 'superhuman will', in which case Einstein was right in asserting that 'God does not play dice'!

Curiously, however, one can find few, if any, traces of these views of Margenau's in his responses with Wigner to Putnam and Shimony's critiques of what the latter took to be the 'orthodox' account of measurement in QM. It

[46] He also argued that the Postulate cannot handle non-ideal measurements, such as those that disturb the system, in the sense that the latter's post-measurement state is not an eigenstate of the observable being measured, and those that annihilate it, as when a photon is absorbed by a detector (see also 1937, p. 358). This rejection of the Projection Postulate as originally conceived has been quite influential although Kronz (1991) has argued that both of the above kinds of cases can be handled through an appropriate generalization.

[47] Again, he had previously insisted that if the ego were to be introduced in this context, the only reply would be to say that 'quantum mechanics does not as yet pretend to be a psychological theory' (1937, p. 367).

[48] By incorporating it, somehow, as a sub-manifold of the relevant Hilbert space or as a feature of the Hamiltonian.

[49] As Wessels (1980) noted, Margenau's argument actually runs contra to Wigner's view.

was Margenau who independently spotted Putnam's initial foray into this area but soon afterwards received a letter from Wigner asking for his opinion and offering to co-author a reply.[50] Let us now consider the background to Putnam's concerns.

3.4 Putnam

Putnam, of course, is one of the major figures of 'analytic' philosophy in the second half of the twentieth century, renowned for his work across a wide range of areas, from the foundations of logic to the philosophy of mind and the philosophy of science. His PhD thesis on the foundations of probability theory was supervised by Reichenbach, who insisted that the claim that a system is disturbed by an observation was 'philosophical mysticism' that has no basis in QM (Reichenbach 1944, p. 15).[51] Thus, he maintained that QM deals only with relations between physical things so that all statements of the theory can be made without reference to an observer.[52] Putnam adopted a similar line, arguing in his seminal paper, 'A Philosopher Looks at Quantum Mechanics' (1965), that measurement should not be regarded as primitive, as he claimed the 'von Neumann axiomatization' took it to be, but should be understood as a physical interaction like any other (and here we see a foreshadowing of the position adopted in Relational Quantum Mechanics, discussed in Chapter 9). He wrote:

> To define a measurement as the apprehension of a fact by a human consciousness, for example, would be to interpret quantum mechanics as asserting a dependence of what exists on what human beings are conscious of, which is absurd. (ibid., p. 147)

[50] In a 1964 interview, Margenau said, 'Last year a man named Putnam, who is a mathematician and logician... published a paper on quantum mechanics in the *Journal of Philosophy of Science*. I saw this paper and I thought it was crazy. A few days later I got a letter from Wigner, saying, "Look. Have you seen Hillary Putnam's paper?" I said, "Yes." He said, "We cannot let that ride. You and I must write a paper in rebuttal, setting him straight." I said, "Well, wonderful. I'd be delighted to do it." Well, I had the same experience again; he practically wrote the whole thing and I deny making any contribution at all.' (Margenau also noted how Wigner insisted on him (Margenau) having his name first, even though his contribution was small); see Margenau 1964.

[51] Moreira dos Santos and Pessoa Jr suggest that Wigner may have attended a seminar given by Putnam while he was at Princeton, after completing his doctorate. Wigner wrote to Shimony afterwards that he had tried to correct a serious mathematical error on Putnam's part (2011, p. 631). However, this was in 1963 and Putnam had moved to MIT by then (perhaps he had returned to give the talk).

[52] Reichenbach himself argued that a form of three-valued logic offered an appropriate framework for describing the structure of the quantum domain.

Returning to the topic forty years later, he remained dismissive, stating:

> one might say—Von Neumann hints at this in his book, and Eugene Wigner famously advocated it—'the collapse occurs when the result of a measurement is registered by a consciousness'. I do not know of anyone who currently advocates this 'psychical' view. (Putnam 2005, p. 626)

However between these two bookends,[53] as it were, von Neumann's approach, and his 'cut' in particular, were the focus of Putnam's 'Quantum Mechanics and the Observer' (Putnam 1981). Here he noted that, given the nature of the cut, in terms of what we take 'the system' to encompass, we might call QM a 'theory of relativity', since 'there is a dependence of truth upon one's perspective' (ibid., p. 197).[54] Indeed, he suggested, one can push this relativity further and place the cut within the brain, or between the brain and mind, concluding (in parentheses), '[p]erhaps the ultimate observer on von Neumann's view is the Kantian transcendental ego' (ibid., p. 197). It is a pity that Putnam didn't take the opportunity to use this passing remark as the basis for reflection on his earlier rejection of consciousness- or ego-based approaches.

What prompted the exchange with Margenau and Wigner was Putnam's 1961-response to a purported resolution of the Einstein–Podolsky–Rosen (EPR) 'paradox' (Sharp 1961).[55] Putnam used this as the opportunity to raise various concerns about measurement in QM and in particular he argued that the theory could not jointly incorporate two conditions: first, that a measurement requires interaction with an 'outside' system, such as an observer and second, that the 'whole universe' can be treated as a quantum system (Putnam 1961). After canvassing various approaches, Putnam suggested, at the end of the note, that we should abandon the first condition and contemplate the possibility that 'macro-observables have sharp values without being measured from "outside"'.

[53] For a detailed comparison of Putnam's 1965 and 2005 papers, see Wüthrich (forthcoming). As Wüthrich has noted, whatever one might think of Putnam's own stance towards the interpretation of QM, he was instrumental in pushing physicists in the 1960s and 1970s to recognize that there really was a 'measurement problem'.

[54] Interestingly—particularly in view of the later Relationist interpretation that we shall consider in Chapter 9—Putnam then extended this 'perspectival' approach, suggesting that when we choose to measure a given property of a system, such as its spin, we choose to 'institute a frame' relative to which the system has the determinate property 'spin-up', say, or the determinate property of 'spin-down', and '*the measurement finds out which*' (1981, p. 209; his italics). He writes, 'Relative to *this* observer, *these* properties are "real"... but relative to a different observer, different properties would be "real"' (ibid.). He concluded by noting that on this view, quantum particles are real entities, '*but which they are is relative to the observer*' (ibid., p. 208; his italics).

[55] Feyerabend (1962) records various interventions from and 'private communications' with Putnam on issues to do with the foundations of QM, including some that originate in discussions from 1957.

This of course is reminiscent of Bohr's approach as Margenau and Wigner noted in their response (Margenau and Wigner 1962, p. 293) but before we get to the debate itself, we should introduce the fourth participant, Abner Shimony, whose work also spanned both physics and philosophy.

3.5 Shimony

Shimony also wrote his PhD thesis on probability, supervised by Carnap,[56] but then went on to complete a second PhD in physics (on statistical mechanics), with none other than Wigner himself.[57] Although it was through his work on probability that Shimony became interested in physics, Born's *The Natural Philosophy of Cause and Chance* also played a role, and reignited his interest in QM (Shimony 2002). He also knew of Margenau's work but didn't take the latter's course in philosophy of physics while at Yale, which he came to view as a mistake. However, he was quite taken with Whitehead's 'process' philosophy, recording that he 'liked Whitehead's mentalism' and the anti-dualist idea that 'there's something in common to mental reality and physical reality, because those entities which we call mental have a kind of experience' (ibid.). This chimed with his 'strong evolutionist' views:

> It seemed to me if creatures like us are evolutionary products, then our mental faculties must be products of evolution, not just our bodies. If our mental faculties are products of evolution, then there must be something mental-like from which the faculties evolve.
>
> (Later in the interview he refers to a form of 'proto-mentality'; ibid.)[58]

At that time, Wigner was beginning to work on the foundations of QM, as we have seen, and Shimony told him about Bell's Theorem (for which Wigner

[56] In contrast with Reichenbach, Carnap did not discuss QM in detail, perhaps because, even as late as 1966, he believed it was too early to draw philosophical conclusions from the theory (Faye and Jaksland 2021).

[57] Shimony initially started his PhD with Wightman, one of the founders of axiomatic quantum field theory. However, he felt that the mathematics he needed to tackle the problems Wightman had posed would take too long to learn and so he changed advisor. Nevertheless, it was Wightman who asked Shimony to study the EPR paper and find out what was wrong with it. Shimony concluded that as an argument it was 'flawless' while acknowledging that one or more of the premises could be false.

[58] The interviewer (Joan Bromberg) noted that the theory of evolution is 'something that is a constant' in Shimony's writings and as we'll see, it features in the debate over the role of consciousness in QM.

produced a simpler proof). It was through Wigner, on the other hand, that Shimony was introduced to London and Bauer's 'little booklet' which he described, of course, as 'a sort of popularization of von Neumann's mathematical foundations of quantum mechanics' (Shimony 2002).[59] Indeed, Shimony thought it would make a useful text for the foundations of the QM class he was then teaching at MIT but as it was in French, he had to translate it. Bauer, apparently, liked the translation (London having died by then) and Wigner felt it should be published with an introduction but the French publishing company (Hermann) refused permission, on the grounds that they wanted to do it themselves.[60]

However, Shimony admitted, '[t]hat booklet was more explicit about the intervention of mentality in the measurement process than von Neumann is, for a very interesting reason. London's first doctorate was in philosophy. He was a student of Husserl. He was interested in physics' (Shimony 2002). And, he continued:

> [a]s a student of Husserl, there were some residues of phenomenology in the little booklet of London and Bauer. Without giving you the details, in the first quantum mechanics paper I wrote, the one called 'The Role of the Observer in Quantum Mechanics', I have a long passage on London and Bauer. That came from reading the book to teach the course at MIT' (ibid.; as we'll see, this point about the phenomenological 'residues' in London and Bauer's work was not appreciated at the time of the debate itself and Shimony's realization came about some years later.)

That paper was first presented at a conference in 1963, organized by Podolsky, with the likes of Bohm, Furry, and Wigner attending and indeed, it was Wigner who asked the organizers to allow Shimony to present.[61]

[59] Shimony also studied von Neumann's book while at Princeton and found it to be very readable.

[60] The Wheeler and Zurek version actually refers to translation*s*: 'English translations—including a new paragraph by Professor Fritz London—done independently by A. Shimony, and by J.A. Wheeler and W.H. Zurek, and by J. McGrath and S. Mclean McGrath; reconciled in 1982' (Wheeler and Zurek 1983, p. 217). As we'll see these reconciliations may have blurred some of the nuances in the original text.

[61] In his interview, Shimony speculated that it was perhaps as a result of this presentation and drafts of his paper being circulated, that he was sent Bell's paper which eventually led to his contribution to the construction of an empirically testable form of Bell's inequality (Clauser et al. 1969). He also subsequently became heavily involved in the production of the informal journal *Epistemological Letters* which was for many years one of the only forums for discussions of Bell's work (see Murgueitio Ramírez 2022, p. 764).

3.6 The Debate

As we've already noted, the debate began with Putnam's 1961 paper,[62] in which he expressed concern about the central role given to the observer when it came to the process of measurement (Putnam 1961).[63] Let's examine his argument in a little more detail.

Putnam began by rejecting the claim that von Neumann's axiomatization gave QM, and the measurement process, 'a logically rigorous formulation' (1961, p. 234). We recall that he saw the central difficulty as arising from the tension between the following two statements:

1. A measurement on some system requires that system to interact with some other 'outside' system, such as a measuring device;
2. The whole universe can be treated as a system by the theory (and hence can be assigned a state function). (ibid., pp. 234–5)

With regard to (2), Putnam added in a footnote: 'For the "whole universe" any suitable closed system which includes the measuring apparatus may be understood in the present discussion' (ibid., p. 235, fn. 1). Now, there is some historical work to be undertaken on where this idea came from, that QM could be applied to the entire universe. Certainly, a hint might be discerned in a 1928 talk by Schrödinger (published in 1935), where he argued that it ought to be possible to eliminate spatio-temporal features, such as electron 'orbits' 'without leading to the consequence that no visualizable scheme of the physical universe whatever will prove feasible' (Bacciagaluppi and Crull forthcoming, p. 73). And it is perhaps not too much of a speculation to suggest

[62] One of the few analyses solely devoted to the debate is given by Moreira dos Santos and Pessoa Jr (2011) who emphasized the 'non-epistemic' factors in play, including political and religious differences and, more significantly perhaps, the fact that Margenau and Wigner, both eminent figures in the physics community, may have viewed Putnam as an 'outsider' (ibid., p. 632; Shimony's contribution is not considered in their discussion). As they have noted, some of the exchanges take on an aggressive tone, from both parties. They also acknowledged that both sides may have misconstrued the London and Bauer piece, citing French (2002) and concluded, 'Like a volley of punches at the end of a close fight, Putnam's arguments against the mentalist interpretation left the philosophical audience with the impression of victory, even if the judges voted for a technical draw' (translated from the original; Moreira dos Santos and Pessoa Jr 2011, p. 639).

[63] Again as already noted, Putnam was responding to Sharp's presentation of a 'new resolution' of the EPR 'paradox' that, the latter claimed, relied neither on Bohr's epistemological presuppositions nor a detailed application of von Neumann's account of measurement. The core of the purported resolution consisted in the claim that EPR err in assuming that separate pure states can be assigned to the parts of the correlated system after measurement. If it is accepted that only the entire system can be assigned a state function, then, Sharp maintained, the paradox clearly dissolves. Putnam, however, used Sharp's analysis to argue that there are serious conceptual difficulties related to measurement more generally.

that the application of the theory to macroscopic systems as in the phenomena of superconductivity and superfluidity, might have encouraged the development of this idea that it might be applied to the universe as a whole.

Certainly, by 1956 Everett felt able to argue that '[t]he theory is ... capable of supplying us with a complete conceptual model of the universe, consistent with the assumption that it contains more than one observer' (Everett 1956, in Barrett and Byrne 2012, p. 152).[64] However, it is not clear whether Putnam was aware of Everett's work at this time[65] but, like the latter, he argued that no 'physically acceptable' version of the then current form of QM could accommodate the above statements. And this is because if the time development of the entire universe is described by Schrödinger's Equation, then, for example, the position of an electron, say, can never be measured, contrary to experience.[66]

He then considered various options for dropping the claim that QM can describe the entire universe, including von Neumann's approach. The last was rejected, not because of any introduction of consciousness (at least, not here) but because it, like many others, implied that systems that are not closed cannot be assigned their own state function.[67] Hence, Putnam argued, if we reject statement (2) above, *there are no systems*, at least not in the usual quantum mechanical sense; that is, that have states representable in terms of vectors in Hilbert space and so on (Putnam 1961, p. 236). Thus, the core issue here has to do with representing the influence of the 'outside' within the formalism and he emphasized in conclusion that treating the universe in terms of two kinds of entities is simply untenable.

[64] Everett defended this claim by appealing to the kinds of factors that are typically found within the philosophy of science, including, in particular, the value of novel predictions which increase our confidence that the theory can be extended beyond its initial domain of application and which generates ' a strong desire to construct a single all-embracing theory which would be applicable to the entire universe' (1956 in Barrett and Byrne 2012, p. 171). We should have more confidence in a theory that is 'universally valid' than one that either restricts itself to microscopic phenomena or includes an arbitrary element such as, for example, consciousness.

[65] As he noted in (Putnam 2005), he does mention it in (Putnam 1991, pp. 17–19; thanks to Matteo Morgani for finding a copy for me), where Wigner's name also comes up, but only in a prelude to consideration of the 'Many Minds' account that we shall touch on in Chapter 9. In his co-authored 1995 paper the 'Many Worlds' interpretation is dismissed as 'incoherent' (Albert and Putnam 1995, p. 22).

[66] Here he drew on Sharp's argument that an observable of the electron, say, will also be an observable of the larger system which in this context will be the entire universe. But it can then be proved that the state function of this larger system cannot be an eigenfunction of this observable at two different times, unless a measurement takes place which of course is precluded (Putnam 1961, p. 235). Sharp himself invoked Wigner's earlier result that, according to von Neumann's account, in a closed system that includes the system of interest and the measuring apparatus, the only quantities that can be measured are those that commute with all conserved quantities; Sharp then offered what he considers to be a simpler argument to the same end (Sharp 1961, pp. 229–30).

[67] Here too he followed Sharp.

3.6.1 Margenau and Wigner's Response

Margenau's immediate response was harsh, as we've seen, and Wigner insisted that they had to co-author a reply.[68] Thus they began by suggesting that the paper's conclusion could not have been intended to have been taken at face value, as a contradiction as glaring as that which was identified by Putnam would surely have already been spotted by physicists. Instead, they viewed it as a challenge to restate the theory of measurement in terms that do not require the mathematical formalism of QM itself. It is at this point that they identified von Neumann's and London and Bauer's approaches, stating:

> According to von Neumann and London and Bauer, who gave the most compact and the most explicit formulations of the conceptual structure of quantum mechanics, every measurement is an interaction between an object and an observer. (Margenau and Wigner 1962, p. 292)

They then rejected the suggestion that 'the object' might be the entire universe, because, bluntly, the observer *has* to be distinct from it. That the object is 'closed', in the sense of being separate, is simply assumed to be the case between measurements but is obviously not the case *in* a measurement since this involves interaction. Furthermore, they continued, it cannot be the case that a larger system, such as that of the object plus measurement apparatus, never undergoes measurement, since to ascertain the result of a measurement requires a further measurement on the measurement apparatus.

The 'chain of transmission of information' from the object to the consciousness of the observer may consist of a number of steps that can be analysed to a greater or less degree. However:

> [o]ne cannot follow the transmission of information to the very end, i.e., into the consciousness of the observer, because present-day physics is not applicable to the consciousness. This point, which may be unpleasant from the point of view of certain philosophies, has been clearly recognized by both von Neumann and by London and Bauer. (ibid., p. 292)

[68] As we've also seen, Margenau's attitude might well be described as 'ambiguous' at best—describing the introduction of consciousness as 'monstrous' at one point, yet subsequently reflecting seriously on the idea and of course, he was far more reflective, philosophically speaking, than Wigner (see also Moreira dos Santos and Pessoa Jr 2011, p. 641).

Thus, the 'cut' must be introduced between the system and observer, with the assumption that 'the observer has a "direct knowledge" of what is on his side of the cut' (ibid., p. 292; here they cite London and Bauer again).

Putnam's specific claim, that an electron subject to external forces cannot be in an eigenstate of position, say, at two different times, was then dismissed as 'obscure' and his conclusion rejected as erroneous. And they also rejected the suggestion that the 'orthodox view' treats the universe as divided into classical and micro-objects, since on that view the former are regarded as 'proper limiting concerns' of the theory describing the latter. Here, of course, the two sides were talking past one another with regard to what should be taken as the 'orthodox' view, with Putnam clearly referring to the Bohr version and Margenau and Wigner, of course, taking the alternative, 'Princeton' line. Their conclusion was that:

> [o]verall consistency of all parts of quantum mechanics, especially when that theory is forced to make reference to 'the entire universe', has never been proved nor claimed. And it is not likely that any expert in the modern developments of logic will demand it. (ibid., p. 293)

3.6.2 Putnam's Counter

Putnam's counter-response was equally blunt, referring to Margenau and Wigner's note as 'a strange document' (Putnam 1964, p. 1) and claiming that they had simply failed to meet his concerns. These he presented again, albeit in more detail, beginning by distinguishing between the system *S*, the measuring apparatus *M* and the rest of the universe, *T*. He then pointed out that although for pragmatic purposes the approximation is made of setting the interaction between $M + T$ and *S* to zero, strictly speaking that can never be the case. As a result, *S* can be assigned neither a state function nor an appropriate Hamiltonian and there can be no rigorous, 'contradiction-free' account of measurement within standard QM.[69]

Putnam then made a series of replies to distinct points in Margenau and Wigner's critique, beginning with the claim that by referring to 'the entire universe', Putnam was trying to apply QM to cosmological issues. However, he

[69] On this point, he drew a comparison with the situation in the foundations of calculus in the eighteenth century (1964, p. 2). In a sense this whole debate exemplifies Vickers' point that whether a given theory is taken to be inconsistent or not depends on how it is characterized (Vickers 2013).

stated, '[n]othing could be wider of the mark' (ibid., p. 2); rather, he insisted, the issue is simply whether the theory can consistently treat measurement as a form of interaction that takes place within a closed system. Whatever that system is taken to be, the point is that *T*, above, should be empty, containing no observers. Margenau and Wigner's assertion, that 'if one wants to ascertain the result of the measurement, one has to observe the measuring apparatus' could then be seen to be 'worthless', as it presupposed that the observer is not part of *M*.

Now, what about the assumption that if measurement involves interaction between *S* and something 'outside', then it cannot also be assumed that the entire universe is a system? Previously, as noted, Putnam had suggested that we might give up this assumption and here he supplied the details: let *T* be empty; then according to von Neumann, when *M* measures an observable *O* in *S*, the state of the system jumps into that of an eigenstate of *O*, as determined by *M*. According to Bohr, *M* can be treated entirely classically, with QM applying only to *S*. However, Putnam repeated, '[t]his is not only implausible on the face of it, but inconsistent since *S* cannot, strictly speaking, have states of its own' (ibid., p. 3). What is consistent, he maintained, is that *S*+*M*—that is, the 'entire universe'—jumps into an eigenstate of *O*.

Having referred to the 'so-called Copenhagen Interpretation' (ibid.), Putnam argued that Margenau and Wigner cannot evade the point by referring to the well-known classical limit theorems or Bohr's correspondence principle, since these only imply that any classical system may be treated as the object under consideration, but then some other system would have to be regarded as the 'observer' and treated classically within this scheme. Indeed, Bohr himself made the point that we are obliged to ignore the quantum mechanical structure of the observer, and the likes of Landau had argued that classical mechanics cannot be taken as reducible to QM precisely for this reason—classical physics has to be assumed on the 'observer' side of the infamous 'cut'. This is where London and Bauer were drawn into the debate again, this time by Putnam: 'London and Bauer would like to reduce the "observer" to a disembodied "consciousness", but Margenau and Wigner admit this is not yet a success' (ibid., p. 3). The alternative sketched above, in which we have a purely quantum mechanical account of measurement, Putnam suggested, is both more direct and 'unmetaphysical'.

As for the Projection Postulate, consider the case of an electron, whose position is measured at t_0 and t_1, and which is free during the interval between. If the position measurements and the free movement of the electron are treated as one 'motion' within a closed system, then the state function of the whole system must be an eigenfunction of the position at t_0 and t_1. However, since the electron is not interacting with the rest of the system between measurements, its state function has to be considered to be 'spread out' during the interval, so that it cannot be an eigenfunction of position at t_1, except by undergoing a discontinuous collapse at that time. If we then drop the assumption of a 'cut' and take measurement to put *M*+*S* into an eigenstate of the observable, then, Putnam argued, although we must still introduce, by fiat, a reduction of the state function at t_1, we can say that 'the whole business' takes place within a single closed system containing the observer.

Of course, a defect of his account, as Putnam acknowledged, is that it does not explain *why* or *how* measurement causes such a reduction but then, he maintained, the 'London–Bauer interpretation' is in an even worse situation:

> On their interpretation the measuring system is always outside the system *S* and includes a 'consciousness'. However, London and Bauer do not go so far as to make it just a 'consciousness'—it must also have a 'body', so to speak. (ibid., p. 5)

But then his main point holds: ignoring the interaction between *M* and *S* before the measurement is not just a useful approximation but is absolutely indispensable here. Furthermore, and critically, measurement, *on this view*, comes down to simply the 'direct awareness' of a fact by a consciousness and so subjective events—namely the perceptions of the observer—have to be taken as capable of causing abrupt changes in physical state—namely, the reduction of the wave packet.[70] Obvious questions then arise:

> What evidence is there that a 'consciousness' is capable of changing the state of a physical system except by interacting with it physically (in which case an automatic mechanism would do just as well)? By what laws does a consciousness cause 'reductions of the wave packet' to take place? By virtue of what properties that it possesses is 'consciousness' able to affect Nature in this peculiar way? (ibid., p. 5)[71]

[70] London and Bauer's treatment is referred to as 'highly subjectivistic', a label echoed subsequently by Jammer, for example (Jammer 1974, p. 499).

[71] Putnam attributed that final question to Shimony (indicating that they had discussed these issues).

With no answers forthcoming to these questions Putnam concluded that there is neither reason for nor plausibility in introducing either a 'cut' between observer and object or a role for consciousness.[72]

3.6.3 Margenau and Wigner's Counter[2]

Margenau and Wigner welcomed the opportunity to reply, but stated that they found it 'difficult', given their feeling that their position had already been clearly stated (1964, p. 7). So, they attempted to point out what they saw as Putnam's error even more clearly than before, focusing on his two premises: that a measurement on *S* requires interaction with some other system (which they found unexceptional); and, the 'whole universe' can be treated as a system for the purposes of applying QM (here, as they added in a footnote, they follow Putnam in taking 'the whole universe' to be any closed system that includes the measurement device). This second premise, they insisted, conflicts with the theory, as it stands, because the latter takes the state function to apply to that which is 'outside' the observer. However, they continued, modifying the premise, as Putnam does, by replacing 'the whole universe' with any system that does not contain the observer, leads him into error.

This centres on his assertion that, given premise 2, as modified, the time development of *S* could obey Schrödinger's Equation *at all times*. And this, Margenau and Wigner claimed, stemmed from his reluctance to accept the impossibility of describing the last part of measurement by Schrödinger's Equation; that is of accepting that the reduction of the wave packet is unavoidable. This latter process, they wrote:

> when properly understood, takes place when the observer interacts with the measurement apparatus and somehow obtains cognizance of its state. The impossibility of describing this part of the measurement process by means of the equations of quantum mechanics was clearly recognised already by von Neumann as well as London and Bauer. (1964, pp. 7–8)

Alternatively, one can simply eliminate the state function and express the predictions of QM directly in terms of probability correlations between

[72] Similarly, Jauch subsequently cited London and Bauer, together with Wigner, as attributing a special role to consciousness in quantum physics, which, he wrote, 'somehow is made responsible for the change of the state vector during the measurement process' (Jauch 1971, p. 42).

observations that one can make on a system. Here Margenau and Wigner referred to some recently published work that we have already touched on. Thus, Margenau had noted that in classical physics, scientists were able to able to map their observations onto the domain of models (which Margenau called the field of constructs, or 'C-field'; Margenau 1963b) and were then able to reason about the phenomena in question following the rules embodied in those models, together with rules of correspondence leading back to observations. In QM, however, the C-field, consisting of states and observables, is connected to observations only via probability relations. Thus, when it comes to the concept of 'state', for example, instead of an isomorphism holding between our models and the phenomena, we have 'a kind of polymorphism which prevents a unique passage from models to Nature and also from Nature to models' (ibid., p. 3).[73]

It is precisely here that the problem of measurement comes to the fore. One can appeal to the Projection Postulate, but its acceptability depends on the adoption of one or other of the following: (a) the state function develops according to Schrödinger's Equation until a measurement is made, whereupon it spontaneously transforms itself into an eigenstate of the measured value, in a way that is 'unaided by procedures which are not part of the measurement act' (ibid., p. 6); or, (b) measurement involves the selection of actual systems from an ensemble of systems so as to ensure the presence of the relevant eigenstate for every system of that sub-ensemble, where this selection may involve operations in addition to the act of measurement.[74]

According to Margenau, (a) is 'literally wrong', whereas (b) holds in most practical circumstances (ibid., p. 7).[75] In those circumstances, the first stage of measurement consists in an interaction between the system and an apparatus but that must be supplemented 'by the act of looking to see what the outcome of the interaction has been, or by some automatic record of the result' (ibid., p. 14). Putnam's concern can now be addressed, insofar as the nature of the interaction cannot be described simply in terms of a time-dependent Hamiltonian, since that will just take the state of the system to another 'stationary' state, for which there would be a definite outcome. The interaction must be such as to 'open' the system up but further, must not depend only on

[73] As we'll see, Everett also referred to such morphisms holding between models and systems.

[74] These two views align with the subjectivist and frequentist interpretations of probability, respectively, and so the debate between their proponents reflects deeper commitments as to whether the state function is a measure of personal knowledge or an objective feature of reality (1963b, p. 7). Of course, by 'objective' here Margenau means in accordance with the best obtainable knowledge (1950).

[75] He states that this argument is presented in section 7 of the paper. There is no section 7.

the coordinates or momenta of that system but on the relevant parameters of the apparatus. It can then be shown that the relevant Hamiltonian will generate an appropriate mixture from the initial pure state of the system (this is presented in Margenau 1963c) and it follows that '[n]o closed physical system can make a measurement upon itself' (Margenau 1963b, p. 15).[76]

3.6.4 The Practicalities of Measurement

In *practical* terms, then, closure is not an issue, since first of all, no finite system is ever really closed and second, for systems of 'superatomic' size, such as a living organism, the relevant stationary states are so numerous within a finite small energy interval, that the best one can do is to assume that they all occur with equal probability—that is, for all practical purposes, one can assume the overall state to be a mixture. Furthermore, it is simply redundant to insist, as the Bohrians did, that the apparatus must be described in classical terms, since QM reduces to classical mechanics in the limit and hence every apparatus that conveys information to us, and in particular our sense organs, satisfies the quantum equations in their limiting form.[77]

In the second half of this two-part paper Margenau presented the proof, mentioned above, that a measurement interaction will take us from a pure state to a mixture, noting that his demonstration follows that already given in London and Bauer (Margenau 1963c, p. 138, fn.7). The upshot is that 'practically', this analysis yields nothing that was not already implicit in the axioms of the theory, at best demonstrating the consistency of these axioms and the 'unique appropriateness' of von Neumann's account (ibid., p. 141).

All of this has only to do with the first stage of measurement, of course, involving the interaction between system and apparatus; '[t]he culminating act is the look one takes and the number one sees' (ibid., p. 141). This act does not affect the state of the system, and is not governed by any law of

[76] Dalla Chiara has compared this issue to that regarding the limits on the semantic closure of any theory as revealed by the paradoxes of set-theory (1977). On her view, the measurement problem arises for logical reasons and thus a 'purely logical interpretation' of von Neumann's thesis could be given, 'which is completely free of any subjectivistic and spiritualistic connotations' (ibid., p. 340). On this account, 'any apparatus which realizes the reduction of the wave function is necessarily only a meta-theoretical object' (ibid., p. 340)—a conclusion that she compares to Gödel's and Tarski's regarding consistency and truth, respectively—but, of course, as she acknowledged, this does not give an explanation of what goes on *physically*.

[77] Interestingly, Margenau suggested that, '[t]he situation in fact is such that if a Compton electron were conscious of its own recoil it could perform a measurement of the energy of a photon with which it collided. But this takes us rather far afield' (1963b, pp. 15–16).

physics, quantum or otherwise. In particular, echoing London and Bauer again, it does not 'cause' any collapse of the wave-function: '[i]t adds nothing to the physical situation' (ibid.). The so-called 'reduction' of the wave-function, according to Margenau, is nothing more than the selection of one component from the mixture produced by the measurement and is no more mysterious than the selection of a monochromatic beam of light from a composite spectrum.

3.6.5 Introspective Orthodoxy

The other reference Margenau and Wigner gave in their reply to Putnam is to Wigner (1963a). This begins in review mode and it is here that we find the comment, 'There is a very nice little book, by London and Bauer, which summarizes quite completely what I shall call the orthodox view' (ibid., p. 7; in the footnote in which he gave the citation to London and Bauer's 'little book' he also referred to Schrödinger's two 1935 papers, in which the 'cat' thought-experiment and the notion of entanglement are presented, respectively—we'll come back to these). Again, Wigner set out here 'the Princeton school' of orthodoxy, in contrast to the Bohrian line defended by Rosenfeld and others.

Thus, in the section titled 'The Orthodox View' he presented von Neumann's two processes, acknowledging that they represent a 'strange dualism' (ibid., p. 7; albeit distinct from the wave-particle variety). As with Margenau, Wigner then demonstrated the consistency of this view, noting that the concern that the processes of the first kind, associated with measurement, might be incompatible with the rest of the theory derives from the apparent impossibility of describing the whole process of measurement in terms of Schrödinger's Equation only. However, he argued, if we analyse the interaction between the object and the apparatus, we obtain a statistical correlation between the states of the two such that the one is mirrored by the other. But then, ascertaining the state of the object reduces to that of ascertaining the state of the measurement device and so the measurement problem becomes that of making an observation on the apparatus. One could of course bring in a second apparatus but in effect the mirroring would continue and, crucially, one would still not have a 'full description' of the measurement since Schrödinger's Equation is deterministic and one cannot recover the probabilistic aspect that is actually observed (ibid., p. 9). And he concluded this section by recalling that 'practically all the foregoing is contained, for instance, in the book of London and Bauer' (ibid., p. 9).[78]

Rejecting views that suggest taking the result of the measurement to be a mixture, from which a particular state vector somehow emerges with the requisite probability, Wigner concluded that there could be little doubt that the orthodox view is correct (ibid., p. 10).[79] And even if the relevant measurement context is made more complicated and realistic, still it is not consistent with the principles of the theory to assume that the end result will be a (proper) mixture of states.[80]

Indeed, Wigner showed, by a straightforward calculation, that in order to obtain a mixture of states as a result of the measurement interaction, the initial state has to be a mixture already.[81] In a footnote he stated that '[t]his point is disregarded by several authors[82] who have rediscovered von Neumann's description of the measurement' (ibid., p. 11, fn. 10). The argument of these authors centred on the claim that the measurement apparatus, as a macroscopic body, must be described by classical mechanics and there are no superposition states in the latter. However, as Wigner pointed out, this runs contrary to QM and here he invokes Schrödinger's 'cat-paradox' (again, to be discussed) to quash the idea that macroscopic bodies must be described in classical terms. The upshot, then, is that it is just not compatible with QM to describe the state of the object-plus-apparatus after measurement as a mixture of states, each with one definite position of the apparatus' pointer. Thus, the orthodoxy, with its dualism regarding changes of the state function, continues to hold sway.

This does not mean that it remains free of conceptual weaknesses. Wigner took seriously the point that most discussions of measurement take place in an abstract and idealized context. Indeed, he showed that no observable that does

[78] As Shimony recorded, 'There are many passages in Wigner's papers... in which this term ['orthodox'] is understood to be the formulation given fully by von Neumann in his *Mathematical Foundations of QM*... and summarized by London and Bauer' (2004, p. 60).

[79] The crucial point, of course, is that one cannot recover from such a mixture the characteristic 'interference terms' that manifest in the behaviour of the beams of particles in, say, the Stern–Gerlach experiment.

[80] Just to recall, a 'proper' mixture is when the system is in one of a set of states, each associated with a definite probability, where it is unknown which state the system is in, and an 'improper' mixture arises when the system is entangled with another so that it is not in any pure state. As Shimony noted, '[t]hat Wigner is fully aware of the distinction between proper and improper mixtures is clear from his citation... of von Neumann's discussion of measurement in Chapter 6 of [*The Mathematical Foundations of Quantum Mechanics*] where the distinction is used extensively but without an explicit terminology' (2004, p. 65).

[81] Shimony described this as '[p]robably the most significant of Wigner's results concerning measurement' (2004, p. 65).

[82] And here he might well be referring to Putnam.

not commute with additive conserved quantities, such as linear or angular momentum or charge, can be measured absolutely precisely and that to increase the precision a very large apparatus is required (ibid., p. 14; a simple proof of this result was provided by his students Araki and Yanase).[83] Most quantities that we are interested in, such as position and momentum, fail to commute with all conserved quantities, so their measurement is actually impossible with a microscopic apparatus. It is in this regard that macroscopic devices may be necessary, in which case, as Margenau also noted, the relevant state vector cannot be distinguished as simply from a mixture as in the case of the Stern–Gerlach experiment. Nevertheless, despite these weaknesses,[84] the point remains, that appeal to processes of the first kind, or 'reductions' of the wave packet, are unavoidable (and he gave another illustrative example in his analysis of a proton-neutron collision).

All of which was to hammer home the point made in response to Putnam:

> a measurement process governed by the standard quantum dynamics ensures that a superposition of eigenstates of the measured object observable is 'inherited' by the composite system consisting of object and apparatus, regardless of how much of the environment is incorporated into the apparatus.... the registration of a measurement result in the consciousness of the subject [is] a definite fact, selected stochastically from the superposition that is exhibited in the final state of object-plus-apparatus.
>
> (Shimony 2004, p. 69)

In this review of Wigner's approach, Shimony both repeated the misleading assertion that von Neumann's conclusion, that a measurement is completed upon registration in the observer's consciousness, was 'formulated explicitly' by London and Bauer[85] and noted Wigner's approving quote of the passage where they refer to the 'characteristic and quite familiar' faculty of introspection, a faculty which, Shimony stated, Wigner 'evidently considers to be a component of the orthodox interpretation' (ibid., p. 61).[86] More generally,

[83] For a summary, see Shimony 2004, pp. 62–3.

[84] And he went on to note the problem of reconciling the orthodox account with relativity theory, given that the observables are typically regarded as instantaneous quantities (1963, p. 14); see also Shimony 2004, p. 64.

[85] Which is a little surprising given his earlier acknowledgment of the phenomenological underpinning of London and Bauer's work as we have seen.

[86] Further revealing his lack of understanding and again surprisingly, given the above point, Shimony wrote, following a brief outline of the 'Wigner's Friend' argument, '[t]hus, the orthodox interpretation, as understood and somewhat amplified by Wigner, is a kind of solipsism' (ibid., p. 62).

Shimony suggested, Wigner's argument above convinced him that 'somewhere in the chain of coordinations linking the physical object of interest to the observer's consciousness there is a breakdown and limitation of the linear, deterministic, unitary dynamics' (ibid.). We'll come back to the nature of such a 'breakdown' after we've examined the London and Bauer text in detail.

Having said that, Shimony questioned whether Wigner's 'solution' to the measurement problem should be identified with the 'orthodox' interpretation (in Princetonian form), noting that elsewhere (as we have already seen), Wigner argued that 'accepting the perceptions of the ultimate subject as the primitive concepts of physics is a drastic over-simplification and flattening of the psychological evidence that points to the deep and murky background of emotions and of the unconscious underlying the sharp conscious readings of apparatus dials' (ibid., p. 62). Shimony presented this as preparation for 'Wigner's remarkably open-minded and judicious exploration of other proposals for solving the measurement problem' (ibid.), but of course, as we have seen, rather than entertain such alternatives, Wigner speculated that we might achieve a unified science of physical and living systems, within which the superposition principle would not be universally valid due to the absolute nature and definiteness of human perceptions.[87]

3.6.6 The Final Word...

Returning (finally!), to Margenau and Wigner's joint response to Putnam, they went on to directly address the concern about applying QM to the 'whole universe':

> Were we to assume that the whole universe of which Professor Putnam speaks includes the observer and that it is meaningful to describe it by the quantum mechanical formalism of states, the conclusions to be drawn would defeat this premise. (Margenau and Wigner 1964, p. 8)

[87] Feyerabend suggested that the indefiniteness in, say, the position of an electron before measurement 'may even reach the mind of the conscious observer making it impossible for him to say that he has received definite information, no matter how certain he himself may feel about it' (1968, p. 318). He objected to Wigner's line on the grounds that the sensory impressions of the observer must have something to do with the state of the electron, 'and here certainty can no longer be guaranteed' (ibid., fn. 24). Of course, this misses the point that Wigner was drawing on from London and Bauer, namely that the faculty of introspection grants that certainty, although, again, to fully respond to Feyerabend's objection requires the phenomenological perspective.

One would have to ask, given such a situation, whether the state in question should be described as a mixture or a pure state. However, if we take it to be a mixture, then the knowledge of the universe would be 'non-maximal'; that is there would be ignorance. But who is it that would be ignorant in this situation? It can't be the observer, since they are included as part of the universe. Thus, it must be a being who is outside of the universe and at this point, 'it is clear that we have now gone far beyond the competence and the intent of quantum mechanics' (ibid., p. 8). Alternatively, we could take the entire universe to be in a pure state, but then the universe cannot make a measurement on itself (and here Margenau's paper sketched above is cited) and so QM does not even apply and Putnam's concerns are moot.

They concluded with the following 'remarkable' and yet-to-be-fully appreciated 'fact': 'present quantum mechanical theory does not recognize any reality independent of an observer' (ibid., pp. 8–9). The choice, as they presented it, is either to formulate the theory so as to refer to observations only, as they indicate in their respective papers, or retain the state function and accept that the changes in this cannot be completely described by Schrödinger's Equation, yielding an unavoidable reduction or collapse:[88]

> We do not say that quantum mechanics is the ultimate physical theory and that all future theories will have a similar character. We do not even maintain that we are glad that the present theory does have this character. However, it does.
>
> (ibid., p. 9)

3.6.7 ... Not Quite

We recall that Putnam subsequently dismissed this apparent dependence of measurement outcomes on human consciousness as 'absurd' (Putnam 1965). In another piece, aimed at a more general philosophical readership, he returned to Margenau and Wigner's position, suggesting that they deviated from the Copenhagen Interpretation 'in a subjectivist direction' by insisting that the observer must include a consciousness and 'treats himself as possessing definite states which are known to him' (Putnam 1965/1975, p. 81). He then noted that the fact that we do not get superpositions on the observer side of the cut is explained by the fact that we have a 'faculty of introspection' (and

[88] Although they do not use these terms, preferring to talk of the changes in the state function containing a 'statistical element unalterably'.

here he cited London and Bauer again) 'which enables us to perform "reductions of the wave packet" upon ourselves' (ibid.)

As Putnam then pointed out, Margenau's own account is actually incompatible with what was presented in his work with Wigner, not least because it abandons the notion of the 'cut' between object and observer and includes the latter in the entire universe whose state, Margenau claimed, could be represented by a statistical mixture (ibid., p. 82).[89] On Margenau's view, we recall, this follows from the claim that pure states can only be assigned to systems on which a precise measurement can be performed. That in turn means that a measurement, considered as 'opening' a previously closed system, must yield a statistical mixture, because after that measurement, the system is in interaction with the rest of the universe, whose state cannot be known exactly and so neither can the system's.

The problem is, Putnam argued, that on this account we cannot be guaranteed that the mixture obtained is the 'right' one. So, jumping ahead to Schrödinger's Cat thought-experiment what is to guarantee that after the box is opened, and the 'measurement' thereby performed, the whole universe, including the cat and me, the observer, will be in a mixture of 'I observe a live cat' and 'I observe a dead cat', rather than one that includes the superposition of 'I see a live cat' and 'I see a dead cat'? Of course, we could always rule out the latter by invoking von Neumann's Projection Postulate but Margenau rejected that, of course, and so Putnam concluded that Margenau's response to the measurement problem is insufficient.

By this point, however, the two sides in the debate had stopped engaging with one another.

3.6.8 Shimony's Additional Concerns

As we also just noted, Putnam referred to related concerns previously raised by Shimony (Shimony 1963).[90] The paper begins with an explicit comparison of the von Neumann–London–Bauer interpretation with Bohr's (Shimony 1963, p. 755; abstract).[91] The former, Shimony claimed, is not supported by

[89] Given Margenau's admission that Wigner took the lead in writing their responses, it may be that he simply decided to let the latter's view have priority.

[90] Moreira dos Santos and Pessoa Jr suggest that Shimony played the role of interlocutor in the debate between Wigner (and Margenau) and Putnam (Moreira dos Santos and Pessoa Jr 2011, p. 631).

[91] I'd like to thank Susann LoFaso of the American Institute of Physics for providing me with a copy of this paper as well as Wigner's 1963 piece (Wigner 1963a).

'psychological evidence' and is 'difficult to reconcile with the intersubjective agreement of several independent observers' (ibid.). The latter may be useful in practical terms but involves a 'renunciation' of any ontological framework in which events, whether physical or mental, macro- or microscopic, can be located. Hence, he concluded, a satisfactory resolution of the measurement problem will not be achievable 'if the present formulation of quantum theory is rigorously maintained' (ibid.).

Thus, the paper is, overall, negative in tone, as Shimony also made clear in the introduction. Here he distinguished two problems with regard to the relationship between physical objects and consciousness: the first is *ontological*, having to do with how two such different kinds of entities can interact; the second is *epistemological* and concerns the issue of how scientific theories can be justified by reference to human experience. Although a complete solution to either requires a solution to the other, Shimony recorded that classical physics had made considerable progress with regard to the second, at least insofar as an understanding of 'the scientific method' is concerned, while leaving the first to languish in obscurity. It was able to do this because fundamental physical concepts could be related to the common characteristics of those objects encountered in daily life, including in the laboratory, which could then be 'directly recognized' by an observer, even though that act of recognition remained poorly understood (Shimony 1963, p. 755).[92] In the quantum context, however, that relation between the elements of theory and experience is no longer extraneous to the physics but is an intrinsic part of the theory itself. (As we'll see, London and Bauer insisted that QM should be regarded as a theory of knowledge.) Here the ontological problem looms large and thus Shimony asked, can it be sidestepped as it was in the classical context? And if not, can a response be given which meshes both with QM, as currently formulated, and psychology? The answer to both, he contended, is negative.

After a brief outline of the theoretical basics, including, yet again, the two kinds of processes, Shimony wrote that, von Neumann's 'most systematic' account of observation was later presented 'more simply (and in some ways more deeply) by London and Bauer' (ibid., p. 757). According to this account, he continued, transitions of the first kind, resulting from a measurement, are understood to be an ineliminable feature of the theory, where 'measurement'

[92] We'll return to this idea of 'direct recognition' of everyday objects when we discuss Husserl's notion of the 'life-world'.

is, in turn, understood to involve the 'registration of the result in consciousness' (ibid.), and here the 'chain' argument was presented again.

Noting, however, that von Neumann himself says rather little about consciousness, Shimony then presented a long quote from London and Bauer,[93] which begins with their consideration of the composite consisting of the system, the apparatus, and the observer and notes that 'objectively', such a consideration seems to be on a par with that of the superposition of system and apparatus. However, they emphasized, from the perspective of the observer, only the system plus apparatus can be considered to be 'objective', since she possesses a faculty of introspection that affords immanent knowledge of her state, thereby cutting the chain. Crucially—and this is something to which we shall return—this passage cited by Shimony includes London and Bauer's rejection of there being any kind of 'mysterious interaction' involved and their insistence that it is 'the consciousness of an "I"' that separates itself from the old wave-function and attributes to the system a new one. His interpretation of these remarks is that here they are presenting 'some important, but incompletely developed, propositions regarding the place of mind in nature' (ibid., p. 759), failing to recognize, at the time, the relevant phenomenological context.

These propositions were extracted and set out by Shimony as follows:

(i) That London and Bauer take the formalism to provide a 'maximal description' of the composite object consisting of the system, the apparatus, and the observer suggests that the last is not given a 'transcendental role' such that either the system or the apparatus could be said to derive their existence from the action of the observer;
(ii) The claim that the observer knows their own state by direct introspection implies that the mind of the observer is included in this maximal description;
(iii) Insofar as at least some of the principles of QM apply to the states of the observer, these states should be taken as capable of entering into a superposition and of supporting meaningful phase relations;
(iv) The laws governing the evolution of the states of the observer are such that the transition to a definite state occurs without any outside disturbance of the composite system, where this transition is effected by a property that only the observer possesses, namely the 'faculty of introspection'.

[93] Shimony later stated that this long passage 'came from reading the book to teach the course at MIT' (Shimony 2002).

As we'll see, although Shimony misunderstood London and Bauer's view, presenting it in terms of the above propositions is useful not least for illuminating the nature of that misunderstanding.

Thus, he took (i) and (ii) above to merely state a kind of ontology, going back to Aristotle, according to which both mental and physical systems exist in nature and interact with one another (Shimony 1963, p. 759). However, leaving aside the fact that London and Bauer explicitly state that there is no such interaction, this fails to grasp their core point about the shift in perspective involved in consideration of a measurement situation: from the outside, as it were, or 'objectively', it appears as if the system, the measurement apparatus, and the observer are on a par, all folded into the description offered by the theory, but from within, again as it were, the observer is in a privileged position by virtue of possessing this faculty of introspection that has to be understood phenomenologically. Given that, it is not at all straightforwardly the case that the observer does not play a transcendental role—something we shall come back to.

It is proposition (iii) that Shimony regarded as 'remarkable', extrapolating as it does, the characteristics of quantum systems to states of mind. (iv) qualifies it, however, by virtue of taking the transition to a definite state to be non-linear and stochastic and thereby not governable by Schrödinger's Equation. This, Shimony agreed, was reasonable, not least because it is difficult to see how we could come up with a Hamiltonian for a mental system—how could energy be expressed via psychological variables?—and also, the process of the observer establishing herself in a definite state, via this faculty of introspection, must involve an element of chance, because 'prior to introspection there were only various probabilities for the observer to be in various definite states' (Shimony 1963, p. 759).

Thus, Shimony interpreted London and Bauer as proposing that the (mental) states of the observer obey the vector relations required by QM, and hence can be in superposition states, but without the usual temporal evolution. Two psychological questions must then be investigated: 'whether mental states satisfy a superposition principle, and whether there is a mental process of reducing a superposition' (ibid., p. 760).[94] He then considered whether a range of psychological phenomena, such as perceptual vagueness, indecision, or conflict of loyalty could be interpreted as instances of

[94] Thus, recalling Margenau's concerns, Shimony viewed London and Bauer as suggesting that QM *does* have some 'competence' in the psychological realm, insofar as it applies to mental states but not, of course, to the ego itself.

superposition, or whether superposition holds in the unconscious, and concluded, in each case, that the answer is 'no'.[95]

Again, however, Shimony missed the point. As we'll see, London and Bauer insist that by virtue of the faculty of introspection the observer can describe her own state 'in an immediate manner'. So, it is not the case that she first has awareness of a kind of perceptual vagueness, say, characterized in terms of a superposition, that then resolves into a definite mental state; rather she can immediately establish her own mental state, thereby snapping the chain and establishing her own objectivity through the attribution of the corresponding state to the system in question. What Shimony overlooked was the crucial shift from the perspective of a second observer, outside the measurement situation, to that of the first observer, within it (it is this that Wigner's 'Friend' argument is concerned with of course).

He did go on to consider the suggestion, which he took to be 'unlike the proposal of London and Bauer' (ibid., p. 763), that the reduction of the superposition takes place immediately at the moment of observation, so that the mind might be regarded as a kind of filter system that selects one definite outcome out of those compatible with the superposition. However, he identified the most obvious weakness of this proposal as 'the difficulty of understanding why there can be no mental states reflecting the states of physical systems in which macroscopic observables have indefinite values' (ibid., p. 763) and concluded that the proposal is nothing but a stratagem for disguising the fact that such peculiar states do not exist. He also raised the objection that from a psychological perspective there is 'probably' no sharp moment at which the observer becomes aware of the given macroscopic variable and hence reduces the superposition. Whether there is or not, this again misses the point, since the crucial element is the awareness by the observer of their own mental state.[96]

Shimony also surveyed other cases of psychological phenomena that could potentially be interpreted as exemplifying superpositions, such as indecision, conflicts of loyalty, and ambivalence. In these cases, the objection above cannot be raised since we do not even know how to begin with the

[95] The concern here can be traced back to Wigner (1961), where he argues that to suppose that a conscious being could enter into a superposition state would be absurd, as it would correspond to a 'state of suspended animation': 'It follows that the being with a consciousness must have a different role in quantum mechanics than the inanimate measuring device... In particular, the quantum mechanical equations of motion cannot be linear' (Wigner 1962, p. 180).

[96] It is typically granted that perceptual experience involves something termed 'presence', in the sense that the experience is immediately responsive to the character of the objects presented to it (see Crane and French 2017—that's not me by the way!).

construction of such an interpretation. Nevertheless, there is no evidence in its favour and, again, the meaning of the relevant phase relations remains obscure.

Given this, Shimony wondered whether an observer might not be consciously aware of the superposition of mental states and whether that superposition could be situated in what Freud called the 'preconscious' (ibid., p. 760). The reduction of the superposition would then presumably occur as the observer's mind moves from a preconscious to a conscious state. However, he noted, there is 'no evidence of causal connection between the superposition of states corresponding to different values of an observable and combination of images in the preconscious' (ibid., p. 761). Hence, he concluded, this suggestion should be dismissed as ad hoc.

3.6.9 Consciousness and 'Q-Shape'

As Halvorson has pointed out, there is an implicit physicalistic assumption that runs through such discussions to the effect that brain states are *identical* with mental states, so that superpositions of the former must yield superpositions of the latter;[97] if this assumption were to be rejected then many of the concerns that have been expressed would fail to get off the ground (Halvorson 2010, p. 157).[98] In addition, the arguments for superpositions of physical states may not carry over to mental states, not least because the explanatory power of the former (with regard to the two-slit experiment, for example) has found little purchase when it comes to the latter (ibid., p. 158). And of course, superposition in QM is a concept that clearly has testable empirical content—the theory tells us which physical states can enter into such superpositions and describes the latter's empirical manifestation. We find nothing similar when it comes to superpositions of mental states, as Shimony also noted. Thus, 'the claim that there are superpositions of mental states cannot be taken to be a serious scientific claim' (Halvorson 2010, p. 159).

Nevertheless, it might be felt that *some* explanation needs to be given as to *why* such states cannot enter into a superposition. Chalmers and McQueen,

[97] For an overview of the inter-relationships between views of QM and those about consciousness, see Atmanspacher 2004. London and Bauer are briefly mentioned, along the well-travelled lines that they went further than the 'cautious stance' adopted by von Neumann and adopted a 'truly radical position' (ibid., p. 60).

[98] In (private communication) Halvorson disavows any support for mind–body dualism as expressed in this paper but states that he nevertheless still finds interesting 'the extent to which we can (or cannot) treat mental states as subject to the same structural laws as physical states'.

for example, have suggested that 'there are special superposition-resistant observables, which as a matter of fundamental law resist superposition and cause the system to collapse onto eigenstates of these observables (with probabilities given by the Born rule)' (2022, p. 20).[99]

One way of cashing out this idea would be to invoke a 'superselection' rule, which would forbid certain observables from entering into superpositions.[100] This would mean rejecting a fundamental assumption made by von Neumann, namely that every self-adjoint operator can be taken to be that of some observable. It was our old friend Wigner who questioned this assumption when he considered the possibility that the operator representing total charge, for example, might commute with all observables, thereby dividing up Hilbert space into superselecting sectors such that linear combinations of states from different sectors are not physically realizable (we recall his comments as noted in fn. 95; for an overview see Wightman 1995).[101]

If consciousness can be regarded as 'superposition-resistant' in this way, then a subject could not be in a superposition of two different conscious states, which would then yield (somehow) the collapse of the physical processes interacting with that consciousness (Chalmers and McQueen 2022, p. 23). There is an immediate problem, however: systems that possess a property corresponding to such a superselection observable will remain trapped forever in one particular eigenstate of that observable. And that may be fine when it comes to observables such as charge, say, but it would be disastrous when it comes to consciousness—we would never wake up from a nap (ibid., p. 27)! One way out would be to take certain observables to be only *approximately* superposition-resistant, so that the superpositions into which they enter tend to collapse over time with a certain probability (ibid., p. 28).

Such an observable might be the *structure* of the integrated information in a system, where this is represented by the property of 'qualia-shape', or

[99] Again, London and Bauer are cited but although they are acknowledged as differing from von Neumann in noting the 'essential role played by consciousness' in the collapse, they are taken only as embodying 'traces' of the view held by Wigner in his '*locus classicus*', (Wigner 1962).

[100] Thus, for example, all particles currently known can be divided into bosons (e.g., photons) and fermions (e.g., electrons). However, we never observe interference resulting from a superposition of bosonic and fermionic states and this is explained via the invocation of such a superselection rule, whose effect is to divide up the relevant Hilbert space into distinct non-combining sectors, corresponding to these different particle kinds.

[101] The context was that of the nature and role of symmetry principles in QM (so, the different particle kinds—bosonic and fermionic—correspond to different representations of the permutation group, with the permutation symmetry operator commuting with all observables). For a nuanced analysis of the different formulations and 'grades' (from weak to very strong) of superselection rules, see Earman 2008 (who also notes similar ideas propounded by Bohm, who was a colleague of Wigner and Wightman's at the time); and for a useful summary of this analysis, see http://www.soulphysics.org/2013/07/what-is-a-superselection-rule/.

'Q-shape' (Chalmers and McQueen 2022, p. 29; see also Tononi 2008; Okón and Sebastián 2020). From a materialist perspective, this 'Q-shape' should be *identified* with consciousness and then we would have consciousness directly yielding collapse. Alternatively, within a dualist framework, 'Q-shape' can be read off the phenomenal character of a given conscious state. So, for example, we could take it to represent the *mathematical structure* of such a state and if it is assumed that there is a phenomenal state with a given Q-shape if and only if there is a physical state with the same Q-shape, then 'phenomenal Q-shape' can be taken to be superposition-resistant and as obeying the same dynamics, thereby again yielding physical collapse (Chalmers and McQueen 2022, p. 32).[102]

Different models involving different (approximately) superposition-resistant properties can in principle be constructed and tested, where these properties can be associated with different systems, from atoms to small organisms, as well as humans and macroscopic measuring devices. The trick, of course, is to get the collapse rate just right: too slow and the model will predict long-lasting superpositions of conscious states that are contrary to our introspective evidence; too fast and when applied to simple systems, the model would bump up against developments in quantum computing. The hope is that further work incorporating both physical evidence and quantum computational simulations will generate constraints that will eventually narrow down the options.[103]

Before we get carried away, however, and just to foreshadow again what is to come, from the phenomenological perspective there really is no problem here: the very act of introspection in effect 'pulls' the 'I' out of the superposition.

3.6.10 Introspection and the Reduction of the Wave-Function

This claim—that it is by virtue of an act of introspection that the superposition collapses—then answers Shimony's second question above, namely whether there is a mental process of wave-function reduction. Of course, the action of this faculty must lie outwith the Schrödinger Equation and hence must be

[102] Chalmers and McQueen suggest that their approach can be generalized to any psychophysical theory linking quasi-classical states to states of consciousness, via some kind of structural isomorphism. The crucial move, they insist, is to combine such a theory, suitably generalized to the quantum domain, with principles governing the collapse of the wave-function, adapted to states of consciousness (2022, p. 40).

[103] Ideally, as Chalmers and McQueen note, a crucial experiment would involve a conscious human being, isolated from environmental effects. But of course, leaving aside any ethical qualms, preparing someone in such a state is technically rather tricky!

stochastic in some sense. In this regard, Shimony acknowledged that there is some evidence for the action of the mind not being governed by causal law and hence it might be reasonable to attribute the reduction of the superposition to that action.[104] However, he objected, no more creativity is felt by an observer when she makes an observation governed by probability, as in the quantum case, than when she makes one that is fully determined and classical.

More decisively, again, there are concerns based on evolutionary theory: if there is such a stochastic factor in play when it comes to the minds of 'higher' animals, then one should expect to see it also when it comes to more primitive entities, unless this factor could be present as a 'structural characteristic' in complex organisms while absent in the components of such organisms. According to Shimony, it is difficult to see how that could be so.[105] However, Chalmers and McQueen's framework offers a response to this concern: we can imagine the emergence of a physical correlate (such as 'Q-shape') that results in a state collapse, albeit with low probability, with consciousness then 'in a position to take hold' (Chalmers and McQueen 2022, p. 55).[106] Having said that, larger issues are obviously in play here, to do with the evolution of consciousness for example, as well as that of whether non-human animals may be said to possess this faculty of introspection (see Allen and Trestman 2020) and if not, what one should say about their 'observations' in a measurement situation (or, to put it bluntly, can Schrödinger's cat reduce the superposition itself?!).

Moving on from what he called the 'dubious extrapolation' of QM to states of mind, Shimony also examined the possibility that the physical system 'is in some sense derivative from the mind or experience of the observer' (1963, p. 762), as maintained by idealism or phenomenalism. As an exemplar of the former he took Kantianism, which he rejected on the grounds that it must face the problem of relating the transcendental mind, from which the universe is derived, with the limited and contingent nature of human beings. Both idealism and phenomenalism have also been less than successful in showing

[104] Here he cited Schrödinger (1958) and Bergson (1944), on how when learning something, for example, repetition leads to it becoming an unconscious activity and how this might be extended to the evolutionary development of such processes as the circulation of blood and breathing.

[105] As he noted, one response here is to take the 'elementary entities in nature [to] have rudimentary mental characteristics'; that is, to adopt panpsychism (see, for example, Goff 2017). In this regard it is worth noting that it is one thing, in this context, to argue that electrons, say, have conscious experiences, even if of a very basic kind, and another to say that they are capable of thought (see Goff, Seager, and Allen-Hermanson 2020)—if the latter is taken to essentially involve psychological attitudes towards certain propositions—such as believing, hoping, and so on—then one might be inclined to insist that a line can be drawn somewhere, if not below 'human animal' then certainly above electrons.

[106] Certain forms of panpsychism could then be ruled out.

how the properties of physical systems can be obtained from combinations of ideas or experiences and Shimony took such difficulties to be indicative of the independent existence of such systems. Finally, he noted that the description of any such combination would be hugely complex, and that stands in stark contrast to the exact laws that govern the behaviour of these systems—a contrast that is rendered less stark if some form of realist stance is adopted (Shimony 1963, p. 763).

Of course, Shimony could only consider such views in the briefest of terms in a paper such as this but it does not take much thought to appreciate how those of an idealist or phenomenalist persuasion might respond—consider, for example, all the recent work on neo-Kantian accounts of science, its laws, and the systems purportedly 'governed' by them (see, for example, Friedman 2013).

3.6.11 Intersubjectivity

Nevertheless, Shimony did raise a significant worry—again, another to which we shall return—namely that of establishing intersubjective agreement in a measurement situation, as when more than one observer opens the box containing the cat, say. It could be argued that the act by which the first observer becomes aware of the value of the relevant macroscopic observable has only a negligible effect upon the state of the apparatus, so that when the second observer takes a look she will note the same value and consequently both observers will come to the same conclusion regarding the state of the system being observed (Shimony 1963, pp. 763–4). Now here Shimony was obviously right in declaring that such an argument is inadequate as it stands, given, of course, the difference between the state described by a superposition and a definite state.[107] Nevertheless, as we'll see, there are ways of ensuring intersubjective coherence here from a phenomenological perspective.

Shimony examined two alternatives: first, that the changes effected by the observers on the state of the measuring apparatus are, for all practical

[107] Likewise, de Broglie, in his consideration of the 'less-admissable' consequences of the London and Bauer approach, argued that the latter's similar explanation is insufficient as it stands, because 'it amounts to confirming the fact that one would like to explain' (1957, p. 31). He goes on to present a dilemma, to the effect that if the wave-function has a subjective character then the 'undeniable agreement' between different observers is only comprehensible if one does not assume the existence of an objective reality; and if such a reality is assumed to exist, then it must be capable of being described by something other than such a subjective wave-function. Of course, the phenomenologist escapes this dilemma with one bound!

purposes, negligible so that it will appear the same to both; and second, that the definite state obtained as a result of the first observer's action is sufficiently stable that it ensures that the second will agree. Option 1 was associated with the likes of Bohm, Danieri, Loinger, and Prosperi and Feyerabend and the core idea is that when it comes to macroscopic observables, there is no practical way of distinguishing the relevant superposition from a *mixture* of states in the right proportions. However, as Shimony pointed out, there are phenomena that reveal 'unexpected coherence' at the macroscopic level.[108] Furthermore, if it is legitimate for the two observers to use the same mixture to describe the state of the system + measurement apparatus before making their observations, this only justifies the claim that they will make the same statistical predictions with regard to an *ensemble* of measurement situations. Crucially, agreement with regard to one specific reading of the apparatus would be a coincidence unless the first observer leaves the arrangement in a specific definite state, thereby effecting a change that is not negligible.

When it comes to the second option, Shimony suggested that this 'appears very reasonable if one accepts the proposal that consciousness is responsible for the reduction of a superposition' (ibid., p. 765) and yet again cited London and Bauer. However, he offered the following thought experiment in response: imagine that both observers observe the joint system + measuring apparatus arrangement by taking photographs of the apparatus. However, although the first observer takes hers before the second, the latter is first to develop and examine her picture. If it is assumed that the reduction of the superposition only occurs when there is 'registration upon consciousness', then for the two to agree it must be that the second observer's action in effect selects a specific image on the film of the first out of a range of possible images compatible with the relevant superposition. But that would require some form of causation in the absence of any physical interaction between the observers.[109]

Of course, the first observer cannot use this to communicate with the second who remains unable to determine whether the reduction of the superposition was due to her action or that of the first. Nevertheless, as Shimony noted, no one seriously maintains that the observations of two separated observers are causally related where this relation is not constrained by

[108] This includes not just the phenomena of superfluidity and superconductivity that London analysed but also, as Shimony noted, the spin-echo and Mössbauer effects.

[109] Shimony sharpened the point by suggesting that the respective observations could occur in different light cones, in which case their temporal order could be reversed, contrary to Special Relativity in which causal relations remain invariant under the Lorentz transformations.

Special Relativity. Granted that the response can be made consistent, he insisted that it remains ad hoc and metaphysically obscure.

Interestingly, Shimony went on to point out that von Neumann also proposed an explanation of intersubjective agreement in terms of his Principle of Psychophysical Parallelism (Shimony 1963, p. 766). We recall that this mandates that it must be possible to describe the observer's subjective perception as if it were part of the physical world, so the dividing line between the observer and the system observed can be arbitrarily drawn and indeed can be 'pushed' arbitrarily far into the 'interior' of the observer. Von Neumann then demonstrated that if the quantum mechanical formalism is applied to the observed part of the world, the predictions obtained are independent of where we draw that line. And, crucially in the current context, he claimed that intersubjective agreement is a corollary of this result, leaving the proof as an exercise for the reader.

According to Shimony:

> von Neumann seems to be asserting that any observer can describe the mental processes of any other observer as if they were physical processes, in other words that one observer can treat all others behavioristically.
>
> (Shimony 1963, p. 766)[110]

The point then, is that as far as the first observer is concerned, the agreement of the second with some result is rendered as equivalent to a 'control reading' using an auxiliary physical device (ibid.). However, Shimony argued, this misses the point about inter*subjective* agreement: if both observers are treated as 'ultimate subjects' and, upon observing a system, independently effect a reduction of the relevant superposition, then their agreement suggests a kind of implausible pre-established harmony! The only alternative is to insist that there can be only one such 'ultimate subject' which is tantamount to solipsism (ibid., p. 767; and here Shimony cited Wigner 1962). As we shall see, there are in fact other options.

Interestingly, Shimony suggested that the 'counterintuitive' conclusions to which von Neumann and London and Bauer were led were the result of a 'rigid distinction' between objectivity and subjectivity (ibid., p. 767). Bohr, by comparison, maintained a certain flexibility in this regard, insofar as he maintained that one cannot talk of the physical attributes of a given system

[110] We'll come across a similar assertion when we consider QBism, in Chapter 9.

without specifying the appropriate measurement context.[111] Intersubjective agreement is then assured by virtue of insisting that the measurement apparatus must be described in classical terms, another aspect that, according to Shimony, clearly differentiates Bohr's approach from that of von Neumann and London and Bauer. This fluidity is revealed in Bohr's application of his notion of complementarity to our own mental activity:

> For describing our mental activity we require, on one hand, an objectively given content to be placed in opposition to a perceiving subject, while, on the other hand, as is already implied in such an assertion, no sharp separation between object and subject can be maintained, since then the perceiving subject also belongs to our mental content.
>
> (Bohr 1934, p. 96; Shimony 1963, p. 770)[112]

From this Bohr concluded that the complete elucidation of a given object requires different perspectives that defy a unique description. Note that this lack of a sharp separation between subject and object follows as a requirement of the possibility of describing our mental activity.[113] Shimony understood this as marking a profound difference between Bohr's view and London and Bauer's, on the grounds that the latter (and von Neumann) insist on the existence of an 'absolute' subject and an 'absolute' object, whereas Bohr does not (Shimony 1963, p. 771).

Nevertheless, it too raises concerns, not the least of which is that if the distinction between subject and object is arbitrary then we appear to have lost an ontological framework in which we can situate the activity of knowing.[114] For Bohr this was a consequence of our dual role as both actors and onlookers

[111] It is this that then comes to be regarded as 'the given', rather than subjective perception, leading Feyerabend to describe Bohr's view as 'positivism of a higher order' (see Faye and Jaksland 2021; for an overview of both the relationship between neo-positivist philosophy of science and the interpretation of QM and also the impact of alternatives to Bohr's account—such as Bohm's—on the development of Feyerabend's methodological pluralism, see Ryckman 2022).

[112] Thus, when one reflects on one's conscious experience, it becomes something other than that which one was reflecting upon (Howard 2013, p. 278).

[113] Bell argued that the fact that QM does not prescribe where and when the subject–object distinction should be made was a serious defect of the theory, rendering it 'intrinsically ambiguous' and 'only approximately self-consistent' (1987). Halvorson has responded by suggesting that this sets the bar far too high in that any theory that could do this would, in effect, 'theorize itself' (Halvorson and Butterfield 2023, p. 306). If there is any ambiguity arising with regard to what the theory is intended to describe, or, that is, in what is to be taken as 'the object', then this is the case for *any* theory and so Bell's concerns here apply to a straw person of his own making.

[114] It is here, again, that Shimony draws a comparison with Kant, suggesting that when applied to the intrinsic characterization of objects, the principle of complementarity generates contradictions analogous to the antinomies of pure reason (Shimony 1963, p. 771, fn. 32).

in the 'great drama of existence' (see Rosenfeld's comment in Jacobsen 2011, p. 387 as noted in fn. 3) but Shimony drew a different conclusion:

> we must try to formulate a view of nature which accommodates all our experience, including our experience of ourselves as onlookers in the world; and we must formulate a theory of knowledge which suffices to provide a rationale for this view of nature. (Shimony 1963, p. 771)

Of course, such a theory is precisely what Husserl offered.

Shimony concluded with a summary: although von Neumann's and London and Bauer's account (singular) can be made consistent, it is *counter-intuitive* 'in the extreme' (ibid., p. 772), faces the *problem of intersubjective agreement* and relies on the mind being endowed with the *power to reduce superpositions* for which there is no empirical evidence. Bohr's account, on the other hand, renounces an 'intrinsic characterization' of fundamental objects in favour of complementary descriptions whose flexibility is bought at the cost of any definite ontology. Perhaps then, he suggested, we should doubt whether a coherent account of observation in QM can be given, without modifying the theory itself.[115] The collapse of the wave-function might then be regarded as a 'small cloud' on the horizon of current physics, akin to the difficulties in explaining black-body radiation, which might 'eventually provide some insight into the mysterious coexistence and interaction of mind and matter' (ibid., p. 773).[116]

3.7 Testing Telepathy

It turns out that some years after this dismissal of London and Bauer's treatment, Shimony *did* acknowledge London's relationship with Husserl's thought and hence that the above criticisms may have been misplaced.

[115] In a footnote he acknowledged that he hadn't, of course, covered all interpretative options and mentioned the Everett interpretation in particular, but took the latter's 'essential weakness' to be its 'extreme violation' of Ockham's principle (ibid., p. 772, fn. 33).

[116] In the acknowledgments he gave thanks to both Wigner and Putnam and also to Howard Stein, with whom he studied QM (among other things) in the 1960s (Shimony undated). Stein also suggested that a 'deep-going' revision to the theory might be required (1972, p. 418), particularly if we allow a role for a 'sentient observer' (ibid., p. 419 and Stein 1982, p. 576; see Pashby 2020). However, ascribing this view to Wigner, he raised concerns about a slide into phenomenalism and insisted that the introduction of the 'contents of consciousness' 'would not clarify the theory, but would make it extraordinarily hard—if not impossible—to formulate' (1982, p. 438, fn. 55).

Surprisingly, this retraction was published in a paper investigating the possibility of a quantum mechanical explanation of telepathy!

The idea that by allowing a role for consciousness, quantum physics could be applied to parapsychological phenomena or situated within certain bodies of mystical or religious thought, is often associated with the culture of the 1960s and 1970s (see, especially, Kaiser 2011). In fact it can be traced back to the very early years of the development of the theory (Marin 2009; see also Zyga 2009 and for a useful overview, Barua 2017). Leading figures such as Bohr, for example, were aware of these connections and explicitly related them to the purported role of consciousness (Marin 2009, p. 809). They also feature prominently in the reflections of Pauli and Schrödinger, as is well known (see Moore 1989, pp. 170–3). The former, for example, maintained that the 'cut' between the subject and object was in fact demanded by consciousness, although where the cut is made remained arbitrary (Pauli 2013, p. 41; Marin 2009, p. 810). Failure to appreciate this, he argued, leads to Western materialism on the one hand, and, on the other, 'Hindu metaphysics' with its 'pure apprehending subject'. The 'Western mind' cannot accept such a 'cosmic consciousness', he claimed, and so adopts this duality of subject and object, which then meshes with the notion of complementarity.[117] Schrödinger likewise rejected the 'Western' view of objectivity that left no place for the mind, arguing that there needed to be a 'blood transfusion' from Eastern thought (1958, p. 130).

Not surprisingly, perhaps, the idea that consciousness might play a role in determining the nature of reality was also appropriated by less mainstream views, such as those associated with 'magical' thinking, for example. So, John ('Jack') Parsons, a founder of the Jet Propulsion Laboratory and, as it happens, adept of Aleister Crowley's 'Thelema' religious movement, thought that 'Crowley's magick teachings seemed to correlate with the work of the "the 'quantum' field folks"' (Pendle 2005, p. 152). In particular, '[t]he illogical nature of the newly coined quantum physics, in which the simple act of observation seemed to affect the physical world... seemed to Parsons to endorse the improbable possibilities of magic and especially the transformative powers of the magician himself' (ibid.).[118]

[117] Pauli's extensive interaction with Jung is well known but for a sketch of their 'dual-aspect' approach to the relationship between quantum physics and consciousness, see Atmanspacher 2004, pp. 67–8 and Atmanspacher 2015.

[118] Parsons' mentor at the California Institute of Technology (which hosted numerous quantum physicists in the 1920s and 1930s including Bohr, Ehrenfest, Einstein, and Schrödinger), was Theodore von Kármán, a leading aeronautics scientist, who studied and taught at Göttingen and was close to Born (with whom he co-authored a series of papers), as well as Bohr and Einstein (Pendle, private communication; see also Born 1971, p. 49; also von Kármán 1962).

The impact of this idea on discussions concerning parapsychological phenomena is even better known. In 1934, Jordan, who, we recall, helped to develop matrix mechanics with Born (with whom he studied) and Heisenberg,[119] submitted to *Die Naturwissenschaften* a paper entitled 'Positivistic Remarks on Parapsychology', urging a reappraisal of this field in terms of his understanding of positivism as taking experience to be the basis of all knowledge, thereby bringing the external and internal, or subjective, worlds together (Gieser 2005, p. 94).[120] In one of the earliest extensions of quantum considerations into biology which foreshadowed Schrödinger's later work (see Joaquim, Freire Jr, and El-Hani 2015), Jordan wrote that, '[o]bservations not only disturb what has to be measured, they produce it.... We compel [the electron] to assume a definite position.... We ourselves produce the results of measurements' (Jordan 1932; translated in Marin 2009, p. 818).[121] According to Marin, '[t]he verb to produce (*hervor rufen*) is the same verb used when a spiritualist group gathers to summon or conjure a dead soul, a "spook", a "phantom"' (ibid.) and this reflects Jordan's burgeoning interest in parapsychology. Marin has also speculated that the likes of Einstein, with his concerns about QM embodying a form of 'spooky' action at a distance,[122] would have read Jordan's writings in the context of the role of the subject (Marin 2009).

Jordan's interest in parapsychology even appears in his 'intuitive' introduction to quantum theory (1936)[123] in which he suggested that the former field had been unjustifiably neglected (Howard 2013, p. 278).[124] In a later

[119] His work on 'transformation theory' influenced von Neumann in his development of the Hilbert space formalism, although the two adopted very different approaches to formulating the theory, with Jordan constrained by the analogy with classical physics, inherited from Bohr (Duncan and Janssen 2013).

[120] The editor asked Pauli for his opinion and the latter urged Jordan to seek a different venue for its publication, suggesting that he get in touch with the psychoanalyst, Jung. The paper eventually appeared in Jung's journal *Zentralblatt für Psychotherapie* (Gieser 2005, p. 94).

[121] Jordan's early interest in biology is noted in an interview with Kuhn; see (Jordan 1963).

[122] This much used (and abused) phrase is the usual translation of 'spukhafte Fernwirkung', although 'spukhafte' might also be rendered as 'eerie'. Although this phrase is now typically associated with quantum entanglement, Hossenfelder maintains that it refers to 'the measurement update', as in the collapse of a single-particle wave-function, which Einstein, in 1927, described as 'peculiar' (Hossenfelder 2021). Howard has noted that even earlier, in 1925, Einstein had referred to a mutual influence between particles of a 'quite mysterious kind', in the context of what is now known as Bose–Einstein statistics (Howard 1990, p. 67).

[123] Norton Wise has argued that it is not so surprising to find such material in an introductory textbook, given that many of the key players in the development of QM looked to psychology to help them understand certain of the concepts they were struggling to grasp (Norton Wise 1994).

[124] Howard has written that, '[w]ith good reason one might say that these last paragraphs of Jordan's *Anschauliche Quantentheorie* [Jordan 1936] represent the *reductio ad absurdum* of his larger philosophical project' (2013, p. 279). Nevertheless, he continues, the book probably did more than any other to establish the connection between Bohr's interpretation of QM and positivism.

note, published in *The Journal of Parapsychology* (Jordan 1951),[125] he pressed the 'chief point', that 'we must abandon the traditional conception of metaphysical reality, as existing independently from any conscious or unconscious mind' (ibid., p. 279). The 'true meaning of reality', he stated, lies in the 'social significance of normal perception', in the sense that by virtue of our perception of a table, we can say that others may see it and from that we may 'induce' that our unconscious mind has similar relations to the unconscious minds of others, yielding a view of Nature and Mind which is 'wide enough to include the empirical facts of parapsychology' (ibid., p. 281).[126]

By the mid-1950s the likes of Pauli also felt able to more freely express their interest in this area, following the work of Rhine in the USA and Soal and Bateman in the UK.[127] And as we noted earlier, Margenau also published on this topic (Margenau 1956, 1957, and 1966), suggesting that researches in parapsychology were comparable to early experimental work on radioactivity, when those physicists interested in this phenomenon had to convince their colleagues that it really existed. In particular, he argued that with recent developments showing how previously unshakeable principles, such as that of causality, had come under doubt, phenomena currently on the fringe could come to be encompassed by the scientific method. And this, he asserted, 'means that the old distinction between the natural and the supernatural has become spurious' (Margenau 1966).

Drawing on his view of consciousness as 'the primary medium of all reality' (ibid.), with the external world a projection that takes on 'ontological existence' after being tested and confirmed in accordance with the scientific method, he argued that even 'ordinary' perception is from the scientific point of view just as mysterious as anything we find in parapsychology. Furthermore, QM has revealed the significance of 'non-material' interactions, such as those involved in measurements, and here he explicitly

[125] I am grateful to John Kurth, Executive Director of the Rhine Research Center for a copy of Jordan's paper.

[126] The well-known Nazi fascination with parapsychology and other 'occult' matters should also be noted, given the influence of Nazi ideology on Jordan's views (see Norton Wise 1994, p. 245; see also Beyler 1996, p. 250 and Gieser 2005, p. 94).

[127] When Jordan acknowledged Pauli's interest in print in 1947, the latter responded angrily, stating that it was no surprise that people who sat in dark rooms started to see things and that he doubted that parapsychological phenomena could ever be proven. In private, however, he expressed a more positive view, if less uncritical than Jordan's (see Gieser 2005, p. 95). The work of both Rhine and Soal and Bateman on parapsychology has since been discredited but at the time, of course, it was widely discussed, even in the august pages of the *British Journal for the Philosophy of Science* (Wasserman 1955).

compared the wave-function with the parapsychologists' 'psi' concept (see Hesse 1961, pp. 295ff.).[128]

Furthermore, the invocation of consciousness to solve the measurement problem by Margenau and Wigner directly inspired members of the 'Fundamental Fysiks Group', who were interested in this connection between physics and parapsychology (Kaiser 2011, p. 169).[129] According to Kaiser, 'Wigner, in turn, commented generously—in public and in print' on a proposed quantum explanation for certain parapsychological 'phenomena', urging consideration of the comparison with quantum entanglement (Kaiser 2011, p. 169).[130] Shimony is also described as having 'dabbled with similar material' (Kaiser 2011, p. 169) but his purported attempt to reproduce telepathy in the laboratory (Kaiser 2011, p. 170), had, in fact, the aim of testing the 'consciousness causes collapse' resolution of the measurement problem.

Thus, the stated intention of Shimony and his co-authors was 'to focus attention upon one of the most radical proposals made by those who take the problem seriously: that the reduction of the wave packet *is a physical event which occurs only when there is an interaction between the physical measuring apparatus and the psyche of some observer*' (Hall et. al. 1977, p. 760). Costa de Beauregard is cited as giving the most explicit statement of this proposal[131] but then they noted that although the same point of view had been attributed elsewhere[132] to London and Bauer:

[128] Margenau went on to team up with a former psychologist, LeShan to write a 'letter to the editor' for *Science*, urging the scientific investigation into ESP (Kaiser 2011, pp. 168–9). The editors' refusal to even acknowledge receipt of the submission, much less publish it, apparently infuriated Margenau and prompted him to co-author a book with LeShan in which they argued that QM offers different possible futures, from which one is selected by consciousness (LeShan 1974; LeShan and Margenau 1982).

[129] This was based on the claim that quantum physics forces us to give up the classical notion of 'objectivity', articulated in terms of 'real physical objects' existing 'out there' independently of our observations (see, for example, Weismann's comments in (Licauco 2014). It should also be noted that the only connection between Margenau and the 'Fundamental Fysiks Group' seems to be that Margenau was editor of *Foundations of Physics*, to which certain members of the group contributed.

[130] The citation given is to 'Discussant's remarks' in the *Proceedings of the 11th International Conference on the Unity of the Sciences* in 1983, on the topic of 'Theoretical and Experimental Exploration of the Remote Perception Phenomena'. However, Wigner was not the discussant here. He did speak but on the subject of 'The Limitations of Determinism' (Wigner 1983) and although there is a brief mention of consciousness there is no discussion of parapsychological phenomena.

[131] This 'explicit statement' is actually just the usual run of the mill association of London and Bauer with von Neumann in holding that the collapse takes place only when the observer takes cognizance of the measurement (Costa de Beauregard 1976, p. 542). Costa de Beauregard studied with de Broglie but came to reject the broadly realist stance associated with the de Broglie–Bohm interpretation, maintaining instead that 'there is no such "thing" as an "independently existing reality", because "observers", human or otherwise, are (largely) generating what they "observe"' (Costa de Beauregard 1992, p. 130; see Dowe 1993). He also claimed that the formalism of modern physics actually postulated the existence of paranormal phenomena such as telepathy, albeit manifesting at a 'liminal' level and hence only perceivable and usable by sensitive and/or trained minds (1992, p. 134; here he cited Wigner as having come to a similar conclusion).

[132] Shimony's 1963 paper as well as Costa de Beauregard's 1976 were cited at this point.

> [i]n view of London's philosophical training as a student of Husserl . . . we now are inclined to believe that the attribution is incorrect and that the passage quoted should be given a phenomenological interpretation. We also believe that it would not be correct to attribute a dualistic ontology to Wigner, since in his most explicit statement he has asserted that the content of the consciousness of the ultimate subject is the only 'absolute' reality.
>
> (Hall et al. 1977, p. 761, fn. 7)[133]

Unfortunately, there is no consideration of how the objections to the role of consciousness made by Shimony himself would fare in the light of such a phenomenological interpretation.

Despite the radical nature of the above proposal, Shimony and Co. insisted there is value in showing how it can be subjected to experimental scrutiny (ibid., p. 761).[134] Their idea was that if the reduction of the wave-function is due to the interaction with consciousness, then with an appropriate experimental set-up in which two observers, A and B, say, interact with the same apparatus, it may be possible for one to send a message to the other via that reduction. However, after running the experiment, the (limited) data they obtained supported the conclusion that 'almost certainly there was no communication between A and B' (ibid., p. 765).

Of course, as Shimony et al. recognized, there are various possible loopholes that could be exploited, including that of simply denying the implicit assumption in their argument, 'that there is a phenomenological difference between making an observation which is responsible for the reduction of a wave packet and making one that is not' (ibid., p. 765). However, they dismiss this on the grounds that it is unconvincing 'without some account of the mind–body interaction which would make it plausible that the psyche can be causally efficacious upon the wave function of a physical system and yet be insensitive to certain gross differences among wave functions' (ibid.).

This paper was subsequently widely cited in further studies of the purported effect of consciousness on physical processes, in which London and Bauer are again presented as adherents of the 'orthodox' interpretation of QM (see in

[133] I wasn't aware of this footnote when I wrote my 2002 piece and as far as I know this and the comments in Shimony's interview (Shimony 2002) are the only acknowledgments in print of the phenomenological underpinnings of London and Bauer's work.

[134] This is of a piece with his attitude towards Bell's Theorem, which was that it provided a rare opportunity for the enterprise that he described as 'experimental metaphysics' (1980, p. 572). This further supports the suggestion that what Shimony and Co. were engaged in here was not an attempt to test telepathy but rather to subject the above 'radical proposal' to 'the same level of control that has been achieved for typical physical hypotheses' (ibid., pp. 572–3).

particular Jeffers 2003, p. 137; for recent summaries see Hu and Wu 2010 and Radin et al. 2012) and has been taken to support the consensus that there is no convincing evidence for any such effect (Jeffers 2003, p. 150; see also Bierman 2003; 2006 and Bierman and Whitmarsh 2006).[135] As it exemplifies, and despite the explicit acknowledgment of London's phenomenological background, the relationship between consciousness and the system continued to be characterized in causal terms within this debate. Thus, in a wide-ranging review of the foundations of QM, in which the possibility of parapsychology is again mooted in the context of the 'problem of wave-packet reduction', and the passage from London and Bauer is given in which they make it clear that there is no mysterious interaction with the wave-function, it was still argued that on such a view, 'changes of human knowledge can modify the physical structure of the system under investigation' (Selleri and Tarozzi 1981, p. 47). This was then characterized as 'a description rather close to parapsychology because of the direct action of thought on the material world' (ibid., p. 48). As a result, it was suggested, the only way to exclude parapsychological effects is to adopt a form of idealism according to which the wave-function describes only the mental state of the observer. However, this results in a kind of *reductio ad absurdum,* as 'the "real world" would become a sort of ghost behind the wall which cannot in any way be known and physics would become only the study of the spiritual activity of man' (ibid., p. 48).

3.8 Conclusion

Perhaps Shimony declined to revise his criticisms following this acknowledgment of London's phenomenological stance[136] because his core concern had to do with how to understand the idea that the mental states of the observer should obey the vector relations required by QM, and hence can be in a superposition. Given the lack of evidence of such superpositions across a

[135] Nevertheless, the debate continues. A useful overview is provided by Okón and Sebastián (2016) who also pointed out the flaws in proposals that purport to show that consciousness is *not* involved in the collapse of the wave-function.

[136] In a comment on Shimony's discussion of realist and idealist tendencies in the quantum context, Ehlers suggested that a Husserlian account of the relation between knowledge and being might be applicable (Ehlers 1971, p. 478). He does not mention the London and Bauer monograph, however. In response, Shimony confessed his ignorance of Husserl's philosophy and drew instead on the work of Merleau-Ponty. However, he claimed that it demonstrates the collapse of phenomenology into either Lockean realism or a form of constructivism. According to Shimony, what Merleau-Ponty exemplified is the fundamental weakness of phenomenology by taking perception as primary instead of—as Shimony preferred—the end-point of evolution (Shimony 1971, pp. 478–80). As we shall see, this is not the most accurate characterization of Merleau-Ponty's view.

range of psychological phenomena, he concluded that the idea should be rejected. Nevertheless, as we have indicated, even without getting stuck into the phenomenological details, the kind of account offered by Chalmers and McQueen suggests ways of meeting this concern. And of course, as we'll explore in more detail, London and Bauer would not have been fazed by Shimony's worries.

For Putnam, on the other hand, the central issue was that London and Bauer's treatment was 'highly subjectivistic' (cf. also Jammer 1974, p. 499):

> *Subjective* events (the perceptions of an 'observer') *cause* abrupt changes of *physical* state ('reduction of the wave packet'). *Questions*: what evidence is there that a 'consciousness' is *capable* of changing the state of a physical system except by interacting with it physically (in which case an automatic mechanism would do just as well)? By what *laws* does a consciousness cause 'reductions of the wave packet' to take place? By virtue of what *properties* [and here in a footnote he acknowledges Shimony as raising this question] that it possesses is 'consciousness' able to affect Nature in this peculiar way? No answer is forthcoming to any of these questions. (Putnam 1964, p. 5)

As we'll see, it is not the case that a phenomenological understanding of London and Bauer's account will provide the answers to Putnam's questions; rather, it will reveal that *these are not the questions we should be asking.*

Despite Margenau and Wigner's protestations, the Putnam–Shimony critique won the day. Indeed, even though London's own biographer covered London and Bauer's account in some detail (Gavroglu 1995, pp. 169–75), and also presented London's philosophical background, the importance of the latter in understanding the former is not considered. Furthermore, too much is conceded there to Shimony's insistence that the account 'rests upon psychological presuppositions which are almost certainly false' (Shimony 1963, p. 772).

Having considered how London and Bauer's 'little book' functioned as a lens through which the von Neumann–Wigner form of orthodoxy came to be viewed, let us now examine the background to their work, before expanding on the correct phenomenological understanding of it.

4
Physical and Phenomenological Networks

4.1 Introduction

In this chapter I want to situate London and Bauer's 'little book' in its historical context. This will cover not only the authors' backgrounds in physics and London's engagement with crucial issues in the foundations of QM, such as Schrödinger's Cat thought-experiment but also his philosophical roots in phenomenological thought.

4.2 Bauer

I shall say comparatively little about Bauer because it appears that there is not, unfortunately, much to say.[1] He wrote his thesis on luminescence and black-body radiation under Langevin in 1912, the first part of which was an exposition of (the old) quantum theory, subsequently published in a volume which included contributions from Bloch, Curie, Langevin, Perrin, and Poincaré (Bauer 1913). Bauer recalled that it was Einstein's work on the photoelectric effect that brought quantum theory to the attention of physicists in Germany (where he went after graduation) and that he himself was convinced of the significance of the theory by Ehrenfest, one of its founding figures (Bauer 1963a).[2]

[1] His entry in the *Dictionary of Scientific Biography* (Massignon 1970) and the obituaries in *Physics Today* (Darrow 1964) and the *Journal de Chimie Physique* (Cauchois 1964; see also Magat 1964) provide only sketchy biographies. He can be seen just behind Pauli's shoulder in the group photograph of the 7th Solvay Congress in 1933 which was on the 'Structure and Properties of the Atomic Nucleus', with Langevin as Chair (see Gamow 1970, p. 125—Bauer's signature is also reproduced).

[2] A useful indication of the state of physics in France at the time is given by Jean Ullmo (Ullmo 1963) who worked with Langevin. He noted that quantum physics only became 'respectable' after the award of the Nobel Prize to de Broglie (Brillouin's presentation of Heisenberg's matrix mechanics in 1925 was apparently received with stunned bewilderment) and recalled the outsider status of Langevin's group, where Bauer was 'tout-a-fait l'animateur du laboratoire de Langevin' ('quite the animator in Langevin's laboratory').

A Phenomenological Approach to Quantum Mechanics: Cutting the Chain of Correlations. Steven French, Oxford University Press. © Steven French 2023. DOI: 10.1093/oso/9780198897958.003.0004

Bauer subsequently wrote a book on Bohr's theory and was the first to lecture on the new quantum physics in France in the 1920s.[3] In 1933 he published an accessible introduction to group theory and its application to QM (Bauer and Meijer 1962/2004; see p. vi; written by Meijer; also Massignon 1970). This was clearly a further common interest of Bauer and London, who was also an early advocate of the 'Gruppenpest', as the likes of Pauli and Schrödinger dismissively called it (Gavroglu 1995, pp. 53–7).[4]

In 1928 Langevin asked Bauer to be 'sous-directeur' of the former's laboratory at the Collège de France in Paris (Massignon 1970, p. 519), where he met London after the latter took up a research position at the Institut Henri Poincaré in 1936 (Gavroglu 1995, pp. 129–35). Although, '[t]hroughout his life, Bauer was keenly interested in the origin and development of the fundamental notions of physics' (Massignon 1970, p. 519)[5] and wrote a number of books on the history of science (ibid., p. 520), there seems to be little evidence that he was particularly interested in philosophical issues, beyond substituting for Langevin in a lecture on the philosophy of quantum theory, where he presented ('something like') Bohr's philosophy (Bauer 1963a).[6]

According to Gavroglu, Bauer never addressed any of the issues raised in the monograph with London either before or after the collaboration (1995, p. 175).[7] Frustratingly, Kuhn mentions that they talked about it briefly but then goes on to focus on the reception of complementarity in France

[3] While visiting Paris the theoretical physicist Peierls met Bauer and found him to be 'a man of exceptional charm' (Peierls 1985, p. 108). He also noted that Bauer managed to flee to Switzerland in time to escape the German occupation—there is actually much more to say about that and about Bauer's and his family's heroism in resisting the Nazis (see 'Bauer, Edmond', *Complete Dictionary of Scientific Biography*. Retrieved 11 June 2018 from Encyclopedia.com: http://www.encyclopedia.com/science/dictionaries-thesauruses-pictures-and-press-releases/bauer-edmond)

[4] According to London's collaborator Heitler, 'the mathematicians had prepared group theory so well for the use of the physicists without knowing it that sometimes I could just copy word for word pages from a group theory paper and use them for my purposes' (Heitler 1963).

[5] Ullmo records that 'Someone like M. Edmond Bauer, who had put before himself the question of understanding the quantum, because he had made his thesis on the question, was absolutely unique. The consequence was that he never made any career. He was perhaps the most brilliant of the young physicists at the time, but they never offered him any chair, and he had great difficulty in having a career at that time. He had to stay as Adjoint de Langevin because his interests were outside the general routine. That was exactly the atmosphere at that time; it was very stuffy' (Ullmo 1963).

[6] Langevin's own view was that QM's real impact was not on the notion of determinism, but on that of *mechanism*: the standard representation in terms of points and forces was just not adequate in the new domain and, instead, Langevin 'thought that the real images were to be taken from membranes tendues [stretched membranes]' (Ullmo 1963).

[7] In support of a non-phenomenological understanding of the London and Bauer piece, Bueno has argued that it is important to offer an account that accommodates the views of both authors and 'Bauer probably would be more sympathetic to a philosophically minimalist account of his work with London than one that adds substantial philosophical assumptions to the approach' (Bueno 2019, p. 134). We'll come back to Bueno's interpretation in Chapter 6 but even granted his point, given that Bauer appears to have had little interest in philosophical issues, that leaves London's stance as even more starkly highlighted.

which Bauer assured him was no different than elsewhere (Bauer 1963b). However, it appears that the crucial section we shall be concerned with ('Mesure et observation. L'acte d'objectivation') was written primarily by London (Jammer 1974, p. 483).[8]

4.3 London: Philosophical Roots

London, by comparison, was an internationally renowned physicist who produced a series of notable applications of QM to a wide range of phenomena, including the nature of chemical bonds and valency in general (with Heitler), inter-molecular forces, superconductivity (with his brother, Heinz London)[9] and superfluidity.[10] In 1953, the year before he died, he was awarded the Lorentz Medal for this body of research. Significantly, however, London brought to this work an acute and well-formed philosophical sensitivity which he had begun to develop prior to his studies in physics (for further details see Gavroglu 1995 and also Jammer 1974, pp. 482–3).[11]

His early essays, written over a period covering his final year of school and the first year of university, reveal Kantian and phenomenological themes concerning the coordinative relationship between an object and its condition of existence and the gap between experience and the laws of physics (Gavroglu 1995, esp. pp. 8–23).[12] It is also here we find early indications of a two-stage methodology that he subsequently applied to his scientific research more generally (Mormann 1991); first, reality must be translated into that which

[8] Darrow quotes the following passage sent to him (by whom? We are not told) from France: 'The work of Edmond Bauer is actually much more important than what has been published under his name. His extraordinary generosity led him to devote a great deal of his time to helping other investigators, some young and some not so young, some of them pupils of his, others who were barely known to him. Also and primarily he was a teacher who strove to pass on to the young something that he had learned from Langevin and Perrin his own masters: the love of work done well down to its least details' (Darrow 1964, p. 87). This may capture an aspect of Bauer's relationship with London, given the twenty-year difference in their ages.

[9] For a discussion of this as an example of theory construction, see French and Ladyman 1997.

[10] For a consideration of the latter see Bueno and French 2018, ch. 5.

[11] Heitler wrote, about London, 'He really started as a philosopher; he studied philosophy first before he changed to physics. And his interest was, even more than mine, on philosophical lines. I remember that his interest in theoretical physics was also perhaps broader than mine… He was very much interested in philosophy, and he took physics, perhaps even more than I did, as a tool to a more philosophical outlook on the world' (Heitler 1963).

[12] His school essays also reveal some interest in physics even then, with indications that he was 'engaged in hands-on experimentation with spark discharges and oscillators' (Heims 1991, p. 179). Heims has speculated that '[h]is father's death may have provided a strong impetus for Fritz London's reflections and his turn away from science toward epistemology and philosophy of science' (1991, p. 180).

we experience and then, second, the latter must be itself transformed to what we choose to express in terms of scientific laws. He also adopted a form of anti-reductionism that rejected explanations of the behaviour of a system in terms of the equations of motion of its constituents. Instead, he argued, it may be the macro-level that is the more interesting, foreshadowing his later work on superconductivity (Heims 1991, p. 181).

Despite these proclivities, London always insisted that he went to Munich to study *physics* with Sommerfeld, rather than philosophy.[13] Nevertheless, while there, he met Pfänder, the leader of the Munich group of phenomenologists and at the time second only to Husserl within the phenomenological movement (Gavroglu 1995, pp. 11–12).[14] London showed Pfänder an untitled essay on the 'logical interpretation of deductive theory', in which he argued that the question 'how is theoretically formed knowledge possible?' should be replaced with, 'assuming theoretical knowledge possible, how is it obtained?' (Heims 1991, p. 182). Pfänder was evidently so impressed that he urged him to write it up and submit it as a dissertation in philosophy.[15] Pfänder was influenced by Lipps' psychological theory of empathy which in turn, 'was influential on London's ideas about the measurement process in quantum mechanics' (Jammer 1974, p. 483). Furthermore, while at Munich, London took classes from Becher who insisted that the mind–body problem was central to metaphysics (ibid.) and advocated a form of mind–brain 'interactionalism' (ibid., p. 484). Thus, according to Jammer, 'London... found in quantum mechanics a field where he could meaningfully apply Lipps' and Becher's philosophy' (ibid.).[16]

[13] Sommerfeld is best known for extending and further developing the Bohr model of the atom. Despite never receiving the Nobel prize himself he was famous for mentoring and supervising four who did, as well as many others who became famous in their own right. Seth has noted how Sommerfeld adopted an explicitly model-oriented approach to quantum phenomena: 'One began with the data, derived empirical laws from them and then sought a model that might produce (or at least reflect) such empirical regularities' (Seth 2010, p. 244). This can usefully be compared with London's two-stage phenomenological approach to theory construction.

[14] Crucial differences between the two emerged after Husserl published his *Ideas*, in 1913, regarded as a major turn towards transcendental idealism which the Munich group resisted.

[15] According to Gavroglu, 'What London was thinking programmatically in 1921 was very close to Husserl's thoughts. In this sense London's problematique was not marginal at all' (1995, pp. 13–14).

[16] Here Jammer cited the much-quoted passage from London and Bauer concerning the 'quite familiar' faculty of introspection (Jammer 1974, p. 484). By giving Becher's interactionalism as a possible source for London's view, Jammer made the same mistake as Wigner in failing to recognize this view's phenomenological origins. Heims, on the other hand, while also recording the emphasis on the 'act of introspection', then shifted his focus to London and Bauer's considerations of the establishment of intersubjective agreement (which we'll come to) and noted that, 'The London–Bauer analysis, including the awareness of a "community of perception", is related to work of the phenomenologists Gurwitsch and Schutz' (Heims 1991, p. 183). Gurwitsch was a friend of London's and we shall say more about him later. The connection with Schutz is less clear (neither are cited in the London and Bauer manuscript) but he became Gurwitsch's closest friend and his book, *The Phenomenology of the Social World*, published in 1932, was praised by Husserl himself (Barber 2022). Heims went on to note that

However, Gavroglu has vigorously rejected these claims (1995, p. 36 and p. 179), arguing that by the time London and Pfänder met, the latter had turned away from Lipps' psychologism and furthermore, there is no evidence that London adopted Becher's 'interactionalism'. Indeed, the explicit philosophical attributions in the London and Bauer monograph are rather different (but, as I shall argue, they are more extensive than even Gavroglu realized). Also, London's thesis was published in the *Jarbuch für Philosophie und phaenomenlogische Forschung*, which was co-edited by Pfänder, with Husserl as editor-in-chief. Hence, there is good reason to conclude that '[t]he dominant features of Fritz London's thesis place it within the phenomenological movement' (Gavroglu 1995, p. 15).[17]

This thesis was concerned with an issue that occupied London throughout his life, namely that of *how we are to conceive of theories*. Here he presented them as mathematical frameworks enmeshing some given 'volume' of fact (Everitt 1996). These frameworks were regarded as closed axiomatic systems that could be compared as to the 'size' of the volume of fact covered and the closeness of the meshing: thus, Einstein's theory of gravitation is better than Newton's, in this regard (ibid.). Interestingly, Mormann considered this to be 'a set-theoretic concretization of Husserl's largely programmatic account of a macrological philosophy of science' (Mormann 1991, p. 70).[18] Within this framework London then developed a 'relational calculus' that allowed him to define the product of relations and 'concatenation laws' by which new relations could be obtained from old ones. This enabled him to characterize theories in terms of the set of their (partial) models, written in modern terms as $\langle A, R_1 \ldots R_n \rangle$, where A is the relevant set of elements and $R_1 \ldots R_n$ are the relations that can be defined over such a set. The content of theories could then be compared, as indicated above, via the set of all consequences of the relevant propositions.[19]

London's scientific research 'would throughout be informed by conscious philosophy' (1991, p. 183). (Meta-comment: although Heims is acknowledged in Gavroglu's book and his interviews (with Edith London and London's post-doctoral fellow Zilsel) are both cited, this 1991 paper is not.)

[17] It is also worth noting the contents of London's personal library, which included the *Collected Works* of Leibniz, Husserl's *Logical Investigations and Ideas*, Cassirer's *Substance and the Conception of Matter* and *The Philosophy of Symbolism* as well as works by Russell and Hegel among others (Gavroglu 1995, p. 36).

[18] Cf. the 'semantic' or 'model-theoretic' approach to theories, regarded by some as currently dominant in the philosophy of science. This concern with how we should characterize theories also appears in Everett's work as we shall see.

[19] Mormann has described London's approach as an 'informal predecessor' of the work of Tarksi and other Polish logicians and compared his analysis of theory content to Popper's later considerations of verismilitude (Mormann 1991, p. 71). However, both Husserl and London over-reached in asserting

The thesis appeared in the *Jahrbuch fur Philosophie und Phenomenologische Forschung*[20] two years after he received his degree from the University of Munich in 1921. After spending a year as an assistant science teacher (Gavroglu 1995, p. 27), London decided to go to Göttingen[21] and work with Born.[22] However, although he wanted to work on the philosophy of the new quantum physics, Born, who at that time was 'very much opposed to philosophizing' (Gavroglu ibid., p. 27; Heims 1991, p. 182), insisted that he do some 'real work'.[23] London balked at this and so he ended up with Sommerfeld who 'persuaded him by the force of his personality to do a very simple and straightforward calculation. I don't know what it was,[24] but he got his thesis and he never became a philosopher again' (Born 1962). As we'll see, that is not quite true!

4.4 London: Physics

After Munich, London obtained an academic appointment at Stüttgart, where he worked on transformation theory,[25] and also tried to cast Weyl's early attempt to unify gravity and electromagnetism into quantum theoretic terms (Weyl 1918). This is significant because Weyl had also adopted an explicitly phenomenological stance towards physics (Ryckman 2005, ch. 6; see also Wiltsche 2021).[26] At the core of this was the claim that '[t]he world exists

the formal completeness of theories (for London the possibility of characterizing the domain of the theory relationally was taken to be a sufficient condition for this)—within ten years of the publication of London's thesis Gödel had produced his incompleteness theorem (ibid., p. 72).

[20] At this time the journal was edited by Pfänder, rather than Husserl. London's work was initially sent to Geiger for revision but as the matter was urgent and Geiger was not available at that time, it was passed on to Pfänder (Alves 2021, p. 455, fn. 6). Husserl went on to inform Ingarden that the new issue of the *Jahrbuch* would soon appear with two 'mathematical-philosophical' contributions, evidently referring to the works by Becker and London (ibid.).

[21] According to Nordheim, London switched from philosophy to physics because 'it was an exciting time' (Nordheim 1962). As part of an exercise regarding the 'social anthropology' of quantum theorists of the time, Heims has noted that 'Fritz London's career is unusual, in that he had received his doctorate in philosophy (not physics!) and turned to physics only after several years' work as a professional philosopher' (1991, p. 179).

[22] Who was a student of London's father, a professor of mathematics (Heims 1991, p. 178).

[23] Born dismissed phenomenological reflection as 'a kind of "a-priorism", not a rational one, like that of Kant but a mystical one... If science stands for anything it has certainly no use for Husserl's philosophy' (Born 1978. p. 96). He was also less than keen on the group theoretical approach which underpinned London and Heitler's work on the quantum mechanical account of chemical bonding (ibid., p. 56).

[24] It had to do with the intensity of band lines in spectra (Gavroglu 1995, p. 27).

[25] See ch.2 fn. 62.

[26] McCoy has argued that this claim is 'seriously in error' and that not only were Weyl's primary intellectual influences drawn from the Göttingen mathematical tradition but also his justification for adopting a pure infinitesimal geometry is actually in conflict with the basic principles of phenomenology (McCoy 2022, p. 191).

only as met with by an ego, as one appearing to a consciousness' (Weyl 1934, p. 1).[27] Furthermore, Weyl maintained, it is through an *act of self-reflection* that the ego comes to realize that it has a function as a 'conscious-existing carrier of the world of phenomena' (Bell and Korté 2016). This yields 'what might be called a *polarized dualism*, with the mental (I, Thou) as the primary, independent pole and objective reality as a secondary, dependent pole' (ibid.).

This stance was evident, at least in embryonic form, in Weyl's now classic analysis of the foundations of space–time (Weyl 1918). Upon receiving a copy of the third (1919) edition of this book, Husserl felt compelled to write to him, exclaiming, 'How near this work is to my ideal of a physics permeated by a *philosophical spirit*' (quoted in Ryckman 2005, p. 112).[28] Here, in his attempt to construct a unified theory of gravitation and electromagnetism (see Ryckman 2005, ch. 6; especially pp. 81–5 for a summary), Weyl argued that characterizing Einstein's theory of General Relativity as representing an objective spatio-temporal reality masks the fundamental issue of how it is possible to assign labels to the points of a continuous manifold, which can serve for their identification, thereby establishing a coordinate system (Weyl 1949). It is only after we have achieved such a labelling that we can 'think of representing the spectacle of the actually given world by construction in a field of symbols' (ibid., p. 75). The solution is to lay down a coordinate system or frame of reference, but this must be 'exhibited by an individual demonstrative act', in effect establishing that the observer is 'here, now' (ibid.). It is because of the necessity of such an act that the objectification inherent in science's representation of the world does not completely succeed and the coordinate system is thus understood as the 'necessary residue' of the elimination of the ego (ibid., see also p. 123; for elucidation of this claim, see Wiltsche 2021).[29]

However, the theory that Weyl tried to construct on this basis was strongly criticized by Einstein and Pauli. Weyl himself eventually realized that it offered little in the way of new empirical results and came to view it as less of a

[27] The standard view is that Weyl began with such a stance but then shifted to a symbolic constructivist view under the influence of Hilbert; for a counter to this, see Baracco 2019.

[28] Weyl's relationship with Husserl is documented in detail in Tonietti (1988). As is well known Weyl's wife was a student of Husserl's although it appears that Weyl was aware of the latter's work before he met her (ibid., pp. 376–7). McCoy (2022) makes no mention of this relationship or Husserl's remark above.

[29] According to McCoy, Weyl's insistence that the intuitions on which physical representation are grounded must be restricted to the 'infinitesimal neighbourhood' of the ego-centre represents a fundamental misunderstanding of Husserl's view when applied to the objects around us, leading to the latter being regarded as 'private mental items' (McCoy 2022, p. 202). However, this criticism confuses the world to which physical representation is appropriate with the 'life-world' and the principles to be adopted for constructing a theory with regard to the former should not be applied to entities in the latter. We shall return to this issue in Chapter 7.

hypothesis and more of a summary or interpretation of our knowledge of field physics, before finally abandoning it altogether (Ryckman 2005, p. 168).[30] Reflecting on this episode, London felt that Weyl was primarily motivated by the metaphysical conviction that nature would not have ignored the 'beautiful possibility' that lay at the heart of his theory (Gavroglu 1995, p. 32), an approach that clashed with London's own phenomenologically driven two-stage methodology of theory construction.[31] In line with the latter, London realized he could correct the flaw in Weyl's work by introducing the quantum wave-function with its non-classical complex amplitude. This foreshadowed the development of what would come to be known as the gauge principle, a fundamental feature of modern elementary particle physics (Gavroglu ibid., p. 33).[32] It was also an early indication of London's application of 'purely' quantum concepts that had no classical analogue (ibid.).[33]

In early 1927 he moved to Zurich to work with Schrödinger and then followed the latter when he was appointed to the Chair of Theoretical Physics in Berlin. It was in Zurich that London collaborated with Heitler to give the quantum mechanical explanation of hydrogen bonding leading the latter to famously and contentiously declare, now '[w]e can … eat Chemistry with a spoon' (Gavroglu 1995, p. 54).[34] This relied on the novel idea of

[30] Referring to Becker's contribution to the phenomenological yearbook, Husserl wrote to Weyl that, '[i]t is nothing less than a synthesis between Einstein's and your discoveries with my Naturphänomenologischen researches. It aims by deep and original means to prove that Einstein's theories, but only when they are completed and recasted through your researches in Infinitesimalgeometrie, represent that form of the "structural lawfulness" of nature (as opposed to the "specific causal" lawfulness of nature) which must be presumed necessary on the deepest transcendental-constitutiven grounds: which therefore (in their form) is unique possible and comprehensible. What Einstein will say, when it is proved that a nature seeks a relativistic structure on the a priori grounds of phenomenology and not on positivistic principles and that only in this way a fully comprehensible and exact science becomes possible' (Tonietti 1988, p. 370).

[31] In this respect London might have agreed with McCoy (2022) that Weyl did not adopt a coherent phenomenological approach to theory construction. This is not to say that McCoy's critique of Ryckman's analysis of the phenomenological character of Weyl's work is correct, however.

[32] Weyl subsequently recognized the significance of the phase factor in a work that made no reference to London, although Pauli cited both authors (Gavroglu 1995, p. 33).

[33] Schrödinger had earlier studied Weyl's book, *Space–Time–Matter* and in a 1922 paper presciently, but rather casually, introduced the imaginary number i into the 'Weyl factor'—which gives the change in length of a vector when subject to congruent displacement—as applied to the Bohr orbit of an electron in a hydrogen atom. Four years later, London wrote a 'playful' letter in which he ribbed Schrödinger for not making more of this suggestion and realizing that he had demonstrated that 'Weyl's theory becomes reasonable … only if combined with quantum theory' (Moore 1989, p. 148).

[34] Heitler learned about quantum theory while still at school by attending evening lectures at the Technische Hochschule in Karlsruhe where it was presented as sitting at the intersection of physics and chemistry (https://www.aip.org/history-programs/niels-bohr-library/oral-histories/4662-1). He too had philosophical interests, attending a seminar on the theory of knowledge, for example. After Karlsruhe he went to Berlin where he took courses on group theory and subsequently met London in Zurich, which he describes as 'a decisive turning point in my career' (ibid.). After being accepted as Born's assistant in Göttingen (where he occasionally lectured on group theory and its application to QM), Heitler became 'very well versed in what other people called the "Copenhagen spirit"' (ibid.).

electron 'exchange' and, as with spin, the lack of any classical analogue was emphasized.[35] This work is generally regarded as marking the birth of the field of quantum chemistry (Gavroglu, ibid., pp. 44–9; see also Heitler 1963).[36]

London then turned his attention to inter-molecular forces and as we've already seen, his ideas were further extended by Margenau, visiting Berlin on a scholarship. While in Berlin, London was invited by Reichenbach to give a talk at the *Gesselschaft für empirische Philosophie* and it was suggested that the resulting paper, 'The Philosophical Problems in Quantum Mechanics' might be published in the first issue of the journal, *Erkenntnis*.[37] However, London declined,[38] apparently reluctant to be associated with the new logical empiricist movement (Gavroglu 1995, p. 61).[39]

With the rise of the Nazis, London and his wife moved to Oxford, where he shifted his focus to low-temperature physics. Here the influence of philosophy on his science became much more apparent and his phenomenological approach to theories effectively shaped the account of superconductivity that

[35] Heitler and London went on to pursue these developments using group theory, with Heitler, in particular, influenced by Weyl and Wigner at Göttingen. Wigner was apparently impressed with London's papers during this time (Gavroglu 1995, p. 51), although he was sceptical that the whole of chemistry could be 'eaten' as Heitler had exclaimed (ibid., p. 54). London himself hinted at a non-reductionist stance in some of his own papers, based on his philosophical inclinations (ibid., p. 74; for the differences between Heitler and London see ibid., pp. 91–2).

[36] As Born noted, with some dismay, it was Pauling who received the Nobel Prize in Chemistry for the quantum explanation of chemical valency and not Heitler and London, although Bauer nominated them both in 1950 (Born 1971, p. 115; see Nomination Archive. NobelPrize.org. Nobel Prize Outreach AB 2022, 20 July 2022 <https://www.nobelprize.org/nomination/archive/show.php?id=11,694>) London was commissioned by Springer to write a book, *Quantenmechanik und Chemie*, on 'the significance of quantum mechanics for chemistry'. Although it was begun in 1929 and was based on lectures presented at the University of Berlin, it was never completed (Gavroglu 1995, pp. 69–71). The draft manuscript emphasizes the role of symmetry and contains an appendix on group theory (ibid., p. 70). London ended up being nominated four times for the Nobel Prize in chemistry and once for the prize in physics, jointly with Landau, in 1954 (the year that Born was finally awarded it), by Robert Marshak, well known for his work that paved the way to the unified electro-weak theory (he also had an undergraduate background in philosophy. And was a dance critic).

[37] Unfortunately, a copy of this paper could not be located in the archives held at Duke University (but I'd like to thank Brook Guthrie, the Research Services Librarian there, for looking for me).

[38] Kojevnikov has recorded that while in Berlin, London followed Schrödinger in embracing wave-particle dualism and that his 'lectures on wave mechanics [in 1928–9] ... opened with a programmatic statement on the dual (wave and particle) nature of quantum objects' (2020, p. 94).

[39] Reichenbach's own contribution to this issue begins with a lament on the '[a]lienation between the world of science and the world of everyday life' (Reichenbach 1930, rep. in Reichenbach 1978, p. 304) that is strongly reminiscent of Husserl's concern in *The Crisis of the European Sciences*. Of course, their responses were very different, with Reichenbach presenting an epistemological critique of apriori concepts and arguing that science requires a shift to a different conceptual framework than that which is appropriate for the 'intermediate dimensions' of the everyday level. He also went on to write a well-known book on the foundations of QM (1944) but as Glymour and Eberhardt note, 'the missing piece in Reichenbach's discussion of the theory is the measurement problem' (Glymour and Eberhardt 2016). Likewise, there is no mention in Reichenbach's earlier writings, although they do offer glimpses into the emergence of QM (see for example Reichenbach 1927 which mentions both matrix and wave mechanics, albeit with more discussion of the latter), as well as tackling issues to do with causality, determinism, and probability (see, for example, Ryckman 2022, p. 780).

he developed with his brother Heinz.[40] The model that they constructed is often described as 'phenomenological' (but not in the sense we are concerned with here; Gavroglu 1995, p. 118).[41] London himself, however, rejected that description and insisted that the model should be described as 'macroscopical' since it goes beyond the data. This was in line with his view of theory construction, whereby the phenomena must first be formulated in a particular fashion and then embedded in an explanatory framework (Gavroglu 1995, pp. 127–8). It was precisely because superconductivity, as a phenomenon, had been represented inappropriately to begin with that it had proven so difficult to accommodate within the framework of quantum theory (for further details see French and Ladyman 1997; Bueno, French, and Ladyman 2002 and 2012).

In a letter to a fellow physicist London described his approach as 'mainly a logical one', in which by adopting 'a new and more cautious interpretation of the facts' the fundamental difficulty of understanding superconductivity could be overcome (Gavroglu 1995, p. 129). In this move the 'macroscopic' nature of the theory was crucial, insofar as superconductivity was seen by London as, again, a uniquely quantum mechanical phenomenon of long-range order (Gavroglu 1995, p. 144).[42] It was this feature which offered ' an entirely new point of view for a theoretical explanation' (London and London 1935, p. 71). And in his 1935 Royal Society presentation (London 1935), he provided a 'sketch' of such an explanation, elaborating on the concluding remarks of the joint paper where crucially, the suggestion is made for the first time that the electrons in a superconducting material are coupled in some way. This then became a 'valuable heuristic' for subsequent developments, with the idea of coupled electrons expressed in the concept of 'Cooper pairs' (Gavroglu 1995, p. 209; see https://en.wikipedia.org/wiki/Cooper_pair).

[40] According to Gavroglu (1995, p. 110), Heinz liked to regard himself as an experimental physicist but in fact made both theoretical and experimental contributions. He was one of the last Jews in Germany to be awarded a degree and soon after also moved to Oxford (for an amusing description of Heinz London's status in the process see Kurti 1968). Neither brother obtained a permanent job there nor until comparatively recently were they commemorated on the hallowed walls of the Clarendon Laboratory (Blundell 2011; photos of them have now been put up (Blundell—personal communication)). Heinz eventually moved to Bristol (where Heitler also ended up, after fleeing from Göttingen) whereas Fritz went to Paris.

[41] The term is often used within the philosophy of science and has different meanings, depending on the context but here it can be understood quite broadly as a descriptive, rather than explanatory, account of the (suitably processed) data. According to Everitt, however, '[t]he Londons' theory was "phenomenological" in both the common and Husserlian sense. It was descriptive, not explanatory; and it involved an apriori leap ([in] setting a constant of integration to zero) justified "an der Sache selbst"' (Everitt 1996, p. 1274). Husserlian phenomenology is also sometimes considered to be 'merely' descriptive, but that is a mistake. And, as Everitt himself goes on to acknowledge, the 'phenomenological' stage of theory construction was only preliminary in London's view.

[42] This emphasis is obviously significant in the context of the measurement problem and the emergent separation between the two forms of the 'orthodox' approach, noted previously.

The idea of the macroscopic model going beyond the (physically) 'phenomenological' and extending into the 'theoretical' but leaving the latter open to further elaboration and development,[43] also explicitly appears in London's classic book *Superfluids* (cf. Gavroglu 1995, p. 143; London 1937, p. 795; cf. 1950, p. 4).[44] This represents the culmination of London's research that began during his last year at Oxford when he became interested in what he later called the 'mystery' of the 'liquid degeneracy' of liquid helium (Gavroglu 1995, pp. 147–8). Again he sought a quantum resolution that had no classical analogue and proposed the application of Bose–Einstein statistics (Bueno, French, and Ladyman 2002 and Bueno and French 2018, ch. 5).[45] Originally developed by Bose in order to obtain a more transparent derivation of Planck's black-body radiation formula that initiated the development of quantum physics, Einstein then applied this novel form of quantum statistics to material gas atoms (Einstein 1924), noting that, '[f]rom a certain temperature on, the molecules "condense" without attractive forces' (Pais 1982, p. 432). He subsequently suggested that the statistics expressed 'an implicit hypothesis about the mutual influence of the molecules of a totally new and mysterious kind' (Einstein 1925; trans. in Duck and Sudarshan 1997, pp. 91–2). For Einstein this mysterious influence could be understood through de Broglie's hypothesis of matter waves, which was an important precursor to Schrödinger's wave mechanics.[46]

London expanded on Einstein's suggestion by not only presenting a proof of this Bose–Einstein 'condensation' but also constructing a 'highly idealized' model that he showed could explain the bizarre behaviour of liquid helium (1938a; 1938b; see also 1954, esp. p. 59).[47] He also argued that the 'condensation' should not be thought of as taking place in ordinary space, since the particles

[43] In a draft of a letter from 1946 London wrote: 'Any macroscopic theory has in general to go beyond the strictly phenomenological data—the same is for instance the case in my "macroscopic" theory of superconductivity which also was suggested by certain molecular ideas, but, of course, not based on them' (Gavroglu 1995, p. 203).

[44] For a review of Vol. I on superconductivity, see Hudson (1951); and for one of Vol. II see Kirkwood (1955).

[45] For an insight on how London was made aware of Bose–Einstein condensation, see Tisza 1988 and fn. 3 in particular; for interesting comments on the macro–micro distinction, see fn. 11. Tisza learned about QM while in Göttingen but was initially more mathematically inclined and, after taking Heitler's course on group theory, also at Göttingen, applied the latter to molecular spectra for his PhD research in Budapest. In 1937, through the help of London and Bauer, Tisza obtained a scholarship to work at the Collège de France.

[46] It is interesting to note that, in a response to Einstein's work which was subsequently abandoned, Schrödinger adopted a 'holistic' approach which attributed quantum states to the body of a gas as a whole, rather than to the individual gas atoms. Given that London was Schrödinger's assistant in 1927, this may be the origin of London's view of superconductivity and superfluidity as forms of holistic quantum phenomena.

[47] There is an intimate connection between Bose–Einstein condensation and the non-classical 'indistinguishability' of quantum particles that was also drawn on by London and Heitler in their

do not mysteriously disappear, but rather occurs in momentum space (1938b, p. 951; cf. p. 39 of *Superfluids* where this is seen as a 'manifestation of quantum-mechanical complementarity').[48] This condensation then generates a 'peculiar omnipresence' in the total volume available to the molecules, which is characterized as a 'macroscopic' quantum effect, just as in the case of superconductivity.[49]

And, as in that case, London's theorizing proceeded in two stages, beginning with a conceptualization of liquid helium as a highly non-standard liquid that was very similar to a gas (with regard to its viscosity, for example; London 1938a p. 643; 1954, p. 37). He then moved to the second stage by applying Bose–Einstein statistics, yielding a model that was only partial, insofar as there was as yet no adequate quantum theory of such liquids (London 1938b, p. 954).[50] Here again we see London's phenomenological philosophy of science in action and as in the case of superconductivity, he remained convinced that the phenomena could only be ultimately explained in quantum mechanical terms (Gavroglu, 1995. pp. 235–6).[51]

This work was undertaken just prior to his collaboration with Bauer, after London had moved to the Institut Henri Poincaré in Paris.[52] While there, he also had the opportunity to hone his philosophical understanding through conversations with Aaron Gurwitsch, who went on to play a significant role in the establishment of phenomenology in the USA (Gavroglu mentions him in passing; 1995, p. 141).[53] We'll come back to Gurwitsch's views after we've

explanation of chemical bonding via electron 'exchange'. As Paty has put it,'Supraconductivity and superfluidity [*sic*] are... properties directly connected to indistinguishability' (Paty 2003, p. 461). For more on quantum indistinguishability, see French and Krause 2006.

[48] In the sense that we get order in momentum space at the expense of order in ordinary space. The notion of 'complementarity' accrued different meanings over the years, but as characterized in terms of causal vs. spatio-temporal descriptions, momentum space fell under the former.

[49] It may be that London's philosophical stance may have led him to be open to the idea of a condensation in momentum space, rather than 'ordinary' space by virtue of the phenomenological (re)construction of the latter, with modern physics driving the extension of such a constructive enterprise to the former.

[50] There is a further similarity here with chemical valence and molecular bonding where there were also no adequate 'microscopic' theories. In the former case, London followed Weyl and Wigner in deploying group theory, sidestepping the well-known difficulties with the many-body problem. Gavroglu has suggested that, 'many years later this difficulty was strangely liberating and helped him to articulate the concepts related to the macroscopic quantum phenomena as a means of superseding such difficulties' (1995, p. 55).

[51] Interestingly, given what we noted regarding Jordan's ideas, London also thought he could apply quantum theory to biological mechanisms (Gavroglu 1995, p. 193).

[52] According to Heims, 'in Paris he quickly became part of the closely-knit group of the inner circle of French physicists, was accepted wholeheartedly, and was given a good position' (1991, p. 187).

[53] Gurwitsch apparently thought that London had a 'deep understanding' of philosophical problems that was related to his understanding of physics (Gavroglu 1995, p. 175, n. 2). Alves speculates that

considered phenomenology itself in more detail. Before we get there, we need to consider a little more physics.

4.5 EPR and Schrödinger's Cat

It may be that it was the conversations with Gurwitsch about the nature of the ego, for example, that motivated London to analyse specifically the role of observation in QM. Gavroglu, however, has suggested that it was simply the desire to write a 'simplified' version of von Neumann's account (1995, p. 171). In support he has noted that there is no mention of either the (now) famous Einstein–Podolsky–Rosen (EPR) or 'Schrödinger's Cat' papers in London and Bauer's work and further posited that 'if anything, the criticism of quantum mechanics expressed in these papers may have had the opposite effect on [them]' (Gavroglu 1995, p. 171). However, at the time neither paper was given the significance that they have today so the lack of mention is perhaps not surprising. Nevertheless, London was certainly aware of both, as Schrödinger and others attested.

Let's begin with the EPR paper. The core argument can be summarized as follows: consider two particles that have interacted, such that their positions are correlated, as are their (linear) momenta. Measuring the position of one particle would then allow us to predict with certainty the position of the other. Likewise, if we were, instead to measure the momentum of one, we would be able to predict the momentum of the other. If we assume that no action performed with regard to the first particle can instantaneously affect the second (on pain of violating the Special Theory of Relativity), and also accept that if we can make such certain predictions without disturbing the system concerned, there must be an 'element of reality' corresponding to the quantity predicted, then we can conclude that the second particle must have a definite position and momentum prior to these quantities being measured. However, that runs contrary to the standard understanding of QM, which should thus be regarded as incomplete (for further details, see Fine and Ryckman 2020).

The germ of this argument can be found in comments made by Einstein at the 1927 Solvay Conference (Bacciagaluppi and Crull forthcoming, pp. 6–7) and was further elaborated in his famous debate with Bohr at the Solvay

'London's acquaintance with phenomenology was to a great extent a result of his contacts with Aron Gurwitsch in Paris in the thirties' (2021, p. 455, fn. 6) but this ignores the previous relationship with Pfänder.

conference of 1930, where, interestingly, the issue of whether QM can be extended to macroscopic systems runs throughout (which itself has a pre-history going back to 1926; see Bacciagaluppi and Crull forthcoming, pp. 8–11). This eventually led to the co-authored paper, published in 1935 (Einstein, Podolsky, and Rosen 1935), but which, according to Einstein, actually buried the main point (Bacciagaluppi and Crull forthcoming, p. 31; Fine and Ryckman 2020). In a 1935 letter to Schrödinger and in subsequent works, Einstein dropped the invocation of 'elements of reality' and the use of simultaneous values of complementary quantities such as position and momentum. Instead, he emphasized what he saw as the incompatibility between the claim that the description afforded by the wave-function was complete and the fundamental principles of *locality* and *separability* (something that also emerged prominently in his exchange of letters with Born; see Born 1971).[54]

Bohr's reaction to the EPR paper is well-known—essentially he argued that the experimental conditions that would enable the measurement of position precluded that of momentum and hence one could not attribute 'elements of reality' corresponding to both simultaneously. However, this failed to engage with Einstein's later versions of the argument (Fine and Ryckman 2020). Others also weighed in, through letters and publications, each misunderstanding the argument in their own way, which together offers an interesting window on the state of play when it comes to the interpretation of QM at the time (Baccigaluppi and Crull forthcoming, p. 41).[55]

Although there is no mention of either EPR or Schrödinger's Cat thought-experiment in the London and Bauer paper itself, London was definitely involved in discussions over both, as Schrödinger's correspondence indicates. Schrödinger himself was also dissatisfied with 'orthodox' QM, insisting that the likes of Bohr and Heisenberg should give up clinging to classical concepts and instead find new forms. Interestingly, he also came to believe that 'the ascription of quantum states to systems is *conditioned by the choice of our*

[54] Whereas locality demands that 'no real change can take place' in one system as a result of a measurement made on the other system, separability embodies the more fundamental notion that spatially separated systems possess their own 'elements of reality' (Fine and Ryckman 2020).

[55] Margenau also responded (after corresponding with Einstein on the matter), first in a short paper in 1935 and then in a longer paper two years later in which he also mentioned Schrödinger's 1935 'cat' paper. In both cases Margenau's focus was on the so-called 'collapse postulate' and he concluded that if that is dispensed with, the paradox is resolved (see Bacciagaluppi and Crull forthcoming, pp. 50–4). Bacciagaluppi and Crull interpret Margenau as maintaining that physical quantities are not real until they are actually measured and thus to be denying the EPR criterion of reality (ibid., p. 54) but even at this time Margenau was beginning to lay the basis for his 'constructionalist' epistemology, discussed earlier.

observations, in such a way that quantum mechanical wavefunctions in fact encode the *relation* between subject and object' (Baccigaluppi and Crull forthcoming, p. 70).[56] In this regard a talk he gave in Munich in 1930 is significant; there he stated, regarding the implications of QM:

> Most of us today feel that this necessary abandonment of a purely objective description of Nature is a profound change in the physical concept of the world. We feel it as a painful limitation of our right to truth and clarity, that our symbols and formulas and the pictures connected with them do not represent an object independent of the observer but only the relation of subject to object. But is this relation not basically the one true reality that we know?' (ibid., pp. 75–6)[57]

This concern with the replacement of the 'purely' objective description associated with classical mechanics with one grounded in the *relation* between observer and system is a prominent theme in London and Bauer's 'little book'.

It is with regard to this relational aspect that Schrödinger saw the conclusion of the EPR argument as even more profound than Einstein envisaged, insisting, again, that statements in QM only deal with the object–subject relation and in a way that seems to be much more radical than any 'natural description' (Bacciagaluppi and Crull forthcoming, p. 85). And it was through these discussions that Schrödinger's own 'cat' paper took shape (Schrödinger 1935a). In a letter to Einstein following the publication of the EPR piece, Schrödinger expressed his happiness at the way Einstein and Co. had 'caught dogmatic q.m. by the coat-tails' (Moore 1989, p. 304). In reply Einstein again applied the theory to a macroscopic system, presenting the example of a pile of

[56] As they go on to state, 'for Schrödinger this relational aspect of the quantum state is an indication of the limitations of the quantum mechanical description and of our inability thus far to construct a satisfactory underlying picture of physical reality' (Bacciagaluppi and Crull forthcoming, p. 70).

[57] There is an all too obvious connection here with Schrödinger's profound interest in Indian thought and the Upanishads in particular (Moore 1989, pp. 168–73), the central claim of which can be summarized as 'the self and the world are one and they are all' (ibid., p. 173). Husserl also expressed an interest in Indian philosophy, although this was focused on Buddhism (Schrödinger also read translations of Buddhist thought; Moore ibid., p. 113) and the level of engagement was much less. Nevertheless, Husserl acknowledged that Buddhism offered an alternative path to the phenomenological attitude, although he maintained that it was inferior to the approach founded on European thought, going back to Socrates, as he regarded this as more consistent and universal (see Schuhmann 1992, p. 31). There is also a historical common cause in play, insofar as Schrödinger was an avid fan of Schopenhauer (Moore 1989, pp. 111–13) and Husserl originally went to the University of Leipzig to study astronomy, where he attended lectures by Zollner, the founder of astrophotometry, who also gave philosophical lectures on Schopenhauer. And Schopenhauer, of course, regarded the Upanishads as one of the most notable achievements of human thought whilst also acknowledging the 'admirable agreement' between his own views and those of Buddhism.

gunpowder that had a certain probability of spontaneously exploding in the course of a year: using an appropriate wave-function, the state of the gunpowder must be described as a superposition of exploded and unexploded and Einstein concluded that '[t]here is no interpretation by which such a function can be considered to be an adequate description of reality' (Moore 1989, p. 305; Fine 1986, ch. 5; but see Uffink 2020).[58]

Likewise, Schrödinger, in introducing his example of the cat (described as a 'burlesque' or 'ridiculous'—depending on translation—case) stated that it shows that 'an uncertainty originally restricted to the atomic domain has become transformed into a macroscopic uncertainty, which can be resolved through direct observation. This inhibits us from accepting in a naïve way a "blurred model" as an image of reality' (Moore 1989, p. 308).[59] As Bacciagaluppi and Crull have put it, ontic indeterminacy might be acceptable at the microscopic level but at that of cats in boxes, it is empirically inadequate (Bacciagaluppi and Crull forthcoming, p. 104).

London was quite clearly aware of Schrödinger's paper. In a letter to Born, from June 1935, responding to the latter's dismissal of the EPR piece, Schrödinger remarked 'at first go you do it pretty much like all the others (e.g., London, Teller, Szilard)' (Bacciagaluppi and Crull forthcoming, p. 348).[60] Here he meant that although Born set out the situation with clarity, he failed to get the point, which was the claim that QM is incomplete. Schrödinger also indicated in a letter to Arnold Berliner (the editor of *Naturwissenschaften*) in August 1935, that both Born and London had read his work, which he had submitted to the journal.

Now, in evaluating the impact of Schrödinger's thought, both on London and others, there are two significant ideas to bear in mind: first, there is the Cat thought-experiment and second, there is the notion of *entanglement*.

[58] Vigier subsequently raised the stakes by substituting an H-bomb for Einstein's gunpowder (in Körner 1962, p. 143). In his contribution to the Born Festschrift in 1953, Einstein deployed a 'bullet' in the box argument that, again, explicitly considered what QM has to say about macrosystems (Bacciagaluppi and Crull forthcoming, pp. 38–9).

[59] The section of the paper in which the 'cat' is presented comes after one in which Schrödinger argues against the possibility of ascribing determinate values to properties of the system. The indeterminacy can't be attributed to an 'actual blurring' either, because 'there are in fact cases where an easily executed observation provides the missing knowledge' (Bacciagaluppi and Crull forthcoming, p. 211)—such cases including that of the cat of course.

[60] Both London and Schrödinger were in Oxford at the time and were clearly in touch on these issues. In another letter to Einstein, Schrödinger wrote, similarly, 'Now I am having fun and taking your note as an opportunity to prod a variety of clever people on this topic—London, Teller, Born, Pauli, Szilard, Weyl' (Bacciagaluppi and Crull forthcoming, p. 363). And in a letter to von Laue in July of that year he remarks that London had not yet sent him an offprint of Laue's note in *Naturwissenschaften* (ibid., p. 370).

4.5.1 The Disappearance and Reappearance of the Cat

It is certainly true that London and Bauer make no mention of the thought experiment itself, but it turns out that almost no one did! Apart from Jordan's later attempt to rule it out (Jordan 1949), there appears to be no mention of it in any work that appeared in the years immediately following its publication. Halpern, in a PBS blog-post marking the 80th anniversary of that publication has written that, '[f]rom the time Schrödinger proposed it, in 1935, to his death in 1961, it was scarcely mentioned in the literature. Even Schrödinger rarely brought it up' (Halpern 2015).[61] Indeed, Halpern has credited Wigner with bringing it back onto centre stage in the context of the debate with Putnam. As for Putnam himself, Halpern has quoted him as recalling, 'I met [mathematical physicist] David Finkelstein at an American Mathematical Society Summer Seminar on "Modern Physical Theories and Associated Mathematical Developments" in Boulder, Colorado in 1960. It may well be that it was from him that I learned about Schrödinger's Cat. I always assumed the physics community was familiar with the idea' (ibid.).[62]

In fact, if we turn to Google N-gram we actually find 0 citations for 'Schrödinger's Cat' until 1957, when the Proceedings of the Ninth Symposium of the Colston Research Society[63] were published (Körner 1957). Organized around the general theme of 'Observation and Interpretation in the Philosophy of Physics', the main focus was actually Bohm's defence of his 'hidden variables' theory.[64] Rosenfeld participated as the emissary of the 'Copenhagen school' on a mission to 'stamp out' this 'new obscurantism' (for a good account of the conference, see Koznjak 2019; also Ryckman 2022, p. 789).[65] As part of this debate, Feyerabend[66] raised concerns about

[61] He failed to mention it in his twin 1952 papers in the *British Journal for the Philosophy of Science* (Schrödinger 1952a and Schrödinger 1952b) where he criticized orthodox QM and compared the infamous 'quantum jumps' to the epicycles of an earlier era.

[62] Putnam went on to present his own variant of the thought experiment, placing the cat in a rocket ship in interstellar space, where it cannot be observed (Putnam 1965/1975, p. 80).

[63] The Society was founded in 1899 using funds obtained from Edward Colston, a Bristol merchant who was involved in the slave trade. From 1948 it supported an annual series of interdisciplinary conferences.

[64] Bohm was a research fellow at Bristol at the time.

[65] The general consensus seems to be that Rosenfeld actually came off worst in the exchanges.

[66] Feyerabend had a three-year fixed-term position at Bristol at the time (having defended his thesis on 'observation statements' in 1951 under Kraft, a member of the Vienna Circle; Popper and Schrödinger wrote letters of recommendation in support; Feyerabend 1995, pp. 100 and 102) and supervised the transcription of the conference discussions. He also read out Popper's contribution, on the propensity interpretation, as the latter was unable to attend (Koznjak 2019, p. 95). Popper argued that one of the advantages of his interpretation was that it avoided the 'interference' of the observer

the orthodox account of measurement, with its distinction between 'ordinary experience' (as described by classical physics) and 'physical theory' (that is, QM; Feyerabend 1957, p. 121).[67]

He rejected as 'too simple a picture' the assumption that both the measurement device and the system each 'jump' into an appropriate eigenstate as soon as the observer consciously looks at the former ('*esse est percipi*'; ibid., p. 125, fn. 11),[68] not least because '[t]he values of macroscopic variables are fixed independently of any account which is given of their observation. The theory does not yield this result, not even as a first approximation' (ibid., p. 125). And it is here that he explicitly referred to 'Schrödinger's cat' in a footnote (ibid., p. 125, fn. 12). Instead he argued that due to decoherence[69] the macroscopic distinguishability of the pointer readings on the measurement device implies that *for all practical purposes* (cf. Bell 1990), describing the system + measuring device as either a pure state or a mixture yields the same results, with respect to the properties of the system. Feyerabend took this to be the first step in the response to Schrödinger's thought experiment, since the relevant mixture of 'cat alive' and 'cat dead' can be substituted for the superposition of the two states, not because of any 'jump' or wave-function collapse, but because under the relevant conditions the difference between the two is 'negligible' (ibid., p. 127).[70]

Feyerabend's conclusion was that that a theory of quantum measurement may be possible in which there is no collapse if appropriate special conditions are introduced, namely that we have macro-observers and macroscopically distinguishable states. Such a theory would incorporate the 'everyday' level as part of the theoretical level, rather than taking it to be something self-contained and independent (Feyerabend 1957, p. 129), a point that resonates with London and Bauer's analysis and with a phenomenological stance more generally.[71]

with the wave function, according to the 'Copenhagen spirit' (Körner1957, p. 69). Feyerabend was awarded a British Council scholarship to study the foundations of QM in London, under Popper's supervision, for the academic years 1952–3 but the scholarship was not extended.

[67] For a summary see also Mehra 2012, pp. 27–30.

[68] In a subsequent paper Feyerabend favourably compared Bohr's methodology to certain principles of dialectical materialism and dismissed the 'disappearance' of the boundary between subject and object supposedly implied by QM as an example of 'secondary' or 'parasitic' philosophy (Feyerabend 1966).

[69] Here he cited Bohm in a footnote (Bohm was an early advocate of the significance of decoherence; see Goldstein 2021).

[70] Mehra has noted that at various points in his argument, Feyerabend in effect begs the question at issue (Mehra 2012, p. 29).

[71] Kuby (2021) locates the origins of Feyerabend's infamous methodological pluralism in part in his reflections on the outcome of that conference and its discussions. In letters to Kuhn commenting on a draft of *The Structure of Scientific Revolutions*, Feyerabend repeatedly rejected the idea that QM should be regarded as a new 'paradigm'. Instead, he argued, from the perspective of the Copenhagen

In the discussion following Feyerabend's paper, it is clear that the likes of Bohm,[72] Groenewold, Landsberg, Mackay, Öpik,[73] Pryce, Süssman,[74] and Vigier, as well as the philosopher Körner, were all well aware of Schrödinger's thought-experiment even though it was not referred to in the physics literature. In that sense, then, London and Bauer were not alone in failing to cite the Cat argument in their work. More importantly, the case can be made that it simply would not have been an issue for them: London, certainly, already accepted that quantum theory could be applied to macroscopic phenomena and hence would have rejected any distinction between macroscopic and microscopic levels in this context. Furthermore, from the phenomenological perspective, the thought-experiment presents no problem, as we'll see.

4.6 Entanglement

Although Schrödinger had clearly aligned with Einstein in adopting a critical attitude towards the 'orthodoxy', their approaches differed (Baccigaluppi and Crull forthcoming). Indeed, Schrödinger himself made one of the more significant contributions to our understanding of QM with his introduction of the notion of 'entanglement' in his 'Cambridge' paper of 1935 (Schrödinger 1935b; see Bacciagaluppi and Crull forthcoming, ch. 3). He famously characterized it in the following terms:

> Maximal knowledge of a total system does not necessarily include total knowledge of all its parts, not even when these are fully separated from each other and at the moment are not influencing each other at all.
>
> (Bacciagaluppi and Crull, forthcoming p. 218)

interpretation, 'quantum theory is...what remains of the classical paradigm when its metaphysical pretensions have been eliminated' (Hoyningen-Huene 1995, p. 379; see also Beller 1999, p. 300 and Ryckman 2022).

[72] Bohm speculated about putting a further human observer in place of the cat, thereby suggesting, in retrospect, a Wigner's Friend type of situation but concluded that any paradox here simply indicates that QM, as standardly understood, is incomplete.

[73] Öpik also mentions the EPR case in his comments.

[74] Süssmann was aligned with Wigner in the so-called 'Princeton school' and in discussion rejected Feyerabend's approach on the grounds that QM could be applied to macroscopic bodies, as exemplified by the phenomena of superconductivity and superfluidity: 'These phenomena are macroscopic phenomena which are simply absurd when seen from a classical point of view; not much less absurd than the extremely "weak" co-existence of dead and alive [clearly referencing the cat here] or the extremely improbable rising of a stone' (Körner 1957, p. 141).

Thus, part of what one *knows* has to do with certain relations or, as he put it, 'conditionings' between the systems. This, he went on, cannot happen if the systems are simply juxtaposed without interaction, but if they 'influence' each other and are then separated, then each leaves a trace on the other, resulting in an '*entanglement* of our knowledge of the two bodies' (Bacciagaluppi and Crull, forthcoming, p. 219).

Schrödinger actually struggled with this idea for some years and realized that the difficulty in interpreting entangled states had become a major stumbling block to the further development and application of (his) wave mechanics. (Bacciagaluppi and Crull, forthcoming, pp. 59ff). In particular, he recorded in one of his notebooks a discussion in one of Zurich's cafes with London (and Heitler and Fues) in which he expressed his concern that if one tried to consider separated systems using forces that act at a distance, then the quantum formalism simply broke down. Clearly then, London was aware of Schrödinger's worries in this regard.[75]

In particular, in their consideration of an entangled system (although they do not use the term), London and Bauer wrote:

> In classical mechanics we are not astonished by the fact that a *maximal knowledge of a composite system* implies a *maximal knowledge of all its parts.* We see that this equivalence, which might have been considered trivial, does not take place in quantum mechanics. There a maximal knowledge of a composite system ordinarily implies only mixtures [note: here they are using the term 'mixture' for what we would now call a 'pure state'] for the component parts—that is, a knowledge that is not maximal. (Wheeler and Zurek 1983, p. 248)

This clearly echoes Schrödinger's statement above which expresses what he came to regard as the characteristic trait of quantum mechanics, but which 'keeps coming back to haunt us' (Baccigaluppi and Crull forthcoming, p. 232). I think we can safely conclude that London, at least, was aware of the relevant context, particularly with respect to Schrödinger's concerns and had a good grasp of this notion of 'entanglement' in particular.

Furthermore, Schrödinger stated that, in the context of taking the measurement process to be a special case of entanglement, from the 'amalgamation' of measurement instrument and object, or system under investigation:

[75] It is perhaps worth noting that even as late as 1959 it was not known whether entanglement would persist when particles separate, although Schrödinger raised the question in 1935 (see Shimony 2002).

> the object can again be separated out only by the living subject actually taking cognizance of the result of the measurement. Some time or other this must happen if that which has gone on is actually to be called a measurement—however dear to our hearts it was to distil the process throughout as objectively as possible.
>
> (Bacciagaluppi and Crull forthcoming, p. 221)

He continued, it is not until this 'inspection' 'resolves [or alternatively, determines; the German word is entscheidet[76]] the disjunction' that there is any discontinuous 'quantum leap'. Now, although one might be inclined to regard this as a 'mental action' since the object itself is now physically out of touch, one shouldn't conclude that the wave-function of the object changes discontinuously because of such an act:

> For it had disappeared, it was no more. Whatever is not, no more can it change. It is born anew, is reconstituted, is separated out from the entangled knowledge that one has, through an act of perception, which as a matter of fact is [certainly] not a physical effect on the measured object. (ibid.)

It is this 'annihilation' followed by a reconstitution that gives the appearance of a 'leap'. What happens in between is, basically, that the bodies become entangled.

What I want to draw from Schrödinger's comments is first, that the wave-function of the system is separated out from our entangled knowledge *through an act of perception,* via which the observer takes cognizance of the measurement outcome. Second, that this does not amount to the mental act inducing some kind of physical effect that collapses the wave-function. Here Schrödinger appears to be anticipating both von Neumann's account and the concerns raised by Putnam and Shimony. In these respects, these passages are strongly reminiscent of London and Bauer's presentation four years later.[77] What they still lack is a comprehensive philosophical framework in the context of which one can understand how the wave-function of the system can be separated out in this way, without that mental act 'causing', in whatever sense, that separation. London's phenomenological account does just that, as we'll see.

[76] This could also be translated as 'decides'.

[77] Obviously, taken as an antedecent this could be taken to undermine my interpretation of the London and Bauer piece since Schrödinger was not an adherent of phenomenology! However, there are other reasons for preferring the former interpretation and I would suggest that, given that London and Schrödinger discussed these issues while in Oxford, London and Bauer drew on this passage from Schrödinger's paper but gave a phenomenological spin to it.

4.7 Conclusion

Although London and Bauer do not mention the Cat thought-experiment, that is not so surprising given how little attention it received in general. And of course, London, in particular, may not have been fazed at all by it, given his extension of QM to macroscopic phenomena—that it should apply to cats would have seemed obvious! As for Schrödinger's introduction of entanglement and its role in measurement, these find echoes within London and Bauer's little book, although whether these come directly from Schrödinger's own thoughts and work or whether the latter were influenced by his discussions with London is difficult to determine. When it comes to the EPR argument, given the emphasis on the reality criterion it is perhaps no surprise to find no mention of this. Furthermore, as we have noted, Einstein and Schrödinger, although typically lumped together as opponents of the 'orthodoxy', had in fact profoundly different views on the underlying issues (see again Bacciagaluppi and Crull forthcoming). Indeed, in a letter to Schrödinger dated 17 June 1935, Einstein wrote that he found 'the renunciation of a spatiotemporal comprehension of the Real idealistic–spiritualistic' (ibid., p. 340) and in a further letter two days later he dismisses the 'spiritualist' or 'Schrödingerian' interpretation as 'insipid' (ibid., p. 342). As we'll see later, London and Bauer utterly reject the suggestion that on their account, the scientific community becomes a kind of *spiritualistic society* which studies imaginary phenomena, produced by the observer (London and Bauer 1983, p. 258). This might have been a direct response to Einstein's allegation against Schrödinger that gives a further reason why London and Bauer declined to address the EPR argument.

As Gavroglu notes, 'London never worked systematically on questions related to the measurement problem in quantum mechanics before, or after, the appearance of [their] monograph, nor did he ever correspond with anyone about these issues' (1995, p. 174). London received the proofs of the manuscript in the summer of 1939 and on the day that Germany invaded Poland, he and his wife sailed for the USA where he took up a position (his last) at Duke University.[78] There he taught quantum theory to chemistry students[79] and,

[78] His brother Heinz stayed in the UK and subsequently took up a position at the Atomic Energy Research Establishment at Harwell where he became the Principal Scientific Officer. He too studied liquid helium in the context of various cryogenic projects during the later 1950s and 1960s (see Shoenberg 1971). Unlike his brother he does not appear to have had any serious philosophical inclinations (although he apparently discussed philosophical topics with his children).

[79] For a sketch of London's teaching at Duke, see Linschitz 1988.

after a hiatus, pursued his work on superfluidity (ibid., pp. 198–206 and 214–17).[80]

Having looked at the historical background to London and Bauer's work in physics and gestured at some of the connections with the work of Schrödinger in particular, it's now time to set out the phenomenological basis of their 'little book', before considering the resolution of the measurement problem that it presents.

[80] Heims notes that in the United States, London's 'two closest friends were philosophers: Aron Gurwitsch, the well-known phenomenologist and one-time assistant to Husserl; and the little-known Ernst Moritz Manasse, also a German-speaking refugee, who had a position teaching in a college for Black students in North Carolina' (Heims 1991, pp. 183–4; see https://de.zxc.wiki/wiki/Ernst_Moritz_Manasse).

5
The *Epoché* and the Ego

5.1 Introduction

There is, of course, an enormous body of literature on phenomenology in general and Husserl's work in particular. Husserl himself wrote a huge amount during his lifetime, much of it unpublished, which has only comparatively recently begun to be examined and analysed. In addition, and not surprisingly, his ideas evolved over time and have been subjected to alternative and, in some cases, conflicting interpretations. Here I shall focus on those features of his thought that I believe help illuminate London and Bauer's account, focusing, on the one hand, on the phenomenological approach to objects, objectivity, and physics in general, and on the other, on consciousness and the nature of the ego. This will cover not only aspects of the *Logical Investigations* and *Ideas I* and *II* but also central themes in Husserl's last, great, unfinished work, *The Crisis of the European Sciences*, in which he appears to decry the mathematization inherent in modern science. This could be interpreted as a decisive rejection of the form of recent physics that is exemplified in QM, particularly as associated with the mathematics of group theory, which London himself favoured, as we've seen.[1] I shall argue, however, that in the context of London and Bauer's work, QM, properly understood, represents, in fact, the scientific completion of Husserl's great project.

Let's begin with a 'preliminary orientation' of Husserl's work as evolving through three stages:

(i) a rejection of the psychologistic understanding of logic and mathematics;
(ii) the elaboration of phenomenology as the 'science' of consciousness;
(iii) the further development of this stance as underpinning intersubjectivity and as extended to cover culture, history, and the 'life-world' in general. (Smith and Woodruff Smith 1995)

[1] This is one of the reasons given by Crease, Kamins, and Rubery for the post-war decline in interest in the phenomenological treatment of QM, the others being the lack of training in physics among the next generation of phenomenologists and the acceding to 'analytic' philosophers of this task (Crease, Kamins and Rubery 2021, p. 407).

A Phenomenological Approach to Quantum Mechanics: Cutting the Chain of Correlations. Steven French, Oxford University Press. © Steven French 2023. DOI: 10.1093/oso/9780198897958.003.0005

Such a framework is useful in the present context since it enables us to situate London's early thesis, for example, in the first stage, whereas, as we shall see, the considerations of consciousness and objectivity that we find in the piece with Bauer span the second and third stages. Nevertheless, it is important to acknowledge that it is typically claimed (although not by all) that a major shift occurred between the publication of Husserl's *Logical Investigations* in 1901 and the *Ideas* in 1913, marking a transformation from a descriptive analysis of intentionality to a form of transcendental idealism, albeit one that was distinct from the Kantian version. This in turn led to something of a schism between the two schools of phenomenological thought that had formed around Husserl in Göttingen and Pfänder and others in Munich (see Salice 2016; Parker 2021). The latter, in particular, continued to maintain a broadly realist stance, having understood the analysis of the *Logical Investigations* as a move away from subjectivism (as manifested in particular by Husserl's rejection of psychologism when it came to logic and mathematics) and towards a form of metaphysical realism. Husserl, however, regarded this as a misunderstanding of his attempts—evolving over the years—to articulate his transcendental-phenomenological idealism which 'seeks to reconcile the empirical reality of the world with the dependence of that reality on consciousness' (Parker 2021, p. 3).[2]

As we'll see, how we should understand the nature of that dependence will be critical to our interpretation of the London and Bauer account. The historical split is also important in this context because, as already noted, London was originally taught by Pfänder, although given when he initially became interested in phenomenology (post-*Ideas*) and also his continued relationship with Gurschwitz, for example, it would be implausible to insist that he was not aware of the developments in Husserl's thought. Finally, although this split between realist and (transcendental) idealist understandings of phenomenology has continued to the present day (see for example Hopp 2020 and the reviews of the latter by Yoshimi 2021 and Woodruff Smith 2021), it has been argued that more continuity than discontinuity can still be found in these developments (see Mohanty 1995; Zahavi 2017; Trizio 2021).[3]

[2] According to Zahavi, the account of intentionality presented by Husserl in the *Logical Investigations* does not support a realist reading (Zahavi 2017, pp. 35–36).

[3] 'Excepting possibly the discovery of the *epoché* in 1905, no major shifts characterize the development of his thought—there is rather a continuous, unceasing attempt to think through the same problems at many different levels' (Mohanty 1995, p. 74; see also Trizio 2021). Having noted that, the significance of Husserl's 'transcendental turn' and the role that the *epoché* acquired should not be underappreciated.

This latter point will be crucial to our understanding of not only the phenomenological treatment of the ego and objectivity but also the stance adopted towards modern science.

So, a quick and crude, ten-words-or-less characterization of phenomenology would be as an investigation into the correlations between mental acts or experiences, the objects that these acts or experiences are about, and the contents or (where appropriate) meanings of these acts or experiences (see Hopp 2020, p. iii). Immediately we might ask: how is this investigation to be conducted? What tools are to be used to achieve its aims? And, of course, what does this enquiry reveal about the nature of our experiences and their relationship to external reality?

5.2 The *Epoché*

Let's begin with the first question and again we can give a quick and crude answer. The method by which such an enquiry is to be conducted has, at its heart, the '*epoché*', according to which one must effectively bracket off, or refrain from positing the existence of, the 'objective' world around us. By means of this device we can investigate the metaphysical presuppositions that underpin our naturalistic attitude to that world and develop a philosophy free from them. Such a 'bracketing off' thus reveals the *essences* of the objects of mental acts, irrespective of whether the objects themselves actually exist or not, allowing us to focus on pure consciousness itself, together with its acts and objects (see, for example, Bell 1990).

As characterized (or, perhaps, caricatured) so bluntly, we can see why the *epoché* has proven to be controversial, leading as it has to claims that it inexorably drives phenomenology into the arms of solipsism (see Zahavi 2017, pp. 51–6). However, such a push can and should be resisted. First of all, it is important to get clear on the aim of the *epoché* which is simply to break the hold on us of a certain dogmatic attitude—that Husserl calls the 'natural attitude' (Zahavi ibid., p. 56)—which is to blithely *assume* the existence of a mind-independent reality. This attitude is adopted not only by scientists, in their scientific work (and, typically, also by philosophers of science, realists, and (many) anti-realists alike) but also by 'laypeople' in their everyday lives. However, even if we all (or almost all) live with this attitude of natural realism on a day-to-day basis we can't simply presuppose it if we are to construct a 'truly scientific philosophy'.[4] We need somehow to break its hold on us if we are to reflect on its fundamental presuppositions, both epistemological and metaphysical. And that is what the *epoché* is designed to do.

It is important to note that this bracketing off does not mean 'denying the existence of' and so the *epoché* does not amount to an endorsement of scepticism. The world and our natural attitude towards it are not denied or negated but rather are now regarded in such a way that they can, in effect, be held up to reflective scrutiny.[5] As Husserl makes clear, the *epoché* should be understood as a 'reorientation' of our attitude towards the world (Zahavi 2017, pp. 57–8): 'what is bracketed is not wiped off the phenomenological board but merely bracketed and thereby provided with a marker' (Husserl 1982, p. 159), where this 'marker' indicates a kind of 'reclassification' within the phenomenological sphere.[6] Again, what is excluded or set aside is not the world itself, but rather a certain naïve attitude towards it.

Such an understanding of the *epoché* then goes some way towards reconciling the realist and idealist stances within phenomenology. For Pfänder, for example, the *epoché* was 'nothing but a safety measure against advance commitments, whether realistic or idealistic, not an opening wedge into a strange new world' (Spiegelberg 1982, p. 113). And more recently, Hopp has written:

> what comes into view in transcendental phenomenological inquiry are the essences of and essential relations among acts, their contents, and their objects... The phenomenological reduction is just a way of making these essential features and relations stand out in relief, and to place everything irrelevant to them in brackets. Thus, there is nothing especially 'radical' about the reduction. (2021, p. 267)

[4] Zahavi has argued that one need not employ the *epoché* in order to apply phenomenology (Zahavi 2021). The issue is complicated somewhat by Husserl's own insistence that phenomenological psychology requires such employment but as Zahavi points out, he then acknowledged (in the *Crisis*, p. 258) that the psychologist could return to the 'natural attitude' having taken the phenomenological turn. Furthermore, as Zahavi also notes, Husserl offered various entry routes into phenomenology, including that of phenomenological psychology but this should not be interpreted in terms of specific instruction on how to conduct empirical research in general (Zahavi 2021, p. 270).

[5] Bitbol refers to the epoché as a 'phase of neutralization' (Bitbol 2021, p. 565).

[6] Here Husserl compared the *epoché* to a mathematical operation that transforms the 'value' of what it operates on. He later compared it to making the transition from a two-dimensional to a three-dimension life (Husserl 1970b, p. 120).

Here Hopp conflates the terms '*epoché*' and 'reduction', as many other commentators have done, but a case can be made for keeping them distinct. First of all, not everything can be put in abeyance—there have to be limits on the universality of the *epoché* else there would be no ground to stand upon at all, in effect, and phenomenology would be impossible. Where the *epoché* stops is at the borders of the domain to which everything is taken to be *reduced.* As Husserl himself makes clear (1970b), the *epoché* can be used to effect a first reduction of the entities and processes of the natural sciences to those of the 'life-world', or the world of everyday life. There is then an additional, 'transcendental', reduction that can be achieved through a further application of this device, bracketing off the life-world itself—where this covers both the world of everyday life *and* that of scientific practices—leaving us with the domain of 'pure' subjective being, or the 'flux of lived experience' (Bitbol 2021, p. 567). This further reduction is particularly problematic, insofar as 'even the standard conditions for mutual understanding and intersubjective agreement about the enduring things that can be shown and manipulated, are suspended' (ibid.). As we'll see, the issue of recovering such agreement will crop up again and again throughout our discussions.

What is important for our purposes right now is that it is by means of the *epoché* that we can grasp the true significance of the role of consciousness and appreciate the nature of that role which is otherwise concealed from us. The *epoché*, then, is the crucial methodological move that opens up the relationship between consciousness and the world to appropriate philosophical analysis. As we'll see, it is precisely that relationship that emerges as central to the whole phenomenological enterprise—indeed, it is only to that relationship that we ultimately have access.

This relationship can then be held up to scrutiny through an act of *reflection*, or 'essential seeing' (Husserl 1982, p. 162). It is important to appreciate that this act is not the same as 'mere' introspection; to claim so would reduce phenomenology to descriptive psychology. Nor is it the same as 'intuition' as typically understood and used these days. The act of 'essential seeing' involves a form of reflective *regard-to* that is different from simply 'inwardly observing' or intuiting what is the case. It comprises a kind of disentangling and explication of those components and structures that are inherent in our pre-reflective experience (Zahavi 2017, p. 23). Again, as we'll see, it is precisely such a disentangling and explication via this 'regard' that London and Bauer argue occurs in the measurement of a quantum system.

As Ryckman has noted, Husserl drew a useful comparison with geometrical thinking here, and indeed with the 'mathematical style of thinking' more generally (Ryckman 2005, pp. 142–3). So, the geometer begins with a particular figure, a particular triangle say, and then abstracts from those particularities to yield an arbitrary example (ibid., p. 143). On that basis she can then freely imagine all possible triangles and indeed, from a basis of a limited number of concepts, such as point, line, angle, and so on, she can obtain all ideally possible shapes in space. In doing so, the geometer has engaged in an act of essential seeing and by conceiving of all possible triangles, she has grasped the essence of the triangle, 'by determining what is universal, *invariant*[7] and pervasively identical, through every imaginable particularization constituted as a *possible* actuality' (Ryckman ibid., p. 143; his emphasis).

Let's pick this comparison apart a little. First, this seeing essentially is associated with a form of giving 'originarily', in accordance with Husserl's fundamental principle that the 'original source' of legitimacy when it comes to our posits about the world lies in the immediate evidence for them (1982, section 141). When a belief is fulfilled by such an originary intuition and there is no counter-evidence or reasons to doubt the belief-formation process, we have knowledge. Now, it is at this point that the split between realist and idealist understandings appears. Does this form of originary givenness yield the thing itself, whatever it is, or rather, some *representation* of the thing? On the realist interpretation, we do not perceive objects as mediating images, like photographs, say. To suggest otherwise, Hopp has argued, would be to commit the 'cardinal phenomenological sin of replacing what we are obviously conscious of with something completely different' (2020, p. 92).

However, the worry here is that this leaves us within the 'natural attitude' and the whole point of performing the '*epoché*' is to take us out of that, with the result that what we have on the 'phenomenological blackboard', as Husserl called it, are *objects as experienced*, or 'noemata'. According to Yoshimi, this does not replace or change the objects but allows us to regard them in a different way (Yoshimi 2021). Furthermore, this shift in regard does not mean that we lose sight of the objects as perceived from within the natural attitude; rather, we shift back and forth between the two attitudes as we perform our phenomenological reflections (ibid.).

The important point to note is that when we 'see' or intuit in the above sense, we grasp or seize that which is fully present and immediately

[7] That which is invariant in geometry was captured group theoretically within Klein's 'erlangen' programme which played a crucial role in underpinning the introduction of group theory into physics (see Bueno and French 2018, ch. 4).

transparent to consciousness: '[t]he datum of eidetic intuition is a pure essence, a structure of transcendental subjectivity that is immediately present to the mind' (Ryckman 2005, p. 142). It is because of this that such a datum can function as the 'ultimate court of appeal of all knowledge' (ibid.) But as already indicated, we should not conclude from this that phenomenology reduces to either phenomenalism or descriptive psychology. As Zahavi has put it:

> Phenomenology is not interested in qualia in the sense of purely private data that are incorrigible, ineffable and incomparable; it is not interested in your specific experience, or in my specific experience, but in invariant structures of experience and in principled questions concerning, say, the presentational character of perception, the structure of temporality, or the difference between empathy and sympathy. (Zahavi 2017, p. 15)

The job of the phenomenologist is to then pin down these essential structures via a combination of successive acts of such essential seeing and a kind of abstraction from the particulars. And crucially, it is by means of the '*epoché*' that we are able to 'tease apart and analyze the structure and dynamics of conscious experiences, studying the laws governing the coherence of sensory data into meaningful profiles, and the separate laws governing the coherence of noemata into increasingly rich internal models of the world' (Yoshimi 2021).

So, in the case of the triangle our geometer friend abstracts away from the particularities of the given triangle to obtain the essence of the universal triangle, now seen as an ideally possible figure that can be particularized in multiply imaginable ways. But again note that there is a difference between an experience as lived through—our experience of the particular triangle say—and our reflection upon it. If reflection upon our experiences is to have any cognitive value it must involve a form of 'stepping back' from the experience as lived through, not least so that, as noted above, the essential structures inherent in the latter can be disentangled (Zahavi 2017, p. 23). And it is by virtue of this 'stepping back', that phenomenology acquires a critical dimension which elevates it above the mere compilation of introspective reports (Zahavi ibid., p. 24; see also Yoshimi 2021).

In the case of the triangle, and geometrical figures in general, this disentangling reveals the triangle's situation in the space of all possible such figures, and their inter-relationships. Thus, the figure is situated within a manifold of ideal possibilities, that are taken to be freely created in imaginative fantasy

(Ryckman 2005, p. 143; Husserl 1982, section 70).[8] We shall come back to this idea of free choice when we consider London and Bauer's work.

5.3 The *Epoché* and Metaphysics

On some accounts, the *epoché* is understood as excluding consideration of the nature of *being* and redirecting our attention instead to issues of *sense* or *meaning* (Carr 1999; Crowell 2001; Thomasson 2007). What it allows us to do, on this interpretation, is to reflect upon the *sense* that the world has for us in its 'givenness' and as a result the manner of that reflection is entirely different from that which is deployed not only in the 'positive sciences' but also in metaphysics, whether 'naturalized', in the sense of being engaged closely with those sciences, or not. However, if we eschew metaphysics entirely, then it is hard to make sense of Husserl's rejection of both the Kantian 'thing-in-itself' and phenomenalism, or of his own affirmations that what phenomenology excludes is 'naïve' metaphysics not metaphysics 'as such' (Zahavi 2017 p. 64).

What form could such a 'non-naïve' metaphysics take? According to Zahavi it would present reality as a 'constituted network of validity' (ibid., p. 68), where that network provides the conditions under which something *appears as real.* Thus, phenomenology should be understood as 'a systematic investigation of the correlation between structures of intentionality and objects of experience' (ibid.) As we'll see, London and Bauer's phenomenological analysis of measurement puts the former, anti-metaphysical, account under significant pressure—after all, that a definite measurement result is obtained when a quantum system is observed doesn't seem to be just a matter of the sense or the meaning that the result has for us in its givenness. On the other hand, although, as again we'll see, their analysis can be nicely situated within Zahavi's alternative 'correlationist' framework, the suggestion that intentionality has anything to do with such a definite result obviously has potentially radical implications. Let's look at this framework in a little more detail.

5.4 Correlationism

The core idea is that for something to be real it must be *presentable* to us in experience and for something to be presentable in that way, there must be a

[8] According to Luft, it is via the *epoché* that we become free to be ourselves (Luft 2004).

relation between the object and a conscious subject, which by virtue of that relation must be understood as interdependent (Zahavi 2017, p. 113). In that sense, then, objects and consciousness are mind-dependent but this does not mean that the former exist 'in' the mind, or are constructed 'by' the mind, nor that they only exist when experienced:

> Objects have their essentially manifest properties even when not being experienced, and can also truthfully possess them before the emergence of conscious creatures and after their eventual extinction. They exist in public space and are intersubjectively accessible and are to that extent given as transcendent; but as essentially manifestable, they do not have a nature that transcends what can be given in experience. (ibid., pp. 113–14)

This interpretation of Husserl as a correlationist has its antecedent in the work of Beck, who wrote, 'Consciousness and world, subject and object, I and world stand in a correlative, i.e., mutually dependent context of being' (Beck 1928, p. 611).

Indeed, Husserl himself talked of the 'transcendental correlation between world and world-consciousness' (Husserl 1970b, p. 151), where 'we only ever have access to the correlation between thinking (theory) and being (reality) and never to either in isolation from or independently of the other' (Zahavi 2017, pp. 174–5). As a result, mind and world should be seen as 'bound constitutively together' (ibid., p. 117) and the conceptual centrality of this notion of correlation implies that *it is the relation that is constitutive of its relata*, rather than the other way around (ibid. p. 118).[9]

The idea here is that once we effect the *epoché* we are no longer in the position of being able to insist that the world is 'clearly' independent of consciousness. That would be to remain within the natural attitude that the *epoché* is supposed to have freed us from. Once we have performed the latter all that we have access to are our intentional acts, which we can conceive of in more or less neutral terms as *correlations* between consciousness and the world. We might then further explicate those correlations as *relations* holding between consciousness and the world but we should not make the mistake of taking that to imply that consciousness and the world somehow exist *prior* to these correlations or independently of one

[9] A useful comparison might be made here with Ontic Structural Realism, which insists that all there *is*, is structure, conceived of in relational terms, and that physical objects are constituted by these relations (Ladyman 1998; French 2014).

another. It is the correlations that *constitute*, in a certain sense, both consciousness and the world.

Here it is worth recalling that, 'the intentional relation is unlike real relations in that it does not require the actual existence of its object term in order to obtain' (Hardy 2013, p. 183). The essence of intention is that the object concerned is referred to or 'aimed at' but such 'aiming' should not be taken to imply the existence of the object. As Husserl himself also put it, we can represent to ourselves all kinds of objects, from physical things, to angels, God, even impossible objects like a round square (1970a, II p. 596, cited in Hardy 2013, p. 183), 'but it makes no difference whether this object exists or is imaginary or absurd' (ibid.). It is the intention that exists and if the object referred to does as well, then that's a bonus, but all it means is that the intention does not exist alone. In particular, if the intentional experience is that of presentation, then it is of the essence of such an experience that the object aimed at is 'intentionally present', where that last phrase means the same thing as 'the intentional "relation" to the object is achieved' (1970a, II p. 558).

Thus, we should not make the mistake of thinking that in every case the positing of such a correlation implies the existence of a relation, which in turn implies the prior existence of the relata. Rather we have something, a web or nexus of intentions of different kinds, that can be described as correlative and thereby understood in terms of relations, but, again, with the caveat that this should not be taken to imply the prior existence of the relata. This amounts to a rejection of dualism and, in particular, of the idea that the nature of this relationship is a *causal* one (Zahavi 2017, p. 118). It must be appreciated that 'the phenomenological investigations of the structure of phenomenality are antecedent to any divide between psychical interiority and physical exteriority, since they are investigations of the very dimension in which any object—be it external or internal—manifests itself' (ibid., pp. 119–20).

This then provides a useful interpretive framework for understanding London and Bauer's analysis of observation in QM. Just to foreshadow a little: what the quantum mechanical formalism captures and represents is precisely that correlation that holds between the observer and the system observed but insofar as it is only the correlation that 'we' have access to, we cannot strictly speak of 'observer', 'system', or 'relation' holding between them, prior to the separation of the first from the second. And as we just noted, this implies a rejection of mind–world dualism and also of the attribution of causal relations in the relevant sense. What we have is a correlation, expressed via the appropriate wave-function, which effectively *constitutes* both relata, mind, and world. We shall, of course, come back to this.

5.5 Objectivity and the Constitution of the World

As should be clear, phenomenology does not fit comfortably within the boundaries of the realism–anti-realism debate as it is currently framed.[10] Again, this is not to say that Husserl eschews the realism of the 'natural attitude'; rather what he rejects is the 'philosophical absolutizing' of the world which is inherent to metaphysical realism (Zahavi 2017, p. 186). In other words, we should not take the world to be as it is given from the perspective of the natural attitude, appropriately refined perhaps, prior to suspension of that attitude and an appropriate reflective analysis.

What, then, is the nature of this constitutive relationship? Answering this will help us get a better fix on the shape of Husserl's position. As Zahavi has insisted, any such answers must be able to accommodate passages such as the following:

> The being-in-itself of the world might make good sense, but one thing is absolutely certain, it cannot have the sense that the world is independent of an actually existing consciousness. The world is in principle only what it is as the correlate of an experiencing consciousness that is related to it, and as correlate of a real and not merely a possible consciousness.
>
> (Husserl 2003, p. 78)

Thus, it is not the case that consciousness and 'real being' should be thought of as simply living peacefully alongside one another, in parallel worlds as it were, but occasionally interacting, perhaps via the medium of representations resembling in some sense external objects (Zahavi 2017, p. 97). Rather they 'belong *apriori* inseparably together' (Husserl 2003, p. 73; quoted in Zahavi 2017, p. 98). And the nature of this belonging together is that of a kind of constitutive binding, in the sense that:

[10] Indeed, according to Trizio, we should resist attempting such a fit, given that this debate, and the philosophy of science in general, have 'relied on philosophical resources of remarkable poverty' (Trizio 2021, p. 305). Ultimately, this is because the philosophy of science, as currently practised, remains within the 'natural attitude' from the perspective of which both realists and anti-realists understand the world as that which lies 'downstream' of scientific knowledge and as a result 'nonsense is inevitable' (ibid., p. 306). Which is a bit harsh, to be honest.

> the objects of which we are 'conscious', are not simply *in* consciousness as in a box, so that they can merely be found in it and snatched at in it; but . . . they are first *constituted* as being what they are for us, and as what they count as for us, in varying forms of objective intention. (Husserl 1970a, II p. 169)

Rejecting claims that this constitutive relationship amounts to a form of metaphysical dependence, whether cashed out in terms of reducibility, or supervenience, or causal relationship, Zahavi has argued that the core issue here is not that of finding a way of metaphysical understanding the relationship between two different kinds of 'stuff', the mental and the physical, but rather that of getting a grip on the nature of *objectivity* (2017, pp. 98–102). Thus, the nature of the relationship is such as to tie the putatively 'objective' to the subjective, in the sense that '[n]o object is thinkable without the actual subjectivity that is capable of realizing this object in actual cognition' (Husserl 2002, p. 277; cited in Zahavi 2017, p. 102). Reducing philosophy to slogans, what should appear on Husserl's bumper sticker is 'No object without a subject and no subject without an object' (ibid.).

In other words, for a feature of reality (to use a broad phrase) to count as 'objective', and hence, for at least certain of such features, to be regarded as an object, it must be able to be cognitively 'realized' and it is of the nature of that realization that the objective will be constitutively informed by the relevant features of the subjective. Furthermore, as noted above, it is not the case that 'mind' and 'reality' are two distinct 'regions of being', as it were, which are somehow related; rather, the world, for want of a better word, exhibits (ditto) two aspects that we refer to as 'consciousness' and 'reality' or 'real being' and that we tend to reify, from the stance of the 'natural attitude', in terms of two kinds of substance. However, the reorientation effected by the *epoché* reveals that such reification is mistaken and that instead what we have is a kind of objectification, and thus manifestation, of one aspect by virtue of its being constitutively informed by the other. As we'll see, this is exemplified in London and Bauer's analysis of the measurement situation.

Nevertheless, and despite the slogan, there is an asymmetry to the relationship insofar as subjects themselves are 'self constituting' in the sense that the flow of consciousness 'constitutes itself as a phenomenon in itself' (Husserl 1991, p. 83). So, when I reflect on my own conscious activity, although the experience I have is a self-experience, it is still an experience like every other; that is, it consists in a kind of directing towards something that was already there, namely a particular cognitive act. But by reflecting on it, I transform it or, better, *constitute* it, as an object of my act of immanent perception (Zahavi

2017, p. 106). Again, this constituting through an act of reflection is central to the London and Bauer account. At the same time, and contentiously, as we shall see, it can be argued that it is by virtue of this self-experience that the ego itself is also constituted.

To say that reality is 'constituted' by consciousness is not to say that objects are created by us. Here, we need to keep the nature of the correlation in mind (Zahavi 2017, p. 108):

> It actually belongs to the essence of the intentional relation (which is precisely the relation between consciousness and the object of consciousness), that consciousness, i.e., the respective *cogitatio*, is conscious of something that is not itself. (Husserl 1973a, p. 170; cited in Zahavi 2017, p. 109)

In other words, if we reflect upon that relation we can see that it is of its essence that it be directed *to* the relevant object *by*, or *from*, the subject. We are given, as it were, the relation and through reflection we come to see the two poles of consciousness and the object of that consciousness, which thus act as the relata of the intentional relation. But that formal description, in terms of relata and relation, should not mislead us into thinking that the relata, consciousness, and the object of consciousness, are given as such, distinct and separate from that relation, by which they can then be related. Just as the object of consciousness is constituted by consciousness, so '[t]he I is not conceivable without a non-I that it is intentionally related to' (Husserl 1973b, p. 244). The two are inseparable, so that between them 'there is no space in which to turn' (Husserl 1959, p. 352); or, put another way, '[b]oth are irreducible structural moments in the process of constitution, the process of bringing to appearance' (Zahavi 2017, p. 111). And that process is one in which both subjectivity and the world are intertwined and necessary, so that we cannot have the one without the other (ibid.).

To repeat and re-emphasize, then, the phenomenological analysis of intentionality doesn't just reveal features of consciousness—and again, in this sense phenomenology differs from psychology—but also aspects of the world, as the two are bound inseparably together. Post-*epoché* we cannot simply assume that what is given to us in consciousness divides along 'natural' lines of subjectivity and objectivity. Hence, we need to reflect upon it in the manner of an essential 'seeing' and in doing so we uncover the fundamental structures of this something that reveal themselves to be intentional acts having certain 'poles' which we identify with 'the mind' and 'objects'—but ultimately, what we have is this 'mutually dependent context of being' as Beck puts it.

The upshot, then, is that by adopting the phenomenological attitude we attend to an object as it is given and by doing so, we reveal what essential structures have to be in place for the object to appear as it does. Thus:

> We analyse the intentions through which we experience things, and we describe the things as presented in those intentions. We thereby also disclose ourselves as datives of manifestation, as those to whom objects appear. The topic of the phenomenological analyses is consequently not a worldless subject, and phenomenology does not ignore the world in favour of consciousness. Rather, phenomenology is interested in consciousness because it is world-disclosing. It is in order to understand how the world appears in the way it does, and with the validity and meaning that it has, that phenomenology comes to investigate the disclosing performance of intentional consciousness. (Zahavi 2017, p. 26)

Within such a framework, of course, the 'ego' occupies a central place. Let us now consider how this ego is regarded from the phenomenological perspective, although, as we shall see, there is a 'twist' in the story.

5.6 The Ego: Lost . . .

For Husserl, the relationship between the experiencing ego and the ego's experience of itself is not in any way phenomenologically peculiar or different from the relationship between the experiencing ego and its experience of any other object. Again, with regard to the latter relationship, it is important to emphasize the distinction between the 'appearance' of a thing as a subjective connection between such appearance and an experiencing ego and as an objective connection between the appearances and the thing or object itself:

> The appearing of the thing (the experience) is not the thing which appears (that seems to stand before us in *propria persona*). As belonging to a conscious connection, the appearing of things is experienced by us, as belonging in the phenomenal world, things appear before us. The appearing of the things does not itself appear to us, we live through it. (1970a, II p. 538)

That is, we don't experience experiences, as such, we *live through them*. What our experiencing ego finds in itself, when it reflects upon the experience, are the relevant acts of perceiving, judging, and so forth (ibid., p. 540).

This analysis can then be applied to our experience of ourselves: the appearing of I myself must not be confused with the 'I' which appears. The relation of myself, as a phenomenal object, to myself as a phenomenal subject, must be kept distinct from the relation of an experience, as a conscious content, to consciousness in the sense of a unity of such conscious contents (which Husserl calls the 'phenomenological subsistence of an empirical ego'; ibid., p. 539). In common discourse the ego is also treated as an empirical object and, from this perspective, however much our scientific understanding of it may change, it will remain 'an individual, thinglike object, which, like all such objects, has phenomenally no other unity than that given it though its unified phenomenal properties, and which in them has its own internal make-up' (ibid., p. 541).[11] If we approach the ego phenomenologically, then it is reduced to nothing more than a 'unity of consciousness' or a 'real experiential complex'. Husserl concluded:

> The phenomenologically reduced ego is therefore nothing peculiar, floating above many experiences: it is simply identical with their own interconnected unity. In the nature of its contents, and the laws they obey, certain forms of connection are grounded. They run in diverse fashions from content to content, from complex of contents to complex of contents, till in the end a unified sum total of content is constituted, which does not differ from the phenomenologically reduced ego itself. These contents have, as contents generally have, their own law-bound ways of coming together, of losing themselves in more comprehensive unities, and, in so far as they thus become and are one, the phenomenological ego or unity of consciousness is already constituted, without need of an additional, peculiar ego-principle which supports all contents and unites them all once again. Here as elsewhere it is not clear what such a principle would effect'. (ibid., pp. 541–2)

From the phenomenological perspective, the ego—as something over and above the complex of conscious contents—has effectively evaporated away.

There is a problem however: how is it, then, that our 'inner consciousness' or 'inner perception' appears to possess what Descartes called a 'self-evident', or, as Husserl prefers, an '*adequate*', quality? A perception achieves adequacy if the object of the perception is actually and *exhaustively* present within it.

[11] Just as we might reject a metaphysics of physical objects that posits some underlying substance in favour of a conception of objecthood in terms of a set or bundle of properties, although there is nothing tying the bundle together.

Doesn't the self-evidence or adequacy of the Cartesian 'sum' restore the ego which has been phenomenologically evaporated? No; what the quality of adequacy ultimately attaches to are only the *judgments* of inner perception themselves (ibid., p. 544). The Cartesian primary and absolutely certain focus is thus constituted only by what is adequately perceived; and this in turn is nothing more than the end result of the phenomenological reduction of the empirical ego. There is no Cartesian, substantial 'kernel' over and above this.

Neither is there any Kantian 'pure ego', understood as the 'unitary centre of relation' to which all conscious content must be referred. This ego, as such a subjective centre, cannot be or resemble such a content. Hence it cannot be described, since any such description would render the ego as an object but to be the pure ego in this Kantian sense is precisely *not* to be an object but to be that which is opposed to all objects.[12] Husserl's opinion here was blunt:

> I must frankly confess...that I am quite unable to find this ego, this primitive, necessary centre of relations. The only thing I can take note of, and therefore perceive, are the empirical ego and its empirical relations to its own experiences, or to such external objects as are receiving special attention at the moment, while much remains, whether 'without' or 'within', which has no such relation to the ego. (1970a, II pp. 549–50)

Thus, the Kantian ego goes the way of the Cartesian ego: reduced to data that are 'phenomenologically actual' (ibid., p. 550) in the sense that all that we have is the 'complex of reflectively graspable experiences' (ibid.). From this phenomenological perspective, the conscious intentional relation between the ego and its objects is simply that between the 'total phenomenological being of a unity of consciousness' (ibid. p. 550) and the intentional experiences, whose object, in this case, is I, myself.

From such a perspective we can better understand Husserl's insistence that it is always questionable to say either that objects 'enter consciousness' or that the ego 'enters into a relation' with such objects (1970a, II p. 557). Such expressions are misleading in two respects: they suggest, first of all, the existence of real events or real relations taking place between the ego, on the one hand, and the object on the other; and second, that there exists a relation between two things—an act and an intentional object—which are both present

[12] Again, drawing on the analogy with material objects, this pure ego is like the substance, famously characterized by Locke as 'something we know not what', which must underlie, support, or whatever, the properties of the thing.

within consciousness in equally real fashion. As Husserl insisted, if we must talk of relations in this context, we should try to do so in a way that avoids the temptation of giving such relations psychological reality.

With regard to the second misunderstanding in particular, Husserl noted that it is suggested by the phrase 'immanent objectivity', used to express the 'peculiarity' of intentional experiences that they are directed towards or 'aimed at' their objects. However, he cautioned, they do so in an *intentional* sense: that is, to say that the experience is aimed at 'the object' means nothing more than that certain experiences are present (which may then differ in character as to whether the object is aimed at presentatively, or judgingly, or desiringly or whatever). The point is, there are not two things present in experience—the object and the intentional experience directed upon it—there is only the intentional experience (1970a, II p. 558). If the intentional experience is present, then, Husserl insisted, '*eo ipso* and through its own essence' (ibid.), the intentional relation is 'achieved', or, equivalently, the object is 'immanently present', as noted above.

With regard, now, to the first misunderstanding, where it is imagined that consciousness or the ego and the 'matter in consciousness' become related in a real sense, Husserl wrote:

> In natural reflection, in fact, it is not the single act which appears, but the ego as one pole of the relation in question, while the other pole is the object. If [*sic*] one then studies an act-experience, which last tempts one to make of the ego an essential, selfsame point of unity in every act. This would, however, bring us back to the view of the ego as a relational centre which we repudiated before. (1970a, II p. 561)

When we simply 'live in the act', when we are absorbed in the perception itself, then the ego, as a relational centre, is 'quite elusive'. The idea of the ego may be waiting in the wings, as it were, ready to appear on stage, or rather, 'to be recreated anew' (ibid., p. 561) but it is only when it is so recreated that we refer to the object in a 'descriptively ostensible' way. In that description what we then have is a complex act which presents the ego on the one hand and, on the other, the presentation or judgement or whatever, together with its relevant subject matter. Of course, in each act there is an ego which is intentionally directed to some object, but this is not to say that there is some *thing*, some 'essential, selfsame point of unity', present in every act. It is only in such a description, *performed after an act of reflection*, that the ego emerges:

> The sentences 'The ego represents an object to itself', 'The ego refers presentatively to an object', 'The ego has something as an intentional object of its presentation' therefore mean the same as 'In the phenomenological ego, a concrete complex of experiences, a certain experience said, in virtue of its specific nature, to be a presentation of object X, is really present'.... In our *description* relation to an experiencing ego is inescapable, but the experience described is not itself an experiential complex having the ego-presentation as its part. We perform the description after an objectifying act of reflection, in which reflection on the ego is combined with reflection on the experienced act to yield a relational act, in which the ego appears as itself related to its act's object through its act. Plainly an essential descriptive change has occurred. The original act is no longer simply there, we no longer live in it, but *we attend to it and pass judgment on it.*
>
> (1970a, II pp. 561–2; Husserl's emphasis)

5.7 The Ego: ... and Found

As we shall see, these passages provide the key to understanding the London and Bauer account of the measurement situation. There remains a further problem, however, and it is summed up in a footnote, inserted by Husserl in the second edition of *The Logical Investigations* and attached to the above claim that he is 'quite unable to find this ego, this primitive, necessary centre of relations':

> I have since managed to find it, i.e., have learnt not to be led astray from our grasp of the given through corrupt forms of ego-metaphysic.
>
> (1970a, II p. 549)

And earlier, in another footnote, he recorded that:

> The opposition to the doctrine of a 'pure' ego... is one that the author no longer approves of, as is plain from his *Ideas*. (ibid., p. 542)

This apparent recantation of his earlier view appears to bring the ego back on to centre-stage. However, we need to consider more carefully that phrase, 'corrupt forms of ego metaphysic'. One interpretation is that what Husserl meant by this is 'the tendency to conceive of a pure ego as a *substantive* res cogitans of some sort, something which is substantial independently of our constitution of it' (Taylor 1998, p. 241). This tendency of course can be traced back to Descartes, who, Husserl maintains, was 'dominated in advance by the Galilean certainty of a universal and absolutely pure world of physical bodies' (1970b, p. 79), which generates the split between the 'material world' and consciousness (we'll come back to this). This split becomes concretized, as it were, in Descartes' philosophy, in terms of *res extensa*, on the one hand, and *res cogitans*, on the other. As a result, Descartes betrays his own *epoché* by identifying the ego with (or in Husserl's terms, substituting it *for*) the individual soul, which is the soul of the naturalistic attitude (Trizio 2021, p. 283).[13] Crucially, such a 'physicalistic' view of the psyche misses the intentional character of consciousness (ibid., p. 283).

Of course, at no point did Husserl posit *that* sort of ego. Rather what he managed to find, or rediscover, is a kind of phenomenological pure ego 'as a descriptive principle relating to the nature of experience' (Taylor 1998, p. 242). There is no substantive core to *this* ego, nothing over and above its relation to the stream of consciousness (see Trizio 2021, p. 143). Indeed, as Husserl himself emphasized, beyond its 'modes of relation', the Ego is empty and 'has no explicable content, is indescribable in and for itself: it is pure Ego and nothing more' (Husserl 1982, p. 191).[14]

Thus, in his later work, (the *Ideas*; Husserl 1982), Husserl presented the Ego as a kind of posit that is required by the 'reflective regard'. Let us consider this mental act in a little more detail, since it will turn out to be an important component of London and Bauer's analysis.

This 'reflective regard' is a form of 'directedness-to' which arises when a mental process is 'actional' in the sense of being effected in the manner of the *cogito*:

> To the cogito itself there belongs, as immanent in it, a 'regard-to' the Object which, on the other side, wells forth from the 'Ego' which therefore can never be lacking. This Ego-regard to something varies with the act: in perception, it

[13] It has been suggested that the combined effect of these substitutions—of 'nature' for the lifeworld and of the soul for the transcendental subject, respectively—'has set the agenda for the endless discussions about realism and idealism that have marked philosophical modernity' (Trizio 2021, p. 283).

[14] However, Husserl subsequently added a warning to the effect that although in the *Ideas* (Husserl 1982) he offered a short route to the transcendental *epoché*, via a kind of immersion in the Cartesian *epoché* of the *Meditations* while critically purifying it of Descartes' prejudices and confusions (1970b, p. 155), this comes at a cost: although it yields the transcendental ego 'in one leap, as it were', it brings this ego into view as apparently empty of content, 'so one is at a loss, at first, to know what has been gained by it, much less how, starting with this, a completely new sort of fundamental science, decisive for philosophy, has been attained' (ibid.). In the *Crisis* (1970b) he presented a new, 'concretely plotted' way, to which we shall return.

> is a perceptual regard-to; in phantasying, an inventive regard-to; in liking, a liking regard-to; in willing, a willing regard-to; etc. This signifies that this having the mind's eye on something, which pertains to the essence of the cogito, of the act as act, is not itself, in turn, an act in its own right and especially must not be confused with a perceiving (no matter how broad a sense) nor with any sorts of acts akin to perceptions. (1970a, I pp. 75–6)

When one is 'living in' the cogito—that is, *not* reflecting upon it—we are not conscious of the 'cogitatio' as an intentional object but we become conscious of it through a 'reflective turning of regard' (ibid., p. 78).

Consider, as an example, a piece of paper, lying in front of me (ibid., pp. 69–71). In perceiving the paper, 'I seize upon it as this existent here and now' (ibid., p. 70) and this seizing-upon involves the singling out of the paper from the 'experiential background' consisting of other objects—books, pens, Pepsi Max cans, etc.—which are not seized upon. Thus, every perception of a physical thing has a 'halo of *background-intuitions*' (ibid., his emphasis) and by a 'free turning of "regard"' (1970a I, p. 71), we can bring our mental attention to bear on these other objects so that they become intended to *explicitly* rather than implicitly. Physical objects cannot be the subject of this regard without being seized upon, but this is not the case with mental processes. This 'regard to' which distinguishes actionality, in the above sense, does not coincide with the 'heeding' of an object of consciousness in which it is seized upon and picked out (ibid., p. 72). Consider the act of valuing, for example (ibid., pp. 76–7): in such an act we have *regard to* the valued, but we do not seize upon it as somehow separate from the thing itself. It is the thing, *as a valued thing*, that we seize upon but only after an 'objectifying turn'.

Using this distinction, we can get a better grip on the different kinds of being possessed by 'immanent' mental processes and 'transcendent' physical objects: when we perceive something immanent, rather than transcendent, this perception, as a reflective regarding of, *guarantees* the existence of its object. Even if what 'hovers' before one is a figment of one's imagination, still the hovering itself, as a hovering, cannot be invented but, as with any other mental process, must exist absolutely (1970a, I p. 101). Hence the perception of something immanent is indubitable, in the sense that there can be no failure of reference. This is not so for something transcendent, of course. This then leads to a further difference between the physical and mental, which bears on the apparent retention of the pure ego: the positing of things in the world is always a contingent positing but the positing of my 'pure ego', as—crucially—

the subject of mental acts—is necessary and absolute in a sense we shall examine shortly.

All of this seems deliberately and explicitly Cartesian[15] but Husserl then added the phenomenological attitude which excludes, or 'parenthesizes', the whole 'psychophysical world of Nature'. What we do when we adopt this attitude in this context is 'direct our seizing and theoretically inquiring regard to *pure consciousness in its own absolute being*' (ibid., p. 113; Husserl's emphasis). Instead of 'living in' our mental processes, with their cogitative positings, as we do when we adopt the natural attitude, we effect acts of reflection directed to them. What we are now living in, when we have adopted the phenomenological attitude, *are these acts of reflection themselves*. These have as their datum, the 'infinite field of absolute mental processes' (ibid., p. 114). It is this, of course, which is the fundamental field of phenomenology and which is left as the 'phenomenological residuum'.[16]

5.8 Reconciliation

The 'reflective regard', then, is a kind of *tool* for exploring the 'infinite field of absolute mental processes' through effecting acts of reflection (Husserl 1982, p. 174). All such acts necessarily have the form of the 'cogito' and it is of the nature of such acts of reflection that they are not only directed towards some object, but that they include a reference to an ego (otherwise how could they be of the form 'cogito'?). If the ego *in this sense* were to be excluded then phenomenology could not avail itself of the very tool it needs; it would be, effectively, impotent.

However, it is not the case that we posit the ego first and then consider the 'welling forth' of the *regard* as some kind of property of it, but rather we start with the *regard*, which is at the heart of phenomenology, and from that, conclude the presence of an ego. But just because we need to refer to it insofar as acts of reflective regard are essential, this does not mean that the ego is not

[15] Husserl asserted that these inferences do justice, 'at least', to a core of Descartes's *Meditations* 'which only lacked a pure, effective development' (1970a, I p. 104). It is this pure effective development that the phenomenological attitude provides.

[16] Of course, as Husserl insisted, anyone can effect a reflection and bring consciousness 'within the sphere of his seizing regard', but effecting a reflection is not necessarily to effect a phenomenological reflection nor is the consciousness seized upon necessarily pure consciousness. 'Radical considerations' such as are involved in the parenthesizing of the natural world and the dropping of the natural attitude, are needed in order to arrive at the cognition that there is this pure field of consciousness which is not a 'component part' of nature (1970a, I pp. 114–15).

relative to such acts, much less some sort of Cartesian substance. This is clear when Husserl writes that when we effect the *epoché*, not only is the whole natural world excluded or parenthesized, but so also is the 'I, the human being' (Husserl 1982, p. 190). What is left is the 'pure act-process' with its own essence which includes, by necessity, the pure ego as the subject of the act.

What is it then? The 'Ego' has no 'explicatable content' and is indescribable 'in and for itself' (ibid., p. 191). It has no properties, 'does not harbour any inner richness' (1982 Book II, p. 111) and is absolutely simple and undivided. As such, it can be understood as nothing but a 'place holder', a 'that which' is intending (Taylor 1998, p. 277). We recall that Husserl described the ego as a pole which stands in relation to the object-pole.[17] Just as the bearer or substrate of the 'exact determinations ascribed in physics' (Husserl 1982, p. 119; see also p. 85) is an 'empty X',[18] so likewise is the ego. The perception of something is 'an empty looking at the Object itself on the part of an empty "Ego" . . . ' which seizes upon the object (ibid., p. 83). Thus, 'To say that all reflected upon experiences are ego related is merely to say that they "appear as" originating from an ego and directed towards an object. That ego qua subject-pole has no properties, no personality, it is simply the putative subject of experience' (Taylor 1998, p. 277).

But what about those dramatic footnotes in which Husserl claimed to have found the ego again? According to Taylor:

> What Husserl means when he says that he has learnt not to be led astray by corrupt ego metaphysics is that he has learnt that to say that there is in fact an ego in consciousness is not to posit some substance which is unknowable in itself. Husserl is in *Ideas* still denying that there is an ego substance or an ego 'in itself'. He is still saying that it is absurd to make claims for an ego which has a certain nature independently of points of view or context. However, in *Ideas* he recognises that to say that every time we reflect on our consciousness we find an ego, just is to say that there is an ego from the phenomenological point of view. The fact that the act of reflection is in part

[17] In a supplement to Book Two of the *Ideas*, Husserl writes, 'Just like any object-pole, the Ego-pole is a pole of identity, a centre of an identity, and is an absolutely identical, *though non-autonomous*, center for affects and actions' (1982 II, p. 324; my emphasis). Again, a comparison with physical objects is made: 'Just as an object has its identity as a pole of relatively or absolutely permanent properties, and just as every property is something identical though non-autonomous (*in* the pole), so the same holds for the Ego' (ibid., Husserl's emphasis), although the Ego is a pole of acts rather than properties. See also Husserl 1970b, p. 171.

[18] This 'empty X' is the bearer of 'mathematical determinations and corresponding mathematical formulae' (1982, p. 85) and exists in the 'objective space' of physics, of which 'perceived space' is merely a sign; we'll come back to this.

> responsible for the appearance of that ego does not militate against its existence. Put another way, to say that the act of apprehending an object is at least in part responsible for the properties that object has is not to say that the object does not really have those properties. The intuition that we really ought to say that the object does not 'really' have those properties relies on a notion of substance, or a notion of the object 'in itself'. If this notion is surrendered the intuition loses its force altogether. (ibid., p. 282)

Hence the pure ego of the *Ideas* should be understood as the phenomenological ego of the *Logical Investigations*; that is, as standing for the unity of a particular stream of consciousness, although reconceived according to the dictates of the *epoché*.

This issue of the status of the ego was also taken up by Aron Gurwitsch, with whom London had many conversations, both in Paris and the USA.

5.9 The Ego: Discarded

Like London, Gurwitsch also had an academic background in both physics and philosophy, studying the former with Planck, no less, as well as learning mathematics with Karatheodory and Schur.[19] His philosophy teacher, Stumpf, sent him to Freiburg to study with Husserl,[20] where the latter echoed Newton in telling Gurwitsch 'perhaps you see further than I do because you stand on my shoulders' (https://web.archive.org/web/20120204102358/http://www.gurwitsch.net/bio.htm). Drawing on a comparison with logic and mathematics, Gurwitsch suggested that phenomenology should be conceived of as a kind of 'mathematics of consciousness', not least because its concern is with 'possibilities, compatibilities, incompatibilities, and necessities', on the basis of which appropriate transformation laws could be established (see Mohanty 1994, p. 937).

[19] Gurwitsch was later an instructor in physics at Harvard, a lecturer in mathematics at Wheaton College, and then assistant professor in the same subject at Brandeis University (https://web.archive.org/web/20120204102358/http://www.gurwitsch.net/bio.htm). He eventually became full professor on the Graduate Faculty of the New School for Social Science Research in New York City, replacing his friend Schutz on the latter's death (Zaner 2010, pp. xvii–xvii).

[20] Stumpf supervised Husserl's habilitation thesis and the *Logical Investigations* is dedicated to him. He also founded the Berlin Institute of Psychology, which was the home of Gestalt psychology and he is perhaps most well known for his work in descriptive psychology although he also wrote on the theory of knowledge and the philosophy of mathematics. In a posthumously published work he was quite sharply critical of phenomenology on the grounds that by bracketing the existence of objects, it precluded any contribution from the physical sciences (Fisette 2019). This of course is a misunderstanding.

He went on to develop a 'constitutive' approach that brought together phenomenology and Gestalt theory,[21] with the latter and its anti-reductionist emphasis on the 'whole' underpinning the 'noematic phenomenology of perception' (see Mohanty 1994 for a useful summary of the extent to which Gurwitsch deviated from the Husserlian framework). Crucially, Gurwitsch rejected the idea of a transcendental consciousness and thus advocated a 'non-egological' account. This too had, at its heart, the idea that consciousness has only a constitutive function, consisting in the correlation between a given act and its intended object (Mohanty 1994, p. 940). Crucially, however, he insisted that 'something is wrong with the transcendental ego' and '[s]omething must be dropped: in you transcendental, in myself and Sartre ego is the drop-out' (letter from Gurwitsch to Schutz, 19 December 1940, in Grathoff 1989, p. 31).

The mention of Sartre is a reference to the latter's paper (Sartre 1936–7), in which he endorsed Husserl's elimination of the ego in *The Logical Investigations* and maintained that the ego-theory presented in the *Ideas* is incompatible with the phenomenological analysis of consciousness (as we've just noted, Taylor has argued that there is no incompatibility here). Gurwitsch then expanded on Sartre's 'non-egological' line, arguing that the 'pure' or transcendental ego of the *Ideas* or *Cartesian Meditations*, is simply not found 'as a datum when acts are considered as they are or have been experienced... unless one adopts the attitude of reflection in their respect' (1941, p. 327). His point here is that our acts are impersonal in the sense that when we are dealing with the object of the act, we are aware of the object, and of our dealing with it in whatever way but we are *not* aware of the ego, much less of the ego's intervening in that dealing (ibid.). As a result, there is simply no function for the ego to assume (1941, p. 328). Indeed, he suggested, adopting an egological conception of consciousness is to effectively substantialize the latter (ibid., p. 329) and just '[a]s in regard to material things, thinking in terms of substantiality gave way to thinking in terms of functions and relations, so... it will have to do in all fields of experience' (ibid., p. 337).

So, the ego only appears when we *reflect upon* our acts:

[21] There is a lot more to say about the connections between Gestalt theory and quantum physics. Born and Einstein, for example, were both friends with some of the central figures involved and both Born and Schrödinger subsequently appealed to notions of 'Gestalt' in reflecting on some of the philosophical implications of QM. On the Gestalt side, Köhler drew on the holistic nature of quantum theory in support of his view and Cassirer brought together group theory and Gestalt theory in his paper 'The Concept of Group and the Theory of Perception' originally published in 1938 and republished in 1944 (Cassirer 1944). For more on the development of Gestalt theory at this time, see Ash 1998.

> As long as we do not adopt the attitude of reflection, the ego does not appear. On the level of non-reflection there is no ego at all. A conscious act, inasmuch as it is free from reflection, does not deal with the ego and is not related to it in any way whatever. (1941, p. 329)

By 'reflection' here, Gurwitsch understands the grasping of some act, *A* say, by another, *B*, in which the former becomes the object of the latter. However, *B* itself is not to be taken as an object grasped by some further act; rather, *B* is experienced non-reflectively, just as an act is when dealing with a non-mental object, as sketched above. It is through such acts of reflection that the ego is manifested, not in the sense that is revealed to be 'there' as that from which *A* emanates in some sense, but as the result of *B*'s grasping of *A*—such grasping conceptualizes *A* as a relation pointing to or having an object as one pole and, necessarily for it to be a relation, the ego as the opposite pole. As he maintained:

> By the mere fact of being grasped by an act of reflection, the grasped act then acquires a personal structure and the relation to the ego which it did not have before it was grasped. *Reflection gives rise to a new object—the ego*—which appears only if this attitude is adopted. (1941, p. 331; his italics)

However, since the grasping act, *B*, is not itself grasped, it has no egological structure, so 'the ego in question is that of the grasped, not of the grasping act' (ibid., p. 331). Likewise, in the absence of such grasping, '[c]onsciousness has no egological structure; it is not owned by the ego; its acts do not spring from a source or center called the ego' (ibid., p. 330).

Furthermore, although through reflection the given act is modified in that it is 'objectivated', this does not mean that the act itself comes into existence due to its being grasped (ibid., p. 332):

> Reflection is disclosing, not producing; the alteration it conveys to the act concerns the mode in which this act is experienced, it concerns the mode rather than the what of the awareness. (1941, p. 332)

Is the ego itself disclosed then? No; through reflection 'the act is brought into relation to an object which did not appear before the act was grasped' (ibid., p. 332).

It is important to note, however, that in being conscious of an object, being *aware* of being conscious of it does not count as 'reflection'—'to know that

I am dealing with the object which, for instance, I am just perceiving, I must not experience a second act bearing upon the perception and making it its object' (Gurwitsch 1941, p. 330). Indeed, it is in this sense that appearing is the same as being when it comes to consciousness and hence it is endowed with absolute character. It is this transcendental character that is revealed by the phenomenological reduction and thus consciousness, so 'reduced', can be conceived of as a 'pure field of experienced acts which are related to objects' (Gurwitsch 1974, p. 187).[22]

Gurwitsch's claim that we can eliminate the ego in this way was almost immediately rejected by Schutz who responded by insisting that if the grasping act *B*, above, deals with the ego at all, then this ego is grasped by *B* as performing (or having performed) act *A*. If a further act, *C*, then grasps *B*, and through it, act *A*, the ego with which *C* deals is grasped as having performed *B* as well as *A* and, furthermore, it is grasped as the same ego, despite all the changes it undergoes 'in and by the flux of the stream of experience' (letter to Gurwitsch 1941, in Grathoff 1989, p. 46). Gurwitsch's response was blunt. In maintaining that the grasping of *A* by an act *B* reveals the acting subject, Schutz had appealed precisely to that which Gurwitsch questioned (letter to Schutz from 1941, in Grathoff 1989, p. 47): 'All that follows concerning the identity of the ego are precisely the things that appear problematic to us' (ibid.).

So, the idea here is the following: we begin with the natural attitude, according to which we conceive of our actions and our dispositions towards others and so on, as emanating from some core sense of self, or 'I', that is substantialized as the ego. However, once we perform the *epoché*, thereby bracketing off that attitude, and undertake the phenomenological reduction, the central device of which is the reflective 'regard-to', we can see that when we apply that device to a given act, *A*, say, there is no such ego to be had, except as the 'pure field' of all such acts of consciousness. The ego as the genesis of our phenomenological exploration of the structure of consciousness is seen to be not 'contained' within that structure.

Here again, in his reply to Schutz, Gurwitsch made a comparison with the progress of science, which he took to consist in the replacement of the category of substance with that of function and relation (Grathoff 1989, p. 48). Likewise with the ego, which gets reconceived as no more than a bundle of experiences.

[22] This idea can be traced back to the lectures he gave at the Sorbonne in 1937 and which were incorporated into the book Gurwitsch was preparing for publication when he left France in 1939 (see Gurwitsch 1974, p. 153).

Of course, there is the question, just as with material objects, of what 'ties' the bundle together but this, he averred, is an issue to do with the empirical ego and so the problem can be hived off to psychology rather than phenomenology.

Of course, someone could protest that this implies a reconceptualization of what an act is but that is precisely what the phenomenological reduction yields. As Gurwitsch noted, the phenomenological 'finding' that results is the 'living through' or self-awareness of an act, in the sense that what is given in perception, say, is not only the thing perceived, but the very perceiving itself (ibid., p. 48). What is not given, however, is the ego, since the awareness of the present perceiving is limited to the given act. In addition, there is a sense of time and duration which contribute to the formation of the thematic field:

> The *grasped* experience finds itself in a new field which consists merely of constituents of the stream of consciousness. If the reflection is carried through ideally in memory, there is a chain of all acts which have ever been lived through, in the ideal case without any gaps.
>
> (Grathoff 1989, pp. 48–9)[23]

Hence, we have a sense of identity, but only in the sense that each act joins the chain, and there is only one chain. And if you want to call that the 'ego', go right ahead but it cannot exhibit any egological activity and cannot be considered a 'source-point'—'*it does nothing and suffers nothing*' (ibid., p. 49). This 'ego' is nothing but the totality of the stream of consciousness.[24]

The significance of the reflective regard-to as disclosing the ego, not in the sense of a substantive something that is always behind the scenes, as it were, but as that which is brought on stage by virtue of that reflection, is something we shall return to in the context of London and Bauer's analysis.

After working with Geiger at Göttingen,[25] Gurwitsch and his wife moved to Paris in 1933, where (as we've already noted) he too taught at the Sorbonne (Grathoff 1989, p. xxii), initially on Gestalt theory and then eventually on his

[23] Grathoff remarks that Gurwitsch mentioned his work on the field theory of consciousness for the first time in 1942, but these letters suggest a slightly earlier date (Grathoff 1989, p. 134).

[24] Schutz went on to reply, insisting that he had not assumed that which was in contention and that he could not understand how Gurwitsch had managed to eliminate the ego and associated egological structures from the grasped act, since such an act can be analysed in terms of the acting and, crucially, the actor (letter to Gurwitsch 1941, Grathoff 1989, p. 53). Of course, it was precisely this sort of analysis that Gurwitsch rejected. Schutz continued by noting that unless we are able to bring the acting ego into view, we could not make sense of talk of a performance being spontaneous. Gurwitsch did not reply in his next letter but we can speculate that his response would have been along the same lines as the above—the difference between activity and passivity may suggest the operation of an ego within the natural attitude but not subsequent to the phenomenological reduction.

[25] Geiger had previously studied and taught at Munich, where he became a member of the Munich Circle of phenomenology, together with Pfander, London's teacher, whom he had known as a student.

'constitutive phenomenology'. Much of the content of these lectures was incorporated into his later work, *The Field of Consciousness*, published in 1964 (in 1957 in French; Gurwitsch 1964). It was in Paris, as we have noted, that he met London and had extensive philosophical conversations with him.[26]

Gurwitsch emigrated to the USA in 1940, where he was instrumental in helping to establish the American 'branch' of phenomenology (see Marcelle 2019; also Zaner 2010). While teaching at the New School for Social Research in New York he had numerous discussions about the philosophy of science, in particular, with Kockelmans, who acknowledged that Gurwitsch 'had an immediate knowledge of [the] sciences themselves, which put him at once on a par with the scientists who devote their entire lifetimes to the study and teaching of these sciences' (Kockelmans 1975, p. 30).[27]

Gurwitsch's phenomenological philosophy of science followed Husserl in grounding the sciences in 'careful phenomenological analyses of the life–world, which analyses then, in turn, were to be given their final foundation in transcendental phenomenology' (ibid.). Thus, he wrote:

> rational, mathematical and finally purely functional interconnections [*Zusammenhänge*] take the place of the regularities and normal sequences of the 'life-world'. (E.g.: in the life-world we know that many things are heavier than others. We experience this in carrying, shoving, etc. Formulated rationally this yields the concept of specific weight.) The problems of physics are to be formulated on this level, and a theory of science which does not begin here, at the genuine beginning, is hopeless, as hopeless as everything which goes under the title 'philosophy of science'.
>
> (Letter to Schutz from 1945 in Grathoff 1989, p. 750; see also Kockelmans 1975, p. 31)

[26] Grathoff records that in Paris 'Merleau-Ponty began attending Gurwitsch's lectures very early and later came to the Gurwitsch home for discussions every other week' (Grathoff 1989, p. xxii). We shall consider Merleau-Ponty's work in Chapter 8.

[27] Kockelmans himself also wrote on the relationship between phenomenology and the physical sciences, arguing that we should understand the character of the latter on the basis of an intentional and constitutional analysis of the practice of modern physics, focusing on the activity of the physicist herself and the intentional correlate aimed at through the attitude adopted (Kockelmans 1966). Although in the Dutch original of this work both relativity theory and QM are discussed, the chapter on the latter was not included in the English edition and a review of the book noted that Kocklemans did not venture far into the thicket of issues concerning the ontological status of physical entities (Kisiel 1967, p. 139). According to van Fraassen (who read Kockelmans book in the original Dutch and took his seminars), 'It is . . . not difficult to see how this view connects with empiricism, for it is at least tempting to read "intentional correlate" as "representation" and therefore to take what is meant by the world of physics as an image, that is, as the scientific representation of the physical world. Or with what I have called empiricist structuralism, which does not focus on the structures in (or constituting) nature but on scientific representation as structural' (private communication). However, he also noted, 'it is also clear that this is not a place to look for a special take on the measurement problem' (ibid.).

This shift, or transformation, is accomplished through acts of consciousness and so, Gurwitsch insisted, the physical sciences do not undermine the principle 'according to which every object and every level of being must be conceived of relative to the acts of consciousness in which they are constituted and of which they are the objective correlates' (1974, p. 182). The objects of physics, of course, are objective correlates of a higher order, since they are obtained from those other objective correlates of consciousness which in part constitute the life-world (as we'll see, this is related to the claims previously made by Husserl in 1970b). And to disentangle these relationships and expose the existential meaning of such objects requires 'appropriate meditations', of the kind that phenomenology can supply. This analysis also applies to QM, of course (Kockelmans 1975, p. 34).

And in this regard, Gurwitsch referred to London and Bauer's work in the context of his 1946 review of a defence of causation even in the face of the 'indeterministic crisis' in current physics (Gurwitsch 1946). Here it was argued that if we grant that observation disturbs the motion of a particle and that we cannot, even in principle, ascertain at a given time the position and momentum of an electron, say, with sufficient precision as to predict its future location, this has to do with the relationship between the observer and the object and not the behaviour of the latter itself which is still causally determined. Gurwitsch dismissed this as resting on a form of 'philosophical realism' that stood in contrast with positivism. However, Gurwitsch continued, these are not the only games in town and although '[t]here does not yet exist a phenomenological philosophy of science' (1946, p. 341), it could safely be assumed that when elaborated it would follow the lines suggested by Cassirer and Brunschvicg, both of whom considered science in terms of 'the constitutive and constructive activities of scientific reason' (ibid.).[28] Gurwitsch then wrote:

> In this connection, I wish to call attention to a work by F. London and E. Bauer [*La Theorie de l'Observation en Mecanique Quantique*] in which the authors with explicit reference to Husserl and Cassirer speak of the act of observation and measuring as of an objectivating act through which a new objectivity is constituted. Quantum mechanics, according to these authors,

[28] Brunschvicg was a professor at the Sorbonne who supervised Simone de Beauvoir's Masters dissertation. In his essay on metaphysics and science he remarked on the development of the 'new physic' and its 'setting up of relations of incertitude' and noted that 'theories of knowledge capable of taking into account the subject and object, in order to put them in relation, are the only ones who survived scientific advances' (Brunschvicg 2006, p. 69).

> rejects the naive realistic idea of objects existing entirely independently of all observation and having their measurable properties, whether the latter are actually measured or not. For the foundation of intersubjective scientific objectivity, naive realism is by no means required. (ibid., pp. 341–2)

Given this, it is perhaps not stretching speculation too far to suggest that while in Paris together, Gurwitsch and London would have discussed their views on a phenomenological approach to physics, including quantum theory and that such conversations may have had a significant influence on the content of London and Bauer's 'little book', especially when it comes to the nature of the ego or the 'I'.[29]

[29] Unfortunately, there do not appear to be any letters between the two in the Gurwitsch archive at Duquesne University (I'd like to thank Jeff McCurry and Sabrina Bungash for looking for me).

6
London and Bauer Revisited

6.1 Introduction

Let us finally now turn back to the London and Bauer manuscript in the light of everything we have covered so far. Just before we do, it is perhaps worth saying something about it as a text. Wigner called it a 'little book', as we noted, and de Broglie referred to it as a 'pamphlet' (de Broglie 1957, p. 31). Such terms are indicative of its ambiguous status as a material object. At only fifty-one pages long, it is too short to count as a textbook, but too long to be considered an academic paper (at least as such things are usually regarded). Furthermore, it did not appear in a scientific journal but was published 'under the direction' of Langevin by Hermann & Co., a Parisian publishing house founded in 1876 by the mathematician Hermann, as part of their series *Actualités Scientifiques et Industrielles*. The series itself was initiated by Enrique Freymann,[1] who was prompted by none other than de Broglie.[2] The end papers list some of the works published in this series, including those by de Broglie himself, as well as Curie and Debye, and also Carnap, Reichenbach, and Tarski (see also: https://www.editions-hermann.fr/la-maison).

The significance of textbooks in helping to legitimate a theory is well documented (see Kragh 2013; also French 2020, pp. 205–7) but Simon has recently emphasized the importance of historical reflection on the form and status of the relevant sources, including but going beyond textbooks, as well as on their intended audience (Simon 2022). So, granted that London and Bauer's 'pamphlet' was not a textbook per se, what role was it intended to serve? Why did it take the form it did? And who was the intended audience? We can begin to answer such questions by considering its opening pages.

They begin with a preface from Langevin, who we recall was Bauer's boss:

> Quantum mechanics has brought an essential advance to science, the finding that in every experiment or measurement there inescapably enters the

[1] Hermann's son-in-law and Mexican cultural attaché, who had taken over the reins.

[2] Freymann had extensive connections in physics and mathematics and through de Broglie, he was introduced to Langevin (and also Einstein).

A Phenomenological Approach to Quantum Mechanics: Cutting the Chain of Correlations. Steven French, Oxford University Press. © Steven French 2023. DOI: 10.1093/oso/9780198897958.003.0006

> duality between subject and object, the action and reaction of observer and system observed, the observer and measuring system being viewable as one entity. (London and Bauer 1983, p. 217)[3]

It continues:

> This present work, where the authors expand lectures given by one of them at the Sorbonne,[4] demonstrates the precision and clarity with which the formalism of quantum theory expresses this representation by the wave function of the information acquired by the observer, and the manner in which each new measurement intervenes to modify this representation.
> (London and Bauer 1983, p. 218)

In this respect, then, the 'pamphlet' is much like many textbooks, based, as it was, on a series of university lectures. However, rather than setting out the basic principles of the theory as a whole, its primary aim was to lay down the fundamentals of *the process of measurement.*[5] Clearly, despite the passage of over a decade since the publication of the foundational works of Heisenberg, Schrödinger et. al., as well as that of the more recent book by von Neumann, there was felt to be a need for such a clear statement. Furthermore, it is worth mentioning that the title of the series in which it was published can be literally translated as 'Scientific and Industrial News'.[6] The idea of the series was to

[3] 'Un des progrés essentiels apportés à la Science par la Physique quantique est la manière fondamentale dont elle fait intervenir dans toute experience ou operation de mesure, la necéssaire dualité du sujet et de l'objet, la présence simultanée et l'action réciproque de l'observateur et du systeme observé, l'observateur pouvant etre considéré comme faisant corps avec ses appareils de mesure' (London and Bauer 1939, p. 3). The translation by Google Translate perhaps brings more fully into view both the correlative relationship between the observer and the system and also the idea of the former becoming one with their measurement device (something we shall return in our discussion of QBism): 'One of the essential advances brought to Science by Quantum Physics is the fundamental way in which it brings into play in any experiment or operation of measurement, the necessary duality of the subject and the object, the simultaneous presence and the reciprocal action of the observer and the observed system, the observer being able to be considered as one with his measuring devices.'

[4] This was presumably London, although he emphasized to his mother that he held only a research position at the Institute Henri Poincaré (such positions were not permanent) and was *not* a Professor at the Sorbonne (Gavroglu 1995, p. 141).

[5] Simon has noted that by the 1920s the rapid expansion of physics had made textbooks covering the whole field simply unfeasible and so there was a shift to a focus on specific subdisciplines, such as QM (Simon 2022, p. 718). It is not unreasonable to suggest that by the late 1930s that subdiscipline itself had grown to the point that texts, albeit smaller than a standard book, were deemed necessary for dealing with specific issues. Furthermore, as mentioned earlier, Simon has emphasized the diversity and differential 'epistemic agency' of such texts (Simon 2022, p. 724) and pamphlets such as London and Bauer's certainly contributed in this regard.

[6] It has also been suggested that a 'characteristic Weimar physics style' became possible, in part because of the extensive communicative interactions between different kinds of texts, from journal

publish the latest developments on a given topic and in this case, these had to do with the act of observation and, as a result, the role of the observer. Publications in the series were expected not just to summarize such developments but also to offer the authors' own perspectives and this, I suggest, allowed London the opportunity to express his particular philosophical leanings.

Langevin went on to emphasize that the core element of this new theory, namely the wave-function, does not just describe the object but also the *state of knowledge of the observer*: 'For a given object, this function, consequently, is modified in accordance with the information possessed by the observer' (ibid., p. 218). Those readers who are clued up on recent work in the foundations of QM might raise their eyebrows at such claims, as it has been argued that the wave-function describes *either* the state of the system under investigation *or* the knowledge of the observer (these are called the 'ψ-ontic' and 'ψ -epistemic' views, respectively) but not both, as Langevin claimed (see Harrigan and Spekkens 2010). We'll come back to this briefly, in Chapter 10, but not surprisingly the proof rests on premises that are problematic from a phenomenological perspective (see also Hance et. al. 2022 for a non-phenomenological response).[7]

Langevin went on to laud London and Bauer for analysing the act of observation in a 'particularly penetrating way' through their two-stage framework in which the system first becomes coupled with the measurement device, followed by the interaction with the observer who 'becomes aware' of the outcome and thereby determines the new wave-function, post-observation, 'by using the new datum to reconstitute his information bank' (London and Bauer 1983, p. 218). And he concluded by stating that the piece does a valuable service by revealing 'the important finding' of this new physics, namely 'how we express our knowledge of the external world' (ibid., p. 218).

This is followed by the authors' own preface, where they insist that most introductions to QM follow a 'rather dogmatic path' in that when they consider the measurement context, intuitions are appealed to, rather than a careful and explicit application of the formalism. As a result, '[a] certain uneasiness arises' (ibid.), in that we are unable to see exactly when and with what justification we can attribute to the system an appropriate state of its own: 'Physicists are to some extent sleepwalkers, who try to avoid such issues and are accustomed to concentrate on concrete problems' (London and Bauer

papers to textbooks and best-selling works in popular science (Simon 2022, p. 725). If we extend this style beyond German-based publications, we can situate the London and Bauer pamphlet within such a nexus.

[7] I am grateful to Philipp Berghofer for reminding me of this distinction.

1983, pp. 218–19). However, it is precisely these issues of principle[8] that interest non-physicists and 'all who wish to understand what modern physics says about the analysis of the act of observation itself' (ibid., p. 219). They concluded that although these matters have already been the subject of 'deep discussions', and here they cite von Neumann's book, a concise and simple treatment has yet to be provided: 'This gap we have tried to fill' (ibid.)[9]

Here the focus of their work becomes clear: it was to provide a 'concise and simple treatment' of the measurement process through a 'careful and explicit' application of the formalism that avoided dogmatism and the appeal to intuitions. As for the audience, insofar as they were considering 'issues of principle' which physicists tend to sleepwalk their way through, they aimed to engage with those non-physicists who were keen to understand what modern physics says about the act of observation. Having said that, they certainly took no prisoners when it comes to setting out the mathematical form of the theory, beginning with the wave-function, its representation in configuration space, its complex conjugate, and so on.

Thus, London and Bauer began their analysis proper by emphasizing that this discussion is not just a matter for speculation but is a 'definite problem' (London and Bauer 1939, p. 6, 1983, p. 219).[10] Recalling our earlier remarks on the history of the measurement problem and the point about the subsequent use of the phrase itself, the emphasis here on the issue as a 'problem' is obviously significant. Furthermore, they emphasized that, '[t]he heart of the matter is the difficulty of separating the object and the observer' (ibid., p. 220). As a result, some of our 'traditional philosophical convictions' must be abandoned. Granted Heisenberg's aim of constructing a theory incorporating only those relations that hold between observable quantities,[11] the formalism implied more relations than had been anticipated which in turn called for interpretation:

[8] '<<questions de droit>>', in the French (London and Bauer 1939, p. 5).

[9] It is this line that perhaps has led to the impression that their work represents merely a simplified version of von Neumann's account. In the French version, the citation is given in a footnote, but in the English translation it is promoted to the text, which has perhaps enhanced that impression.

[10] They also follow the standard line by insisting that the completeness of QM is testable, insofar as its structure could not be reproduced by 'hidden parameters' lest some 'battle-scarred' results be given up.

[11] According to Bitbol, Heisenberg's move here 'irresistibly evokes the dynamics of phenomenological reduction' in that he first dismisses the semi-classical ontology of the old quantum theory and then 'redirected attention towards the epistemic acts of measurement and symbolization' (Bitbol 2021, p. 564). Thus, we have something akin to the *epoché*, followed by reflection on the reduced domain. We might add to this Heisenberg's conclusion that QM should not be regarded as a theory of individual objects, echoed by London and Bauer.

In this way, the discussion of the formalism taught us that the apparent philosophical point of departure of the theory, the idea of an observable world, totally independent of the observer, was a vacuous idea. Without intending to set up a theory of knowledge, although they were guided by a rather questionable philosophy, physicists were so to speak trapped[12] in spite of themselves into discovering that the formalism of quantum mechanics already implies a well-defined theory of the relationship between the object and the observer, a relation quite different from that implicit in naïve realism, which had seemed, until then, one of the indispensable foundation stones of every science. (London and Bauer 1983, p. 220, 1939, pp. 7–8)

6.2 The Analysis

The subsequent two sections present a brief overview of the principles of QM, with the reader directed to works by Bloch,[13] de Broglie, Dushman,[14] and Kemble for more detailed expositions. Beginning with 'Schrödinger's wave-function', what are now regarded as the standard features of the formalism are covered, including the role of the Hamiltonian, Schrödinger's Equation, the identification of the spectrum of eigenvalues with the allowed values of energy in Bohr's theory, and the recovery of the 'stationary states' of the latter (although they go on to note that the new formalism allows one to obtain statistical predictions for any physical quantity, not just the energy of the system). Born's 'statistical' interpretation of the wave-function is also given, with the note that in the context of Schrödinger's attempt to rid the theory of discontinuities, this 'may be considered to be a particularly conservative attempt to maintain the picture worked out by Born and Einstein and to embody it in a coherent theoretical system' (London and Bauer 1983, p. 223). The usual vector notation is then presented, encompassing (infinite

[12] The French word here is 'entraînés' which could be translated as 'trained' or, perhaps better, 'drawn into'.

[13] Bloch was in Zurich when Heitler and London conducted their research on the chemical bond and although he was only a young student he recalls that they formed a strong friendship (Bloch 1964; see also Bloch 1981). At that time, around 1927, Bloch stated that no one he was in contact with, including Heitler and London, seemed at all excited by issues of how to interpret the theory, putting this down to a lack of penetration of the 'Copenhagen spirit'. Interestingly he was another 'early adopter' of group theory, which he described as 'the fashion' at the time and which von Neumann had emphasized to him was something 'tremendously important'. It was while staying with Bohr in Copenhagen in 1931 that he began to understand 'the whole problem of measurement, that one cannot show a sharp line of distinction between the observing subject and the object to be observed' (ibid.).

[14] Dushman is perhaps the least well known but he did significant work on thermionics and wrote *The Elements of Quantum Mechanics* for chemistry students.

dimensional) Hilbert space, Hermitian operators and their representation via matrices, thereby relating the formalism to the matrix mechanics of Heisenberg, Born, and Jordan (ibid., p. 230).

Having laid down the central features of the formalism, London and Bauer then brought the focus of attention to the measurement process via the 'classic' dilemma inherent in Born's 'statistical interpretation': on the one hand, we might take the wave-function to represent our lack of knowledge of the system, in which case an observation leads to an 'enrichment' of our knowledge (ibid., p. 232); on the other, we may assign it an 'objective' character and take it to be a complete representation of the state of the system (again, this dilemma has more recently arisen in terms of the distinction between 'ψ-epistemic' and 'ψ-ontic' interpretations). But in that latter case, of course, it is difficult to understand the wave-function's statistical character and, in particular, the question arises, on what grounds can we add the new knowledge gained through observation to the supposedly complete knowledge that we already had (ibid., p. 233)? Here they follow Heisenberg in resolving the dilemma: 'it is the *process of measurement* itself which introduces the element of indeterminacy in the state of the object' (London and Bauer 1983, p. 233).

With that out of the way, they went on to explain the difference between pure states and mixtures, in particular emphasizing that the former cannot be reduced to the latter (after giving further formal details regarding projection operators). As an example, they take the case of spin, demonstrating straightforwardly that it is impossible to decompose the statistics associated with the pure state into a mixture of definitely oriented spins (ibid., p. 244).[15] They then noted that they would come back to this example to show how the observer is able to pull off the 'juggling trick' of extracting a definite outcome from such a state.

Before they reached that point, however, London and Bauer excavated the feature responsible for the introduction of probabilities in the first place. As they remarked, if we consider only a single system, we appear to have in the above a theoretical framework that is analogous to that of classical physics, with no reason to add any 'foreign' statistical structure. However, when two

[15] It is noteworthy that they use spin as their example in a publication of this type, and at that time. Today spin is typically the 'go to' quantity when it comes to illustrating quantum phenomena, but, for example, when Einstein, Podolsky, and Rosen wrote their famous paper four years previously, they used position and momentum (perhaps because they were responding to Bohr's stance incorporating complementarity which was usually articulated in these terms). Nevertheless, and despite Pauli's initial concerns, spin as a 'purely' quantum property appears to have been quickly accepted as such and featured in the early textbooks of Darwin (1931, pp. 154–6) and Dirac (1930, pp. 129–33). A useful discussion of the history of spin can be found in Morrison (2007).

systems are in play, each of which is taken to be in a pure state and which are brought into interaction, then we obtain what Schrödinger had only a few years before called an 'entangled' state (see our discussion in Chapter 4). Now, although London and Bauer do not use this term, as we recall, they do paraphrase Schrödinger's point that with such states a maximal knowledge of the composite system does not imply maximal knowledge of the component parts, as we have noted (London and Bauer 1983, p. 248). And this is because the wave-function for the former 'contains still other relations, to wit, *statistical correlations between* the components' (ibid.). It is this loss of knowledge with regard to the components, as represented by the description that we obtain for each, that is expressed by the appearance of probabilities.

As a result, London and Bauer continued, we must make a 'characteristic distinction', that has no classical counterpart, between two different modes of evolution of a system (here they are following von Neumann, of course): the first is reversible, causal, and of constant entropy and applies, of course, to isolated systems; the second is irreversible, acausal, and leads to an increase in entropy and occurs when one system interacts with another: 'Once thus degraded, the system has no chance in and by itself ever to regain its initial degree of determination' (ibid., p. 249).

And of course, it is just this kind of transformation that a measurement brings about. However, a necessary condition is that the state of the system be disturbed as little as possible. There also needs to be a 1-1 coordination between the relevant values on the scale of the measurement device and those of the quantity under consideration. Thus they consider the measurement of some quantity $F(x, p_x)$ of a system in the state $\psi = \Sigma_k \psi_k u_k(x)$ where u_k is an eigenfunction corresponding to the value f_k of F (ibid., p. 250). The system is then coupled with an apparatus capable of measuring F, where $G(y, p_y)$ is the coordinate specifying the position of the apparatus 'needle', $g_0, g_1 \ldots g_\rho$ its eigenvalues, with corresponding eigenfunctions $v_0(y)$, $v_1(y) \ldots v_\rho(y)$. The values of the g_k must then be set in a 1-1 relationship with the f_k, so the index $\rho_{(k)}$ can be replaced by k. After the measurement, then, the wave-function of the combined system + apparatus will be:

$$\Psi(x, y) = \Sigma \psi_k u_k(x) v_k(y)$$

However, London and Bauer wrote, such a coupling does not yet a measurement make. 'A measurement', they write, 'is achieved only when the position of the pointer has been *observed*' (London and Bauer 1983, p. 251). But then, they continued:

> It is precisely this increase of knowledge, acquired by observation, that gives the observer the right to choose among the different components of the mixture predicted by theory, to reject those which are not observed, and to attribute thenceforth to the object a new wave function, that of the pure case which he has found. (ibid.)[16]

The sense of this curious phrase, 'the right to choose', will become clear shortly.

It is at this point that London and Bauer noted 'the essential role played by the consciousness of the observer in this transition from the mixture [that is, the superposition] to the pure case' (ibid.). They now consider the ensemble of three systems composed of (*object* x) + (*apparatus* y) + (*observer* z), described by a global wave-function analogous to the above:

$$\Psi(x, y, z) = \Sigma \psi_k u_k(x) v_k(y) w_k(z)$$

where the w_k represent the different states of the observer. They wrote, 'Objectively—that is, *for us* who consider as "object" the combined system x, y, z—the situation seems little changed to what we just met when we were considering only apparatus and object' (ibid.) The function $\Psi(x, y, z)$ represents a maximal description of the ensemble such that we do not know in what state the system x is. However—and here we have that much-cited passage:

> The observer has a completely different impression. For him it is only the object x and the apparatus y that belong to the external world, to what he calls 'objectivity'. By contrast he has with himself relations of a very special character. He possesses a characteristic and quite familiar faculty which we can call the 'faculty of introspection'. He can keep track from moment to moment of his own state. By virtue of this 'immanent knowledge'[17] he attributes to himself the right to create his own objectivity—that is, to cut the chain of statistical correlations summarized in $\Psi(x, y, z) = \Sigma_k \psi_k u_k(x) v_k(y) w_k(z)$ by declaring 'I am in the state w_k' or more simply, 'I see $G = g_k$' or even directly, '$F = f_k$.' (ibid., p. 252)

[16] It was standard practice at the time to use the term 'mixture' or 'coherent mixture' to refer to what we now call a superposition. I am grateful to Jeremy Butterfield for pointing this out.

[17] As they noted at the end of this section, there might be some 'restrictions on the immanent knowledge of the observer' (London and Bauer 1983, p. 252). However, these will have nothing to do with quantum indeterminism, of course. As they say, 'it is not ordinarily required for a discussion of the measurement process that one should have an all-encompassing knowledge of the observer; for example, there is little chance of making a big mistake if one does not know his age' (ibid.).

In a typed note inserted by London in his own copy of the monograph, he wrote:

> Accordingly, we will label this creative action as 'making objective'. By it the observer establishes his own framework of objectivity and acquires a new piece of information about the object in question. (ibid.)

Furthermore, London and Bauer insisted that:

> it is not a mysterious interaction between the apparatus and the object that produces a new ψ for the system during the measurement. It is only the consciousness of an 'I' who can separate himself from the former function $\psi(x, y, z)$ and, by virtue of his observation, *set up a new objectivity* in attributing to the object henceforward a new function $\psi(x) = u_k(x)$. (ibid.; their emphasis)

6.2.1 Adopting the Phenomenological Stance

These are the passages that featured centrally in the debate between Margenau and Wigner, on the one side, and Putnam and Shimony, on the other. As we've discussed, they've been interpreted as mere summaries or, at best, more explicit presentations of von Neumann's account (with the exception of Shimony's later acknowledgement, of course). *However, the reference to relations of a 'very special character', the phrase 'immanent knowledge', the role of the 'I', or ego, and the emphasis on the free creation of a new objectivity, all clearly demand a phenomenological reading.*

Note, first of all, that at the beginning of this characterization, the observer is not set outside the domain of QM. She too is represented by a wave-function *within* the superposition. But by virtue of possessing this characteristic faculty of introspection, she obtains 'immanent knowledge'—that is, absolute and indubitable knowledge—of her own state by virtue of which she can, on the one hand (namely in terms of the ego), *separate herself* from the superposition and, on the other (namely in terms of the object in question), 'create' (in the French original it is 'constituer' or constitute[18]) a 'new objectivity'.

[18] Again, this is an unfortunate translation since 'constitute' is phenomenologically preferable insofar as it better expresses the relevant dependence relation (Alves 2021, p. 462, fn. 25). Book Two of Husserl's *Ideas* is subtitled, 'Studies in the Phenomenology of Constitution' and the first section is concerned with 'The Constitution of Material Nature' (Husserl 1982).

However, we recall from Chapter 5 that the ego should not be thought of as 'there', prior to this separation. As London's friend Gurwitsch emphasized, the 'structure of directedness' (Trizio 2021, p. 185) that is consciousness should not be conceived of as somehow 'stretched between the opposite poles of the pure ego and the intentional object'; *rather, it is through reflection on the acts that form that structure that the ego appears.*

Thus, the above separation should not be thought of in terms of consciousness, that is the ego, 'causing', in whatever sense, the wave-function to collapse, but rather as that of the *mutual separation* of both the ego-pole and the object-pole through the characteristic act of reflection. That yields a relational act, in which, as Husserl put it, 'the ego appears as itself related to its act's object through its act'. It is of the essence of such an act that the ego should appear but, again, this is not to suggest that the ego is something substantial, over and above this act. It is merely an empty, non-autonomous centre of identity or subject-pole engaged in a likewise 'empty looking' at the object. The latter is then objectified, or 'made objective', in the sense of having a definite state attributed to it, by this objectifying act of reflection. It is precisely through such a reflection that the 'chain of statistical correlations' is cut (an obvious allusion to the 'von Neumann chain', of course).

It is also important to remember that this act is not equivalent to that of introspection within the natural attitude, for the reasons already covered as to why phenomenology is not merely a form of descriptive psychology (Zahavi 2017, pp. 6–29).[19] The reflection is not just 'on' the experience, qua mental object. That would require some form of separation or sub-division by means of which the experience could be isolated and then reflected upon. If that were the case, then in the context considered here a definite state would already have had to have been achieved, via the familiar 'reduction' of the superposition, such that the act of reflection or introspection would be that of mere reportage.[20] And that, of course, would open the door to all the objections raised by Putnam and Shimony. This misses the object-oriented character of

[19] Bitbol has argued that the term 'introspection' is ambiguous insofar as it may connote the inspection of some 'inner realm', thereby implying the adoption of a dualist attitude. As an alternative, the phrase 'knowledge by acquaintance' is offered (2022, p. 272) but even this, I feel, presupposes an acquainting ego.

[20] Gutland has argued that Husserl's phenomenological method can be regarded as a kind of introspection (Gutland 2018). Specifically, he relates certain elements of this method to six common features that can be identified in introspection (Schwitzgebel 2016), including, crucially and unfortunately, that pertaining to the detection of a pre-existing mental state. However, as Gutland himself notes, Husserl emphasized that to claim that there exists a clear distinction between pre-reflective and reflective experience begs the question as to how the definite awareness of the former is achieved (Gutland ibid.).

experience; or as Zahavi puts it, '[i]t is by intending the object of experience that I can attend to the experience of the object' (2017, p. 26).

And that experience, of course, is structured in a certain way, so by attending to it we attend also to that structure and thereby to the constitutive role played by consciousness. As Husserl put it:

> The task that now arises is how to make this correlation between constituting subjectivity and constituted objectivity intelligible, not just to prattle about it in empty generality but to clarify it in terms of all the categorial forms of worldliness, in accordance with the universal structures of the world itself.
> (Husserl 1977, p. 326; cited in Zahavi 2017, p. 26)[21]

As we saw, it is this 'bringing to light' (ibid.) of the constitutive functions of consciousness that distinguishes the task of phenomenology as different from that of 'all positive sciences', including psychology. Furthermore, and crucially, it is by engaging in this act of introspection referred to by London and Bauer that the relevant correlation in the measurement context is made manifest, with the ego-pole and object-pole established as relata and a 'new objectivity' constituted.

And as we have noted, this objectivity is 'freely' created. In his Paris lectures of 1929, Husserl insisted that:

> we persistently *create for ourselves* new configurations of objects... which have for us lasting reality. If we engage in radical self-examination—that is, return to our ego... then all these forms are seen to be creations of spontaneous 'I'-activity... There we also find all the sciences, which, through my own thinking and perceiving, I bring to reality within myself.
> (Husserl 1964, p. 30; my emphasis)

However, it should not be thought either that these acts of creation are unconstrained—at least not across the board—or that, on the other hand, they are subject to our will. With regard to the former, we have already emphasized that phenomenology does not collapse into solipsism and that Husserl explicitly acknowledged the existence of 'external' objects. With regard to the latter, and relatedly, it is not the case that we can choose, in

[21] Husserl subsequently referred to this notion of correlation as the 'first breakthrough' that occurred during work on the *Logical Investigations* and which affected him so deeply that his whole subsequent life-work was dominated by the task of systematically elaborating it (1970b, p. 166; cited in Zahavi 2017, pp. 26–7).

effect, how things will appear. As Husserl stated above, the 'configurations' of objects are the result of *spontaneous* activity. Thus, in the case of quantum measurement, we cannot choose, by an act of will as it were, which outcome will be observed; that is, we cannot choose whether the result is 'spin up' or 'spin down', say. When London and Bauer talk of the observer 'setting up' a new objectivity, or establishing their own framework of objectivity, they do not mean this via some conscious choice as might be comprehended within the 'natural attitude'.[22]

There is no absolute or prior given framework of objectivity residing in some 'I' which is somehow apart from the whole process of observation and which then, by reflecting on 'its' mental states, collapses the superposition of these states. Rather the very act of observation itself involves a creative construction of objectivity by which the observer, as an 'I' and the object being observed come to be separated. The state of the ensemble as a composite object is correctly described 'externally', via the formalism of QM, in terms of a superposition but from 'inside' that object, as it were, the observer in reflection upon, and keeping track of, her own state creates her own objectivity In the double sense of constructing the 'I' in the first place and in doing so, separating this 'I' from the composite and thus gaining 'the right to choose' among the different components of the mixture predicted by the theory.[23]

However, as just emphasized, that choice is not one made by an 'I' surveying the terms of the superposition from the 'outside' as it were, and picking one. Once the 'I' has separated, there is of course no longer any superposition! This 'right to choose' is a feature that manifests after the separation, since as a 'right' it can only be possessed once there is an 'I'. And the sense of rejection of the other possibilities is not that of consciously preventing them from becoming actual but rather that of determining, post-separation, that terms in the

[22] As we shall discuss in Chapter 9, in 1962 a conference on Everett's so-called 'many worlds' interpretation was held at Xavier University, chaired by Podolsky of EPR fame and featuring as discussants both Shimony and Wigner. In the discussion, Kaiser Kunz (a theoretical physicist best known for his work on electrodynamics) emphasized that 'the basic thing' is that 'Quantum mechanics gives us multiple values, so to speak, and our problem philosophically is, when do we pick the solution. We make it. We simply force it to agree with what we have observed. So this observation is taken as the correct solution' (Barrett and Byrne 2012, p. 276). We might compare this with Dirac's suggestion at the 1927 Solvay Conference that it is 'nature' that makes the choice (see Barrett 1999, p. 25; as Barrett notes, although Dirac appears to take the reduction of the wave-function to be a physically real process, he does not say that the choice is made when a measurement is performed but rather when the components of the wave-function are no longer able to interfere with one another, thereby foreshadowing the process of decoherence). Heisenberg, on the other hand, insisted that it is the observer who makes the choice, in the sense that it is the act of observation that forces nature to make a particular choice from among the relevant set of eigenstates (ibid., p. 26).

[23] Bitbol has characterized the transition from the superposition to the 'reduced' state vector as a change in perspective (2022, p. 272).

quantum mechanical description that would be applicable 'from outside' as it were, do not correspond to what has been observed. In other words, as London and Bauer go on to say, this right to choose and thereby create their own objectivity can be attributed to the observer in virtue of their immanent knowledge of their own state. Again, it is not a 'right' that they possess whilst 'in' the superposition, but can only be attributed post-separation, when they have that certain knowledge of their own state.

Furthermore, between 'living in' the observation, as an experience, and describing it (as in the Wigner's Friend scenario, set out in Chapter 3), 'an essential descriptive change has occurred', as Husserl put it. In making such a description we are no longer 'living in' the observation, but 'we attend to it and pass judgment on it' and in doing so we cannot avoid reference to an ego or 'I'. In such a description, performed after an 'objectifying act of reflection', the ego is 'inescapable' since, as Gurwitsch argued, it *necessarily* appears as related to the object of the act of observation. It is important to be clear about what is going on here: the reflection that takes place in the measurement situation is not itself a *phenomenological* act, in the sense that one must first undertake the *epoché* in order to perform it.[24] I am not suggesting that physicists have to be phenomenologists when they make observations! The reflection is a 'characteristic' act that we perform all the time, from moment to moment, as we observe the world around us. Normally we do not explicitly 'keep track' of our mental states, in the sense of making a note of them, say, but what the argument involving Wigner's Friend illustrates is that we do possess this 'characteristic faculty' and can say what our state is, if needs be. What phenomenology provides is an analysis of this act and the uncovering, as it were, of this separation. Further 'radical considerations', such as the 'parenthesizing' of the natural world, are required in order to generate the phenomenological attitude.[25]

6.3 Revisiting the Debate

We can now appreciate just how the London and Bauer analysis has been misinterpreted, not just in the debate from the early 1960s, but throughout the literature on the measurement problem. We recall that Shimony interpreted it

[24] We recall Husserl's point above that effecting a reflection is not necessarily to effect a *phenomenological* reflection.

[25] I am grateful to Oliver Pooley for encouraging me to be clearer on this point as expressed in my 2002 paper.

as proposing that the (mental) states of the observer obey the vector relations required by QM, and hence can be in a superposition. The question then is whether mental states actually do satisfy a superposition principle, and, further, whether there is a mental process of reducing it. And after surveying a range of psychological phenomena, such as perceptual vagueness, indecision, and conflict of loyalty, Shimony concluded that the answer was 'no' in both cases.

However, as should now be clear, the entire basis of this criticism is wide of the mark, laying as it does in the claim that when 'I' observe my mental states, no superposition can be found. According to the interpretation outlined above no superposition can be found because an 'I', as a consciousness which is 'in' a certain state, can only be posited after the separation has occurred. That might suggest that, likewise, the question whether there is a mental process of reducing a superposition is inappropriate since, again, this presupposes some feature of consciousness 'outwith' the superposition, capable of effecting the reduction. Here we do need to tread carefully as the characteristic act of reflection might be seen as precisely performing that sort of role. First of all, however, the function of that introspective act is to produce a *separation* of the ego-pole from the object-pole. That might seem like semantic quibbling over the difference between 'separation' and 'collapse' but the terminology is crucial in this context. It is through such an act in general that the ego appears and thus the basis for the observer being able to assert that they are 'in' a definite state is established. Second, although this act is relational, it is through it that the states of the relata—ego and object—are manifested. As we have discussed, it is this relational act of introspection that is primary, rather than the relata, and its introduction is not an ad hoc move to solve the measurement problem but is phenomenologically fundamental.[26]

Shimony also acknowledged the aspect of creativity in London and Bauer's account, when it came to the role of introspection, but again failed to grasp its (phenomenological) nature. The issue as to whether there is any more 'creativity'—understood in its typical, non-phenomenological sense—in quantum situations as compared with classical ones is irrelevant. In both cases, it is not a matter of whether the observer *feels* more creative when making an

[26] Heidegger referred to '[t[he way in which in the most recent phase of atomic physics... the subject-object relation as pure relation... takes precedence *over* the object and the subject, to become secured as *Bestand* [resource]' (Heidegger 1977, p. 172; reproduced and translation modified by Sacco 2021, p. 517). As a result, this relation possesses a 'pure relational character' in the sense that the relata—subject and object—are 'sucked up as *Bestände*' (ibid.). That doesn't mean, he continues, that the relation has somehow vanished with the loss of the relata—on the contrary 'it now attains its most extreme dominance' (ibid.).

observation in a quantum context or not, since from the phenomenological perspective, the very act of objectification is a *creative act.*

And, again likewise, the point is missed when it is asked how irreducibly stochastic behaviour could occur in complex organisms and not in the 'primitive entities' of which they are composed. The relationship between the 'I' and the object cannot be causal, not because it is stochastic yet still physical, but rather because this relationship cannot be described in physical terms at all. There is no causal relationship because the mental and the physical are different modes of being which are not akin and which cannot be set side by side, as it were. Any relationship that there is can only be a phenomenological one. Having said that, there is a further question that arises here, which we touched on in Chapter 3, namely: at what point in the evolution of animals does this ability to effect a separation of ego- and object-poles and establish a definite state for the latter occur?

6.3.1 Can Animals Adopt the Phenomenological Stance?

Husserl himself havered somewhat when it came to extending the status of personhood to animals (Vergani 2021, p. 67). So, on the one hand, we can recognize that a dog, say, has a perceptual system similar to ours, that it feels heat and cold, hunger and thirst, and on that basis we can construct 'a whole series of analogies that are structuring, through passive synthesis, our relation with the animal, showing that the animal is in some way the subjective pole of its acts. As pole of its acts, it is in some way an "I", an ego' (Ciocan 2017, p. 183; see also Vergani 2021, p. 72). Nevertheless, Husserl insisted, we cannot attribute to a dog, say, a 'personal I' because personhood involves belonging to a community of persons via which a human being becomes the subject of a cultural world.

Having said that, he also acknowledged that animals may relate to one another via empathy, performed in the sphere of their own immanent experience. As a result, a kind of hierarchy may be established, based on the constitutive capacities of different species and also, consequently, the differing determining contents associated with the relevant form of world-givenness. Such a hierarchy must of course be based on the relevant scientific research, conducted within the natural attitude. Indeed, it may not only be revised but also re-formed as the differences between the cognitive apparatuses of different species become apparent (see Birch, Schnell, and Clayton 2020 and Birch et. al. 2022). And here of course we come to the nub of the matter for our discussion: given these differences, the series of analogies starts to stretch to breaking point as we consider species with radically different bodily schemas, such as jellyfish or insects or even amoeba (Ciocan 2017, p. 186; also Vergani 2021, p. 73).[27]

Crucially, Husserl argued that animals live in a restricted temporality and thus fail to recognize an infinitely open world. This in turn, he claimed, is linked to their inability to use tools, including that of language (Vergani 2021, pp. 74–5). And without that, he insisted, it would not be possible to constitute an ideal object; thus, he wrote, 'the animal itself has no generative world' (Husserl 1973, p. 181; cited in Vergani 2021, p. 75). So, although we can glean from Husserl's writings that he took the threshold between humans and animals to be plural (ibid., p. 67) and that the analysis of human society and culture can be applied by analogy—but only so far—to animals and animal societies, he remained hesitant in extending the notion of personhood (Vergani 2021, pp. 81–2).

Where does that leave us? Well, first of all, we can say that Husserl's pluralist approach to the dividing line between humans and animals gives some scope for affording certain kinds of animals the ability to effect the kind of separation that London and Bauer envisage and create their own objectivity, whilst also allowing that in certain cases, we may not be in a position to extend the underpinning analogy. Second, as we noted, his countervailing suggestion that animal consciousness may lack certain constitutive features is undermined to some extent by recent studies that demonstrate tool-use among an increasing number of different kinds of animals. If we allow that this then implies an extension of animal temporality we might admit that at least some animals have a generative world in terms of which a separation of ego- and object-poles can be accommodated.[28]

Hence, we can at least give a partial response to Shimony's concern by acknowledging that at least some animals may be able to effect the objectification of the world in a phenomenological sense. Where one should draw the

[27] Interestingly, one of Husserl's principal sources in the field of ethology was Köhler's book, *The Mentality of Apes* (Köhler 1927)—as we have already noted, Köhler was also one of the leading Gestalt theorists (Vergani 2021, p. 79, fn. 11). Husserl went on to distinguish between 'higher' and 'lower' orders of animals but also between wild and domesticated and between different examples of the last. He even went so far as to suggest that domesticated animals belong to our familiar world more than other humans who are strangers (ibid., p. 76).

[28] According to Morris, the tendency to take the lone animal as the unit of comparison has distorted not only comparative studies of human and animal behaviour in general but those of phenomenologists such as Heidegger and Merleau-Ponty in particular (Morris 2005). If we accept that 'animal life is pervasively a group phenomenon' (ibid., p. 50), then we can accommodate a sense of animals' 'being-in-the-world', whilst acknowledging that it still has a different temporality to ours.

line, within or across species, and also, therefore, in evolutionary terms, is a fluid matter, however. Certainly, insofar as we also acknowledge that the relevant analogy breaks down at some point, we must accept that we just cannot say how certain kinds of animals perceive the world. Perhaps it may even be suggested that some do not perceive it in terms of definite states, if that notion even makes sense in this context.

6.3.2 Back to the Complaints

Returning to Shimony's concerns, then, his complaint that London and Bauer's approach is 'counterintuitive in the extreme' obviously hinges on a particular understanding of what counts as 'intuitive'—as far as the phenomenologist is concerned, it is Shimony who was relying on an inappropriate understanding of this notion.[29] Likewise, his ontological concern as to how two such different entities—consciousness and the world—can interact carries no weight insofar as, again, it presupposed that which is phenomenologically denied. As for the epistemological issue that he also raised, regarding the justification of theories by human experience, this must also be recast. We shall return to this in Chapter 7 when we consider Husserl's attitude to science more generally.

It should now also be clear just how wide of the mark Putnam's criticisms are. First of all, to say that 'London and Bauer would like to reduce the "observer" to a disembodied "consciousnes"' (Putnam 1964, p. 3) is clearly mistaken. Granted the role of consciousness in their account as manifested in the separation of the ego-pole, from the phenomenological perspective the engagement of consciousness with the world is via embodiment, something that Merleau-Ponty subsequently emphasized, as we shall see in Chapter 8. Indeed, Putnam himself acknowledged that 'London and Bauer do not go so far as to make [the measuring system] just a "consciousnes"—it must also have a "body", so to speak' (ibid., p. 5). Furthermore, the interaction between the measurement apparatus and the system is *not* ignored by London and Bauer's treatment.

Second, and more importantly, this does not involve 'subjective events' causing abrupt changes of *physical* state; hence, Putnam's series of questions

[29] One might respond similarly to Shimony's criticisms of Wigner in his 'telepathy' paper, discussed previously, in which he acknowledged the phenomenological underpinnings of London and Bauer's account.

that we covered in Chapter 3 (ibid., p. 5) is entirely beside the point. We recall Husserl's insistence that the ego and the given object are not related in a 'real' sense, so again, asking for evidence of a 'physical interaction' is to demand something that was never in the frame to begin with. More interestingly, perhaps, the questions asking by what laws and in virtue by what properties does consciousness yield the 'reduction' of the wave-function are more directly answered in phenomenological terms—indeed, there is a sense in which answering these kinds of questions was Husserl's main focus from the start!

Furthermore, if we go back to Putnam's original concern, then it is just not the case that, within this treatment, the observer is excluded from consideration. As we have seen, the observer is included in the superposition—at least from the external perspective. Internally, as it were, the observer does become separated and in that separation, is no longer described by the formalism but this does not set the observer outside QM, as a physical object which should be, but isn't, described by the theory. From this internal perspective, the observer, as an 'I' or ego, is not a 'natural' object at all, but rather a phenomenological one. There simply is no possibility of describing the observer *in this sense* in quantum mechanical or any other physical terms—indeed, there never was. Thus, the phenomenological reduction has not somehow taken the observer outside the purview of QM. We recall that this reduction is not to be conceived of as some sort of abstraction from the natural world but as a more radical and entirely different sort of process. The further concern that this solution to the measurement problem somehow blocks the application of QM at a cosmological level is therefore also misplaced.

However, despite dismissing this whole approach as 'absurd', Putnam is correct in labelling it 'subjectivistic', albeit perhaps not in the sense that he intended. Likewise, there is indeed a sense 'which enables us to perform "reductions of the wave packet" upon ourselves' (Putnam 1965/1975, p. 81). Of course, everything hinges on what is meant by 'perform' here as it is indeed the exercise of the faculty of introspection that effects the above separation which then yields what Putnam, following common practice but mistakenly, thinks of as a 'reduction'.

6.4 Intersubjectivity, Community, and... Telepathy Again

Granted these misunderstandings, concerns remain. In particular, if this separation of the ego yields an act of objectification, as London and Bauer claim, how is intersubjectivity established?

Husserl's answer was to invoke empathy, understood not as a *sui generis* kind of knowledge, as Lipps maintained, but as 'other-experience' or 'other-perception', in the sense of an immediate experience of 'the other' (Zahavi 2018, p. 735). This involves co-attending to the other's experience in the sense that when comprehending their experiencing, ours passes through that experiencing and reaches all the way through to what they experience (paraphrasing Husserl, cited in Zahavi 2018, p. 736). As a result, the other is given as an alternative perspective on the world and 'our consciousness in empathy transcends itself and is confronted with otherness of a completely new kind' (ibid.). It is on the basis of such empathy, according to Husserl, that communities are formed.

However, some felt that this put the cart before the horse. Another student of Pfänder's was Gerda Walther, whose thesis, *On the Ontology of Social Communities* was published in the *Jahrbuch* in the same year as London's. Influenced by the Marxist view that we are by our very nature 'socialized beings', she argued that there was a direct inner connection between us of which everything else is the external expression (Lopez McAlister 1995, p. 197).[30] In particular this gives rise to a feeling of togetherness, of belonging together—an example being that of a group of scientists working on the same problem (Walther 1923, p. 20; Zahavi 2018, p. 742; see also Lopez McAlister 1995, p. 198).[31]

Like Gurwitsch, later, Walther also had concerns about the concept of the ego.[32] Unusually, however, hers arose as a result of her interest in the phenomenology of mysticism and parapsychology,[33] which was shaped by her understanding of community as grounded in a 'dyadic empathic encounter

[30] Intending to become a 'Socialist agitator', Walther enrolled in Pfänder's class 'Introduction to Psychology' simply in order to fill in an hour in her schedule (Lopez McAlister 1995, p. 190). She then took his 'Introduction to Philosophy' and sat in on his lectures in logic, which prompted her to read Husserl's *Logical Investigations*, followed by the *Ideas*, which then led her to move to Freiburg to study with Husserl in person. While there she not only attended Husserl's classes but also took courses in analytic geometry and set theory. However, she was put off by Husserl's prescriptive approach towards PhD supervision and returned to Münich to be supervised by Pfänder.

[31] Gurwitsch, on the other hand, subsequently argued that communities can persist even in the absence of positive feelings and suggested that it is a shared tradition or heritage that is the crucial factor (Zahavi 2018, p. 748). This, of course, has an unfortunate resonance with Heidegger's insistence that a community could not be identified with a 'multitude of separate Is' but should be conceived of as 'an ethnic-cultural unity rooted in the forces of blood and soil' (Zahavi 2018, p. 750).

[32] At the inaugural meeting of the Freiburg Phenomenological Society, chaired by Heidegger, Walther gave the opening lecture, 'On the Problem of Husserl's Pure Self' (Parker 2017, p. 49; see also Pellegrino 2018, p. 22).

[33] And we might see this as a phenomenological parallel to the considerations sketched in Chapter 3, section 3.7.

(ibid., p. 749; see also Parker 2017, pp. 54–5 and Pellegrino 2018, pp. 20–1).[34] This resulted in perhaps her most well-known work, *The Phenomenology of Mysticism* (Walther 1923; see also Ales Bello 2018) in which she undertook to analyse the essential features of mystical experiences, through the application of the 'epoché' in order to suspend our prejudices about such phenomena (Ales Bello 2018, p. 139).[35] As a result, she came to the view that the 'inner connection' between people that underpins a community should be understood not as *empathy* but as a form of *telepathy* and, not surprisingly, her colleagues in the philosophical community looked askance at this shift (Lopez McAlister 1995, pp. 202–3; see also Parker 2017, pp. 56–60).[36]

It was views such as this that helped lay the foundations of a spiritual or mental 'Geist' that Einstein, and others, feared would creep into science via the introduction of subjective elements into QM. London and Bauer reflected this fear when they wrote that, '[i]t might appear that the scientific community thus created is a kind of *spiritualistic society* which studies imaginary phenomena—that the objects of physics are phantoms produced by the observer himself' (ibid.).[37] Indeed, their 'little book' raised the concern that 'the novelty of its language shared many terms in common with mystical language' (Marin 2009, p. 818). However, they went on to argue,

[34] Walther also grappled with the issue of where to situate Husserl in the realism-idealism debate. This informed her preparation of the index for the second edition of Husserl's *Ideas*, where she listed under the entry for 'phenomenological idealism' passages that might be deemed 'pro' and 'con'. Unfortunately, her version was subsequently replaced, because, she believed, 'she had fallen out of Husserl's favour, likely due to her work in parapsychology, though she speculated that it may have also been due to her entries on "phenomenological idealism"' (Parker 2017, p. 51).

[35] Although Pfänder took Walther's 'plunge into another world' seriously, as a religious experience, he subsequently urged her not to pursue the study of parapsychology (Parker 2017, p. 52). Husserl, likewise, did not object to Walther approaching such experiences in phenomenological terms but expressed doubts as to whether they could serve as the appropriate basis for consideration of the purported object of such experiences (Ales Bello 2018, p. 138). Heidegger, however, was less than complimentary, presenting it as an example of how phenomenological research had 'sunk to the level of wishy-washyness, thoughtlessness, and summariness, to the level of the philosophical noise of the day, to the level of a public scandal of philosophy' (Parker 2017, p. 57).

[36] She had to leave academia and work as a secretary/assistant for Dr Albert Freiherr von Schrenk-Notzing who was engaged in parapsychological research (Lopez McCallister 1995, p. 192), and believed that his investigations were supported by 'the abandonment of the materialistic conception of the universe' (von Shrenk-Notzing 1920, p. 18). However, the mediums who von Shrenk-Notzing studied were subsequently exposed as frauds by Hans Thirring and others. Thirring, who became well known for his work in General Relativity (Thirring 1963), was a fellow physics student at the University of Vienna with Schrödinger, with whom he had discussions about parapsychology, in particular telekinesis (Pietschmann 2020).

[37] This passage perhaps explains the exchange London and Bauer had regarding the title of their final section, with London suggesting 'spiritisme' to follow 'scientific community', Bauer responding with 'solipsisme' (but given the insistence that phenomenology does not lead to a form of solipsism that never had a chance of being adopted!) and London pressing for 'Communité et realité scientifique' before settling on 'Scientific Community and Objectivity'.

such fears could be seen to be unwarranted once the new understanding of objectivity established by QM was properly grasped. Let us now consider that in a little more detail.

6.5 Anonymity and Objectivity

First of all, London and Bauer's claim that it is only through the activity of an observer that we achieve objectivity would seem to be undermined by the following: consider a Stern–Gerlach apparatus that effectively splits a beam of particles into those with spin 'up' and those with spin 'down'. As they noted, such an arrangement can be used as a kind of 'filter', yielding—via a suitably placed slit—a secondary beam all with a definite property, either spin up or down (London and Bauer 1983, p. 257).[38] Doesn't this restore the classical form of objectivity, which is grounded in the possibility of associating with a system the set of all its measurable properties, in a unique and continuous way, even when it is not being observed?

The answer is a firm 'no': they pointed out that such a filter can never put an *individual* object into such a state but only into a superposition (by virtue of the slit interacting with a given particle in the beam). We can only attribute the relevant wave-function associated with the definite state in question (spin up/down) 'at the expense of the individuality of the object, as one does not know in advance *which* are the atoms that have the property in question' (ibid.). In the absence of a 'supplementary check' by an observer, we cannot tell whether a given particle has gone through the filter or not. Thus, the filter can be said to produce such cases in an 'absolutely *anonymous* form' (ibid.).[39]

This is not really a worry since the aim of most experiments in particle physics is not to measure the properties, say, of an individual particle, but rather of the *kind* or species of that particle—physicists are typically interested

[38] The Stern–Gerlach arrangement was first proposed by Stern (who was an assistant to Born) in 1921 and performed by him and Gerlach in 1922; it involves sending a beam of particles (silver atoms in the original experiment) through an inhomogeneous magnetic field and observing their deflection. It has become an exemplar of quantum measurement, possibly, in part, due to the influence of London and Bauer but also because it so clearly demonstrates the observation of a single value of a given physical property (spin) and thereby exemplifies the transition from a superposition (for a useful summary, see: https://en.wikipedia.org/wiki/Stern–Gerlach_experiment; as Franklin and Perovic note, the experiment was initially viewed as a crucial test between the classical theory of the atom and the 'old' quantum theory of Bohr and Sommerfeld and remained robust through the transition to the 'new' QM (Franklin and Perovic 2021)).

[39] This idea of anonymity came to be expressed through the analogy of '[p]ounds, shillings and pence in a bank balance' by Hesse (Hesse 1961, p. 273). Hesse was the only woman who attended the 1957 conference organized by Feyerabend, discussed in Chapter 4.

in the spin of electrons in general, rather than the value of the spin of *that* particular electron. London and Bauer went on to note that '[q]uantum mechanics, truly a "theory of species", is perfectly adapted to this experimental task' (ibid.). However, they continued, 'given that every measurement contains a macroscopic process, unique and separate, we can hardly escape asking ourselves to what extent and within what limits the everyday concept of an individual object is still recognizable within quantum mechanics' (London and Bauer 1983, p. 257). Here they were echoing what Born and Heisenberg had earlier identified as the 'loss' of individuality of quantum particles that was implied by the new quantum statistics (see French and Krause 2006).[40] What is interesting is that this issue of the limits on the everyday concept of an individual object and the claim that the 'new' physics is a 'theory of species' can also be found in Husserl's concluding reflections of his career, as published in his *Crisis of the European Sciences* (Husserl 1970b).[41] We shall consider these in due course but first, let us consider this idea of measurement being a 'macroscopic' process, each one 'unique and separate'.

This is taken up in the final section, which begins by acknowledging that 'At first sight it would appear that in quantum mechanics the concept of scientific objectivity has been strongly shaken' (London and Bauer 1983, p. 258), and it may appear as if we are driven towards solipsism. However, they insist, '[n]o physicist has retired into a solipsistic isolation' (ibid.) because of QM and, furthermore, there remains a 'community of scientific perception' in the sense of agreement as to what constitutes the object of the investigation.

How is such agreement achieved? The answer hinges on their apparently blunt claim that '[i]t is easy to recognize that the act of observation, that is, the coupling between the measuring apparatus and the observer... is truly a

[40] Alves has misunderstood this as resulting from the observer's ignorance about the state of the system (2021, p. 461), thereby failing to grasp the crucial implication of quantum statistics. As Hesse noted, 'electrons are not indistinguishable but separate individuals, *they have no self-identity*' (1961, p. 273; italics hers).

[41] Forman claimed that, '[a]mong the lessons of QM none is clearer or surer than the denial of individuality to subatomic particles' (Forman 2011, p. 211), as noted by Heisenberg. Yet, Forman continued, not only did Heisenberg never again mention this denial but Bohr insisted on precisely the opposite, claiming that such particles had an 'indestructible individuality' (ibid.). The explanation for both these observations, according to Forman, is the emphasis on individuality as a cultural value in Weimar Germany (ibid., p. 212)—which supports his overall conclusion that the quantum physicists allowed themselves to make the theory out to be 'whatever their cultural milieu obliged them to want it to be' (ibid., p. 214). However, that this milieu was not so homogenous or dominant as Forman assumed is clear from a consideration of not only these remarks by London and Bauer and Husserl but also those of Heidegger who wrote that with the development of quantum physics 'even the object vanishes also' (1977, p. 172; in Sacco 2021, p. 517; for more on (non)individuality and QM, again see French and Krause 2006).

macroscopic action and not basically quantal' (ibid., p. 258).[42] Now, this might look suspiciously Bohrian,[43] particularly if we were to read it as 'the coupling... is truly a macroscopic action and *hence* is not basically quantal'. However, such an interpretation would fly in the face of London's own research—just because something is macroscopic does not imply that it is not quantal!

To understand this claim, then, it is important to consider the very next sentence: 'consequently one always has the right to neglect the effect on the apparatus of the "scrutiny" of the observer' (ibid.). The original French text is rather more revealing here as the word 'scrutiny' is actually a translation of '<<regard>>', where the placing of this phenomenological term between << >> indicates its significance.[44] We recall that it is in the reflective *regard-to* that the ego emerges as one pole of the relationship with the object and insofar as this regard is an act, conceived of as an essential phenomenological device, it cannot of course be described in quantum terms, where these are situated within the natural attitude. We further recall that when the *regard* is directed to a physical object, that object is 'seized upon'. This is not so for mental processes whose very *existence* is guaranteed by the regard. The existence of physical objects is not guaranteed in this way, of course, and hence the effect of the observer's 'scrutiny' can be neglected. Furthermore, this turning of the 'mind's eye' on something must not be confused with an act of perception, such as an observation. Thus, the regard or 'scrutiny' does not change or affect the apparatus, as an object, in any way, and so a 'collective scientific perception' can be created in which a second observer, looking at the same apparatus, will make the same observations.

This obviously bears on the 'Wigner's Friend' thought experiment. As we have noted, Wigner himself drew on London and Bauer's appeal to

[42] Again, there is a mis-translation here, as the English version states 'after the coupling is turned off' (1983, p. 258), which makes no sense in this context (Alves 2021, pp. 459–60, fn. 19).

[43] However, Bitbol has suggested that we can also find something akin to phenomenological elements in Bohr's writings since, although he retained an ontology of quantum objects represented in complementary ways, he emphasized the inescapable role of the measurement set-up and intersubjective communication, both of which, of course, are crucial features of the physicists' 'life-world' (Bitbol 2021, p. 564). Bohr also gave due importance to human experience, describing conceptual frameworks in general as merely logical representations of relations between experiences (ibid., pp. 567–8).

[44] 'on a par conséquent toujours le droit de négliger la réaction sur l'appareil du <<regard>> de l'observateur' (London and Bauer 1939, p. 49). Another translation of 'du regard' is 'the look' and here one can draw an obvious connection to the work of Sartre, who argued that it is through 'the look' of another that I become aware that I am an object for them and hence can adopt a third-person perspective regarding myself (Sartre 1943/1956, pp. 340–400). 'The look' in effect breaches my subjectivity and situates me in an intersubjective context. Alves also replaces 'scrutiny' with 'look' but does not remark on the phenomenological significance (nor on my earlier note to this effect in French 2002, p. 487; Alves 2021, pp. 459–60, fn. 16).

introspection in this context to conclude that consciousness plays a decisive role in measurement. Indeed, London and Bauer went on to state that 'nothing prevents another observer from looking at the same apparatus; and one can predict that, barring errors, his observations will be the same' (London and Bauer 1983, p. 258). Intersubjective agreement is thereby established but it is important to emphasize, again, that this is underpinned by a phenomenological understanding of that 'looking at'.

This then ties in with the final line of this passage: 'The possibility of abstracting away from the individuality of the observer and of creating a collective scientific perception therefore in no way comes seriously into question' (ibid., p. 258). As we discussed above, the difference in emphasis between a personalistic and collective interpretation of phenomenology runs as a thread throughout its development. London and Bauer appear to lean towards Husserl's approach to the creation of a community, rather than Walther's (not surprisingly perhaps). Indeed, this idea of 'abstracting away' from the individual was a significant element of Husserl's thought, although the details as to how this 'collective scientific perception' is actually created then need to be spelled out (see Alves 2021, p. 475).

There also remains a further question: even if intersubjective agreement is reached, how do we establish that the objects of physics are 'objective' in the requisite sense? As London and Bauer pointed out, and as indicated above, in classical physics the proof that we are dealing with something 'real', in the sense of existing—at least in principle—independently of all observers, is grounded on the possibility of continuous connection between the properties of an object and the object itself, even when it is not being observed. When it comes to QM there is no such possibility:

> In quantum mechanics an object is the carrier, not of a definite set of measurable properties, but only of a set of 'potential' probability distributions or statistics... referring to measurable properties, statistics which only come into force on the occasion of an effective, well-defined measurement. If one abstracts away from all acts of measurement, it is meaningless to claim these measurable properties as realized; the very mathematical form of the statistics does not allow it. (London and Bauer 1983, pp. 258–259)

The central idea here obviously resonates with Heisenberg's suggestion that the probabilities of QM be regarded in terms of 'a new kind of "objective" physical reality. This probability concept is closely related to the concept of natural philosophy of the ancients such as Aristotle; it is, to a certain extent, a transformation of the old "potentia" concept from a qualitative to a quantitative idea' (Heisenberg 1955, 12; see also Heisenberg 1958, p. 41). But I also want to highlight the last claim in that passage from London and Bauer, that QM itself prevents us from regarding these measurable properties as instantiated, once we are outside the measurement context. This is usually understood as a dismissal of the so-called 'ensemble' interpretation of QM (initially advocated by Einstein for example), which takes the description of the quantum state to refer to an ensemble of similarly prepared systems, rather than representing individual cases. However, I want to suggest that, in the context of London's phenomenological stance, here they were also arguing that in this regard the theory exemplifies an important feature of Husserl's view—which we will discuss in the next chapter—namely, that once we abstract away from the everyday 'life-world' and consider the objects of physics as represented in terms of the appropriate mathematical manifold, we cannot ascribe properties associated with the former to the latter. As we shall see, it is not that the entities involved are different but rather that we are operating with distinct descriptions in each case.

Nevertheless, we are still able to interpret or predict experimental results and '[i]t is enough, evidently, that the properties of the object should be present at the moment they are measured and that they should be predicted by theory in agreement with experiment' (London and Bauer 1983, p. 259). However, we have lost the earlier 'guarantee' of the objectivity of an object, understood classically in terms of the above possibility. Hence, '[i]n present physics the concept of "objectivity" is a little more abstract than the classical idea of a material object' (ibid.). As we noted earlier, understanding this concept involves 'the determination of the necessary and sufficient conditions for an object of thought to possess "objectivity" and to be an object of science' (ibid., p. 259).[45] This problem, London and Bauer remark, was perhaps first

[45] In the English translation the scare quotes around this second use of the term 'objectivity' have been dropped. Furthermore, as Alves has noted, there is a further difference between the English translation and the French original (2021, p. 469, fn. 44). In the former, London and Bauer state, 'Par sa cohérence interne et par la portée de ses applications la théorie nouvelle montre qu'ill n'est pas vrai que <<*l'objectivité*>> *d'un objet* doive etre garantie par la possibilité formelle de lui attribuer ses proprétés mesurables de *façon continue* aux époques où il n'est pas soumis à une observation. Il suffit évidemment que ses proprétés soient présentes au moment de leur mesure et qu'elles soient prévues par la théorie en accord avec l'expérience' (1939, p. 50), whereas in the English, it reads 'Is it not a guarantee of "the objectivity" of an object that one can at least formally attribute measurable properties to it in a continuous manner even at times when it is not under observation? No, as this new theory shows by its internal consistency and by its impressive applications. It is enough, evidently, that the properties of the object should be present at the moment they are measured and that they should be predicted by theory in agreement with experiment' (1983, p. 259). Alves speculates that the English translation follows a revised text, possibly 'by London himself?' that he could not find (Alves 2021,

posed by such 'mathematicians' as Malebranche, Leibniz, and 'especially', Bolzano.[46] More recently, however, they note, 'Husserl ... has systematically studied such questions and has thus created a new method of investigation called "Phenomenology"' (ibid.). The reference here is to both the *Logical Investigations* and the *Ideas.*

They also cite Cassirer's 1910 *Substance and Function* and his 1936 work, *Determinism and Indeterminism in Modern Physics* (Cassirer 1956) at this point. It is curious, however, that although in the French original Cassirer is mentioned only in a footnote, in the English translation he is elevated to the text alongside Husserl.[47] Although this was perhaps done for purely editorial reasons (again, this might be attributed to Shimony), it dilutes the significance of the Husserl citation. Cassirer, of course, was no phenomenologist but he also emphasized that it is not the case that first there is subject and object in terms of which experience is understood but rather that 'in one and the same process of objectification and determination the whole of experience comes to be divided for us into the "spheres within and without," into "Self" and "World"'quoted in Kaufmann 1949, p. 810).

Since it is an empirical science, physics cannot enter into such issues 'in all their generality', London and Bauer continued. Nevertheless, it both uses philosophical concepts 'sufficient for its needs' (London and Bauer 1983, p. 259) and abandons those that come to be seen as unnecessary and as containing elements that are 'useless and even incorrect, actual obstacles to progress' (ibid.) Such obstacles are represented by the classical notion of objectivity, whereas it is the phenomenological conception which is now sufficient for physics' needs.

6.6 Criticisms of the Phenomenological Interpretation

This phenomenological reading of London and Bauer's work has been disputed. Bueno, for example, has offered a 'minimalist interpretation' according

p. 469, fn. 44) but it is more likely that this is due to Shimony (as discussed previously in Chapter 3). That would explain the rhetorical shift that takes the coherence of the theory and the significance of its applications from 'pole position' to secondary place in the answer to the fabricated question.

[46] In general, all three were, of course, concerned with the relationship between our ideas and that which they are taken to represent. However, Alves focuses in particular on their common background with regard to 'the doctrine of the continuous as the *link and the point of unity* between mathematics and physics' (Alves 2021, p. 470). As we'll go on to discuss, the relationship between mathematics and physics was also a major concern of Husserl's.

[47] Alves and others have failed to note this (Alves 2021, p. 469).

to which the reference to 'immanent knowledge' above should not be understood phenomenologically (2019; see also Alves 2021, pp. 456–7). On his understanding, these passages concern the series of inferential steps, taking the observer from their own state, to that of the apparatus and thence to that of the system, which are ultimately grounded in the statistical correlations between each of the various states. This is:

> a process of making an inner state objective in virtue of the correlations between the inner state and those that are not, namely, the states of the apparatus and of the system being measured. And this entire process is prompted by the new piece of information that the observer obtained as the result of the measurement and which was provided by the apparatus that interacted with the quantum system in question. (2019, p. 136)

Crucial to this minimal interpretation is the assertion that if consciousness were responsible for the production of a new wave-function of the system, then there would have to be a 'mysterious interaction' with the latter. However, that seems to be precisely what London and Bauer deny. Nevertheless, Bueno understands the 'separation' of the 'I' referred to here as simply the post-observation removal of the term referring to the observer from the description of the ensemble (ibid., p. 137; see also Alves 2021, p. 467).

The problem with such an interpretation is that it acquires its minimalist character at the expense of dismissing not only crucial elements of the text, such as the role of the 'regard' for example, but also London's explicit commitments to the phenomenological stance, which, as we've seen, Shimony himself eventually acknowledged, together with his background, his discussions with Gurwitsch, and so on. As a result, I do not think that it can be plausibly maintained that 'it is not consciousness, in any significant sense, that is relevant to the outcome in question' (Bueno 2019, p. 138).

Alves, on the other hand, has acknowledged these commitments, but has argued that London's primary concern was to do with establishing objectivity (2021). Thus, he has presented what he calls my 'hermeneutic strategy' as leading to a dilemma:

> Either the 'birth' of the 'I' is the cause of the collapse of the wave function, and in this case London is committed to a mentalistic account that is open to Putnam's criticisms, or the 'I' does not collapse the wave function, and, in such a case, the superposition will remain, at the sub-'I' quantum level, while

> at the 'I'-level there will be a definite subjective state that has no correspondence with the quantum objective state. (2021, p. 468, fn. 41)

However, I regard this as a false dichotomy. It misses the core point that it is the appearance of the 'I' as the subject-pole of the act of reflection that leads to the separation of subject and object, yielding a definite state for the latter. Part of the problem here has to do with the use of the notion of causation in Alves' rendition, which is clearly inappropriate. There is of course a transformation of the wave-function, from one that describes a superposition, to one that describes a definite state, but to present this in causal terms is to misunderstand the nature of the separation articulated by London and Bauer. Furthermore, but relatedly, there is an underlying presumption of a substantivalist framework in Alves' account, whereas what I am proposing is a relational, or better, *correlational*, alternative, following Zahavi.

Thus, Alves has argued that insofar as phenomenology avoids the slide into Berkeleyan idealism by taking the relevant dependence relation to hold between a consciousness and an *object-for-consciousness*, understood as a unity of sense and validity, rather than the object per se:

> [t]his undermines . . . the overall strategy of making phenomenology productive for quantum theory *from the observer's side* by establishing a dependence (supposedly phenomenologically construed) of any kind of being on a consciousness subject and particularly on a transcendental 'I'.
>
> (Alves 2021, p. 472)

Accordingly, what is constituted by the intervention of the observer is a new *object of knowledge*, 'in a "framework of objectivity" determined by the theoretical presuppositions that guided the relevant sense-bestowing and positing acts' (ibid., p. 473); in particular, the observer does not create the reality she observes (ibid.; see also p. 478).

This represents a misconstrual of my interpretation, as I hope the discussion in the previous section indicates. Indeed, the contrast that Alves repeatedly draws is between the conscious observer 'physically producing' a new state and constituting new knowledge (ibid., p. 460) but I do not claim the former, of course (at least not in the usual sense of 'physically'). This misconstrual is further deepened by Alves' continued use of the phrase 'collapse of the wave function' in his considerations of both London's interpretation (he refers to the latter's 'reliance' on the concept of collapse; ibid., p. 478) and my analysis of it. However, as I have tried to make clear, there is no collapse on this interpretation (at least not in the usual sense) and although Alves does mention in passing London's remark about cutting 'the chain of statistical correlations' (Alves 2021, p. 466), no analysis of it is offered.[48] Indeed, Alves has situated London firmly within the Copenhagen Interpretation (ibid., p. 477 and p. 478) but as again I hope to have indicated, that is inappropriate.

According to Alves, then, 'London's most fundamental epistemological issue is the *objectivating act*' (2021, p. 457). In defence, he has offered London and Bauer's description of the Stern–Gerlach set-up, arguing that 'in the mathematical description [of that set-up], the quantum system progresses from a pure case to a mixture and then to a set of new pure cases *without the intervention of a psychological observer.* Indeed, the set {apparatus + screen} counts as an *observer* in a pure physical sense: it triggers the so-called "collapse" of the wave function into a set of several eigenstates' (ibid., p. 460). We have already covered London and Bauer's understanding of this arrangement and what it yields, so let me just say that it is implausible to argue that the apparatus + screen could 'trigger' the collapse of the wave-function, given not only von Neumann's point that they would also become incorporated into the superposition but also London's own emphasis on the extension of QM to macroscopic phenomena in general.

Rather than producing a pure case, consciousness:

> constitutes a new objectivity as correlated with a new piece of knowledge: the fact that the quantum system (and the apparatus, correspondingly) is in such and such a state. (Alves 2021, p. 460)

On this view the role of the observer is entirely epistemic (ibid., p. 462). However, the problem with such a claim is how to ensure it does not run afoul of the usual concerns regarding epistemic interpretations of QM more generally. On my interpretation, it is by relating the objectifying act to the

[48] With regard to the 'cutting' of the chain of statistical correlations and engaging in the creative act of 'making objective', van Fraassen has argued that, '*[t]here is nothing in the observer's experience that corresponds to this.* It is not true to his experience. The observer is not aware of the correlations in the entangled state of the situation as a whole, and certainly not aware of any action on her part that would be, or even feel subjectively, like cutting anything, let alone cutting statistical correlations. Nor is the observer aware of engaging in a creative act—the assertion that he is, cannot be a deliverance of experience, and cannot be something revealed by phenomenological analysis' (private communication). However, as noted above, the act in question is not the same as introspection within the natural attitude but rather should be understood in terms of spontaneous 'I-activity' that makes manifest the relevant correlation, qua such, that then has the ego and the object as distinct poles. It is only once this activity has taken place that we can speak of the observer, or rather the 'I', as being 'in' a certain state and so she cannot of course be aware of the correlation itself.

reflective consciousness—which Alves insists on keeping distinct (ibid.)—that we can avoid such concerns.[49]

As we noted, London and Bauer argued that it is the phenomenological understanding of objectivity, as systemically studied by Husserl, which was sufficient for physics' needs.[50] However, more needs to be said about the nature of that understanding and the relationship between the 'needs' of physics, comprehensively mathematized as that science is, and those of the everyday 'life-world'. This relationship was the focus of Husserl's final, and uncompleted, work, *The Crisis of the European Sciences* (Husserl 1970b) and in the next chapter I shall suggest that London and Bauer's 'little book' can be regarded as contributing to the completion of Husserl's project.

[49] Relatedly, Alves has also argued that although the wave-function, as a mathematical object of 'pure thought', 'fulfills *objectivity* conditions that allow it to become a *physical object* of natural science' (2021, p. 476), what is still missing is any reference to a 'determined object' (ibid.), since the wave-function encapsulates a whole set of possibilities (this is taken to be why QM is unsuited to an ontological interpretation). But of course, my contention is that this is missing, for Alves, precisely because of his insistence on keeping the objectifying act distinct from that of reflective consciousness in observation (ibid., p. 462). It is via the latter that we secure the reference that is sought for.

[50] According to Alves, 'if London's references are coherent rather than disparate, they suggest that his deep concern was about the mathematical treatment of exact and formal essences as a framework for the science of nature, delivering a transcendental account of the way an "object of thought" can be endowed with *objectivity* and then become an object of (natural) *science*' (2021, p. 470).

7
Completing the *Crisis*

7.1 Introduction: Husserl on Objectivity

The following passage from Husserl is helpful in reaching an understanding of how the concept of objectivity should be understood from a phenomenological perspective:

> A purely Objective science aims at a theoretical cognizing of Objects, not in respect of such subjectively relative determinations as can be drawn from direct sensuous experience, but rather in respect of strictly and purely Objective determinations: determinations that obtain for everyone and at all times, or in respect of which according to a method that everyone can use, there arise theoretical truths having the character of 'truths in themselves—in contrast to mere subjectively relative truths'. (Husserl 1978, p. 38)

So, objectivity is grounded in those determinations that are established by science for all subjects for all times. However, if those subjects are taken to include only human beings, then this sense of objectivity will fall short of what science aims for—it will not 'transcend the subjectivity of the human species as a whole' (Hardy 2013, p. 136).[1]

Hence, Husserl insisted that this sense of objectivity must be expanded to embrace *all* possible subjects, whatever their sensory constitution: it must thereby be established via what 'any possible subject of the pre-delineated ideal community can bring out and determine in rational experiential thought on the ground of his [*sic*] "appearances" and the communications of others concerning their "appearances"' (1982 III, p. 55). Given this, we need a language that goes beyond the specificities of the human race (Hardy 2013, p. 136). This would be mathematics, of course, and so objective nature comes to be 'exclusively determined by "exact" mathematical-physical predicates,

[1] Cf. van Fraassen's response to the suggestion that someone fitted with electron microscope eyes would be able to observe electrons, namely that such a person could no longer be considered part of our epistemic community (van Fraassen 1985).

A Phenomenological Approach to Quantum Mechanics: Cutting the Chain of Correlations. Steven French, Oxford University Press. © Steven French 2023. DOI: 10.1093/oso/9780198897958.003.0007

absolutely not intuitable, not experiencable' (Husserl 1982 III, p. 56). And so we arrive at a 'unique physical nature, with the one objective space and the one objective time, consisting of nothing but physical things that are characterized purely by concepts having the exactness described in physics' (ibid.).

We'll come back to the role of mathematics shortly but first let us consider the relationship between these 'physical things' and perceived objects (see Hardy 2013, pp. 136–46). In his initial considerations of the absolute character of consciousness, Husserl explicitly excluded 'the whole of physics and the whole domain of theoretical thinking' (1982 I, p. 86), insisting that we must remain within 'simple intuition' and its associated syntheses, including perception. Here the name of the game, as already noted, is to establish that those objects that appear in perception are nevertheless *transcendent*, in the sense that they cannot be reduced to elements of our consciousness, whether those be sense data or whatever.

But then we must face the question: are the objects as determined by the physical sciences *doubly* transcendent, in the sense of being so not only with regard to our consciousness but also with regard to the object as it appears in our sensuous intuition (Hardy 2013, p. 137)? Could the object as it appears be the appearance of some 'hidden' object? Husserl answered no, giving an argument from infinite regress: the putative 'hidden' object must be perceivable, if not by us, then by some member of the ideal perceiving community, perhaps with different sensory features. But then, it could only be perceivable by virtue of *appearing* to that member of the perceiving community, where that set of appearances would be subject of course to the same phenomenological analysis via the *epoché* and all which that entails as the first set. And of course, that second set of appearances necessitates the postulation of a further 'hidden' object, and so on. Hence, Husserl concluded, the relationship between the object as it appears and the object as determined by the physical sciences cannot be that of appearance-hidden.

7.2 Phenomenology, Realism, and Empiricism

This suggests that Husserl's stance does not 'fit neatly into the space defined by the axes of the realism-antirealism debate'[2] (French 2020, p. 217; however, see

[2] 'no form of metaphysical realism or subjective idealism are compatible with transcendental idealism, for which reality is no less real for being ontologically relative to transcendental consciousness' (Trizio 2021, p. 301).

Hardy 2013 and 2021 for a realist perspective and Wiltsche 2012 and 2015 for that of the empiricist).[3] And this is because is not a matter of the relative sizes of things but, rather, has to do with the different intentional acts involved, namely *perceptive* and *idealizing* respectively.[4] Atoms are unobservable not simply because they are very small but because, as categorial unities of thought they cannot be the correlates of an act of perception at all (Trizio 2021, p. 130; see also Wiltsche forthcoming).[5] They are not posited to causally explain our perceptions, nor should they be understood as mere theoretical constructs that represent some 'hidden' reality, or features thereof—rather they *are* the very things of perception *but as characterized by physics*. As Husserl put it, 'the perceived physical thing itself is always and necessarily precisely the thing which the physicist explores and scientifically determines following the method of physics' (1982 I, p. 119; trans. in Trizio 2021, p. 112).

This might seem an astonishing claim, given the apparent differences in the properties attributed to the thing that is perceived and that of physics (Husserl 1982 I, pp. 120–1). The clue to understanding it lies in Husserl's dismissal of the distinction between 'secondary' and 'primary' properties as one of the unfortunate consequences of the mathematization of nature. Instead, he reformulated the relation between that which we perceive and that which is the focus of physics in terms of different forms of '*givenness*'.[6] The former provides the 'mere "This"' or 'an empty X' which acts as the bearer of the relevant 'mathematical determinations' (Trizio 2021, p. 104). This '"This"'/'empty X' is given to us in perceived space which is merely a 'sign' of 'Objective space', understood as a three-dimensional Euclidean manifold, representable only symbolically.[7] This objective space, in turn, is non-intuitive and hence the relationship between the two spaces cannot be one that allows

[3] More recently, Wiltsche has acknowledged that the realism debate narrows down the question of our ontological commitments to the positions adopted in different sub-sections of the naturalistic attitude and that it is more productive to take a genuinely transcendental stance (forthcoming).

[4] What Husserl meant by 'idealization' is not what current philosophers of science mean: 'The essential feature of idealizations in Husserl's sense is not simplification but *exactness*' (Trizio 2021, p. 136).

[5] Thus, Husserl would not be fazed in the least by the aforementioned thought experiment involving replacing human eyes with electron microscopes because what would be intuitively given to such beings would be by necessity, things of perception: 'Even if God could have perceptions *corresponding* to individual atoms... he would perceive them *as things* and would have to theorize about them exactly as we do, replacing the secondary properties with mathematical, primary properties' (Trizio 2021, p. 130).

[6] Gurwitsch likewise argued that the distinction between 'primary' and 'secondary' qualities should be abandoned (Kockelmans 1975, p. 30).

[7] Husserl still has a bit of problem here: although at the beginning of his career, Husserl took the Euclidean nature of space to be an empirical matter, from the *Logical Investigations* on he insisted repeatedly that insofar as space and time describe invariant features of any possible material nature, they belong to the a priori ontology of nature (2021, p. 192). As a result, he could not accommodate the

perceived space to be eliminated, since it is through the latter that the 'empty X' is given that provides the 'substrate of the judgements formulated by physicists in their theoretical language' (ibid., p. 104). Crucially, it is a matter of 'eidetic necessity' that we cannot sidestep this realm of perception and simply walk into the space of mathematized nature.

Returning to Husserl's claim above and the issue of the relationship between these two 'things', we should not envisage this in terms two distinct realities, set side by side, as it were. There is no way to 'literally point' to 'the thing of physics' and no way to individuate it by means of grasping its own 'thisness'.[8] Thus, to talk of it is to use what is only a figure of speech, because what there *is*, is only the thing of perception and that thing *as characterized by physics* (Trizio 2021, p. 113; cf. Hardy 2021, p. 449). And this has nothing to do with the purported dimensions of the latter; rather it is because the kind of theoretical thinking engaged in within physics can only emptily intend an object, whereas perception can fulfil the *relevant meaning intentions.*

As idealized 'objectivities' these objects are given to us as intentional poles, that are, crucially, immanent to consciousness and which cannot be grasped by the senses but only by the intellect. It is precisely for this reason that scientific models and theories come to be constructed around them.[9] However, this is where care must be taken, as it is all too easy to move from thinking about models in paleontology, for example, to thinking about those of physics and taking the latter to be 'insufficient' representations of hidden realities just as the dinosaur skeletons in the Natural History Museum are imperfect representations based on whatever paleontological evidence is available (Trizio 2021, p. 121). Such a slip can be attributed to the failure to pay due attention to the 'constructional' nature of these categorial determinations of realities that are produced by thinking which prevents their sensuous fulfillment.

To put it rather simplistically then: there is just the one nature, namely that which is given to us via sensuous intuition, but scientists are rationally compelled to 'determine' that nature through their theoretical frameworks.

non-Euclidean space–time of relativity theory, never mind its tight inter-relationship with matter. Indeed, it has been noted that Husserl maintained an 'embarassing' silence on this issue even after his famous exchange with Weyl and his disciple Becker's phenomenological interpretation of space–time physics (Trizio 2021, p. 193). The option would be to liberalize the core phenomenological account of idealization but whether this would be sufficient to accommodate General Relativity remains to be seen (ibid.; see also Wiltsche forthcoming).

[8] Thus, the issue of the individuation of objects goes beyond the implications of QM.

[9] 'One can think of the several models of atomic structures that were developed in the years in which *Ideas I* was written' (Trizio 2021, p. 121).

There is no 'reaching beyond the world which is there for consciousness' on this account (Husserl 1982 I, p. 121) and hence the realist supposition that these theoretical posits represent some unobservable reality beyond the appearances gets no purchase. Indeed, the implicit supposition that the latter are causally related to the former was rejected by Husserl as nothing but a myth of causal depth (Hardy 2013, p. 145).

Of course, that is not to suggest that physicists should stop positing elementary particles and the like and attributing imperceptible properties to them—that is all part of the natural attitude (Husserl 1982 I, p. 212). However, as the phenomenological reduction reveals, '[t]he transcendency belonging to the physical thing as determined by the physics is the transcendency belonging to a being which becomes constituted in, and tied to, consciousness' (Husserl 1982 I, p. 123). It is this constitutive tie that underpins the correlation between consciousness and the material world, understood as a kind of psychophysical conditionality. This in turn bears directly on the issue of how to understand London and Bauer's proposal in that Husserl rejected the assumption that 'consciousness is just the endpoint of a causal relation, the existence of which is wholly contingent with respect to its source, physical reality' (Trizio 2021, p. 118).[10] From the phenomenological perspective, then, both the perceived 'world' and that described by physics 'are just constitutional layers of *the* world, they are both transcendent constituted poles' (Trizio 2021, p. 139).[11]

7.3 The 'Crisis' of Modern Science

Unfortunately, however, it was precisely the creation of such 'idealized objectivities' that has generated a crisis in human thought. In his final work (Husserl 1970b; for a useful introduction, see Moran 2012; also Trizio 2021), Husserl argued that the 'life-world'[12]—in which resides our 'natural', pre-theoretical understanding of things—has been overlaid with the above 'mathematization',

[10] It is puzzling, then, that Alves should claim, in his alternative analysis of the London and Bauer work, that 'the most fundamental lack in quantum theory... is the lack of *reality*' (2021, p. 464) and that 'the sheer quantum mathematical approach is by itself incapable of producing proof of its objects' (ibid.). His reason, essentially, is that the world presented by QM is one that is utterly unintuitable but phenomenologically of course that is just what we should expect (see also Alves 2021, p. 469, fn. 41).

[11] Atoms and particles 'exist *qua* endpoints of the constitution of material nature in transcendental consciousness' (Trizio 2021, p. 170).

[12] One explanation as to why Husserl introduced this term rather than just 'world' is that he saw the latter term as effectively corrupted by the tendencies that led to the very crisis he was concerned with (see Trizio 2021, p. 304).

as initiated by the likes of Galileo.[13] As a result modern science has lost its meaning or significance 'for life' and consequently is no longer able to address what is crucial for human existence.[14] Husserl's aim was not only to draw our attention to this loss but also to make us aware of the implicit presuppositions underpinning this mathematization, which have shaped the 'mindset' of modern physics ever since (see Berghofer, Goyal, and Wiltsche 2021, p. 417).[15]

The way forward, then, is to recognize that the life-world is the 'only real world' (Husserl 1970b, pp. 48–9), where this is understood not just as the 'coherent universe of existing objects' (ibid., p. 108) but as the world which is 'valid for our consciousness' (ibid.), as existing through all of us living together. The emphasis here is on its *shared* nature: the life-world is framed for us in spatio-temporal terms but to get to this from our initial 'solipsistic' conception we need the crucial further step of intersubjective experience, which is achieved, for Husserl, through *empathy*. As we have seen, this allows us to move away from what is fundamentally a 'first-person' standpoint by putting ourselves in someone else's shoes and not only incorporate their viewpoint but, in an iterated shift, incorporate their consideration of *our* viewpoint (Beyer 2018).

This in turn yields the space that appears as the form of all possible things and which:

> is an ideal necessity and constitutes an Objective system of location, one that does not allow of being grasped by vision of the eyes but only by the understanding; that is, it is 'visible' in a higher kind of intuition, founded on change of location and on empathy. (Husserl 1982 II, p. 88)[16]

Here 'change of location' may be taken as shorthand for 'a complex series of perceptions correlated to a likewise complex series of kinesthetic data, to

[13] A great deal has been written about Husserl's focus on and fascination with Galileo, including his apparent lack of appreciation of the latter's experimental work. As Berghofer et al. note, Galileo is in the frame here because it was his distinction between primary and secondary qualities that drove the wedge between the pre-scientific life-world and science (2021, p. 416). For our present discussion we can simply take Galileo as an exemplar of the physicist applying mathematics.

[14] It is important to recognize that, as Husserl emphasized, this crisis 'does not encroach upon the theoretical and practical successes of the special sciences' (1970b, p. 12), even though 'it shakes to the foundations the whole meaning of their truth' (ibid.).

[15] In addition to the 'prescriptive role mathematical models play in the physical constitution of reality', Berghofer et al. add '[those] presuppositions regarding the non-perspectivity and the subject-independence of physical knowledge' (Berghofer et al. 2021, p. 417).

[16] Empathy plus mutual linguistic understanding as the geometrical concepts become shared 'cultural acquisitions that by virtue of being sedimented in the "sensible embodiment" of speech and writing, are graspable by all' (Trizio 2021, p. 225).

which the transformations from the "there" to the "here" corresponds' (Trizio 2021, p. 161). The 'there' corresponds to the position of another subject's body whose point of view is then adopted, either in reality or imaginatively, taking us from the oriented, subjective, and sensuous space of the perceiver to the non-sensuous and uniform, objective space in which every body and all perceived things have their place.

It is within this objective space that the 'empty X' of the object-pole is posited, together with the differently embodied subjects, also posited, to whom this 'identical something' (Trizio 2021, p. 161) appears endowed with the relevant secondary qualities. This then provides the link between what is given to us in perception and the idealized language of physics by virtue of the relevant spatial forms admitting of being grasped in 'geometrical purity' and being exactly determined. All the features of some thing which is deemed to be objective are so by virtue of their connection with that which is 'fundamentally objective', namely space, time, and motion (Husserl 1982 II, p. 89). 'Real' properties are those mechanical properties that express the lawful dependencies of the spatial determinations of bodies. The objectivity of this spatial form is inherited by mechanical properties, which can be directly mathematized, whereas non-mechanical properties, such as those associated with colours or heat, can be so only indirectly via the relevant causal explanations. As a result, the 'thing of physics' is objective in the sense of not being relative to a given subject's body and as a 'mathematical ideality' comes to be regarded as the same for all possible subjects. Thus:

> we come to an understanding of the physicalistic world-view or world-structure, i.e., to an understanding of the method of physics as a method which pursues the sense of an intersubjectively-Objectively (i.e., non-relative and thereby at once intersubjective) determinable sensible world.
>
> (Husserl 1982 II, p. 89)[17]

We see, then, that the incorporation of the viewpoints of other subjects involves the assumption of a common 'world', at least in certain respects, if the metaphor is to be taken at all seriously.[18] Hence, we must presuppose that

[17] Here I have been broadly following Trizio in his claim that there is continuity between Husserl's *Ideas* I and II, with the differences having to do with their different aims rather than a shift in his thought (Trizio 2021, pp. 168–74).

[18] Schutz, whose book *The Phenomenology of the Social World* was praised by Husserl, as we noted previously (Chapter 4, fn. 16), argued that we should distinguish between 'Consociates who share the same time and spatial access to each other's bodies, Contemporaries with whom one shares only the same time, and Predecessors and Successors with whom one does not share the same time and to whose

the objects that we perceive can be situated within this common world in such a way that they go beyond, in some sense, *my* particular experiences. As Husserl emphasized, they may surprise us by revealing features that had hitherto gone unperceived, although this does not signal a slide into the naïvely realistic presupposition of a completely mind-independent world, of course.

This emphasis on the constitutive role of 'the other' is obviously significant when it comes to understanding the Wigner's Friend argument in the context of London and Bauer's approach to the measurement problem. However, an appeal to empathy is not going to be enough to ensure a common framework of objectivity in this case[19]—what we need is some assurance that not only will the observer and her friend agree that there is an object 'there', but that they agree as to its *state*, as given by the measurement. Bringing the matter even more into focus: there is nothing in Husserl's account that can ensure that when the observer records a measurement of 'spin-up', say, her friend does as well. For that, we need to appeal to something else but fortunately QM itself can come up with the goods, as we'll see.

7.4 The Completion of the Crisis

What the mathematization of physics yields is a psychophysical splitting of the two constitutive layers of material nature, namely the intuitive and the idealized.[20] Nature then comes to be seen as a 'really self-enclosed world of bodies' (Husserl 1970b, p. 60), with a concomitant idea of self-enclosed causality in virtue of which everything is determined unequivocally and in advance (ibid.).[21] And that remains the case even given the developments in QM:

> In principle nothing is changed by the supposedly philosophically revolutionary critique of the 'classical law of causality' made by recent atomic physics. For in spite of all that is new, what is essential in principle, it

lived bodies one lacks access' (Barber 2022). The first, who are present to one another physically, partake of each other's 'inner time' and thereby 'grow old together', whereas the experiences of the other kinds can only be inferred.

[19] Neither is the point about geometrical concepts becoming shared 'cultural acquisitions'.

[20] Where, again, it must be remembered that the notion of idealization here (*Mitidealisierung*) does not match straightforwardly with current notions found in the philosophy of science (Moran 2012, p. 68). Nevertheless, as Wiltsche has argued, as a central notion in Husserl's account it demands further elucidation in the context of modern physics (Wiltsche forthcoming).

[21] 'The objectivation of the world through idealization is . . . the theoretical performance that overcomes the subjective–relative character of the life-world' (Trizio 2021, p. 218).

> seems to me, remains: namely nature, which is in itself mathematical; it is given in formulae and it can be interpreted only in terms of the formulae.
> (Husserl 1970b, p. 53)

Here it is helpful to draw a comparison with the neo-Kantian attitude towards QM as expressed by Cassirer (1956) who argued that the above characterization of quantum physics as undermining the 'law of causality', although understandable, was mistaken, and that the true impact is on the notion of *object*. The apparent loss of the individuality of particles associated with the new quantum statistics (see French and Krause 2006) was taken to imply that the object 'constitutes no longer the self-evident starting point but the final goal and end of the considerations: the *terminus a quo* has become a *terminus ad quem*' (Cassirer 1956, p. 131). Objectivity is no longer secured via object-hood but through laws and symmetries, represented in terms of the mathematics of group theory as adopted and applied by the likes of Weyl, Wigner, and, of course, London.[22]

For Husserl, however, this encapsulated the whole problem in a nutshell. The mathematizing move that he associated with Galileo was manifested in that particular historical context by the use of geometry to represent motion. But then geometry itself is 'arithmetized' via the application of algebra, yielding a realm of shapes that can be conceived of in their 'pure exactness' as measurable, with the very units of measurement taking on the meaning of spatio-temporal magnitudes.[23] 'This arithmetization of geometry', he wrote:

> leads almost automatically, in a certain way, to the emptying of its meaning. The actually spatiotemporal idealities, as they are presented firsthand [originär] in geometrical thinking under the common rubric of 'pure intuitions' are transformed, so to speak, into pure numerical configurations, into algebraic structures. In algebraic calculation, one lets the geometric signification recede into the background as matter of course; indeed drops it altogether. (1970b, p. 44)

[22] The rise to prominence of group theory itself was largely due to Klein's 'Erlangen' programme as already noted, in which it was applied to geometry, thereby characterizing the differences between Euclidean and non-Euclidean forms.

[23] Geometry is 'the sense-foundation ... for modern exact physics as a whole, for ... the mathematization of nature that founds modern physics is, at bottom, a *geometrization of nature*. Furthermore, the geometrization of nature is an idealization of nature, because geometry is a discipline dealing with ideal space and figures' (Trizio 2021, p. 223).

This process substitutes a 'symbolic' meaning for the original one, yielding, ultimately, the formal-logical idea of a 'world-in-general' (ibid., p. 46). Thus, whereas for Cassirer it was the prominence given to laws and, in particular, symmetries, captured via group theory, that represented the decisive advance of modern physics, for Husserl it was precisely this that contributed to the crisis.

However, the apparent tension here can be dissipated and the crisis overcome through London and Bauer's analysis. The key lies in Husserl's acknowledgment that although classical physics represented the world in terms of indivisible elements moving around in space and time (and here we might recall that (mistranslated) passage in London and Bauer where they reject this classical conception and the notion of objectivity based upon it), modern physics offers an alternative conception according to which the idea of individual elements is dropped and nature is taken to be determined in terms of groups and types: '[t]he new physics conceives of the world as a hierarchy of typicalities, not as a universe of atoms' (Moran 2012, p. 85). Thus, in an (unpublished) appendix, from 1936, Husserl stated that:

> Determinate nature can be univocally calculated according to groups, and to corresponding types, but not according to the individual elements of the group, that is, with respect to the movements and other alterations of such elements. Since nature's universal conformity to laws deductively includes only types as universally calculable—in other words, since the nature of natural science is only a nature typical in itself—the alterations of the ultimate elements are predetermined only with probability, after the type to which they belong, and which predetermines a certain margin ('*Spielraum*')[24] and nothing more. (given in Trizio 2021, p. 190)[25]

We can read this in terms of causality only applying at the collective level and not at that of the distinct particles. Of course, this does not mean that the behaviour of such individual elements is chance-like but rather that such behaviour is determined on the basis of the group to which the element belongs. Husserl then concluded that 'the new physics is the physics of a nature conceived in an individual-typical way' (translated in Trizio 2021, p. 191), suggesting that the new quantum physics stood in contrast with the

[24] A slightly better translation might be 'leeway' or 'latitude'.
[25] The original can be found in the Husserl Archives.

'constructible from the bottom up' nature of classical mechanics (Trizio 2021, p. 191).[26]

Two significant conclusions can then be drawn: first, that this lack of bottom-up constructability means that the fundamentalist aim of obtaining an objective, non-relative description in terms of the world's ultimate constituents cannot be met. Instead, we must accept a certain relativity to the level of description, albeit one that is distinct from and does not replace the in-built relativity of intuition (and which still relies on idealization; Trizio 2021, p. 191). Second, QM offers a non-reductionist view of nature that meshes better with Husserl's overall outlook than does classical physics, in that the physical world is no longer considered to be a simple sum of atoms (ibid., p. 192).[27]

Now, Husserl did not touch on the issue of the interpretation of QM, nor did he consider the measurement problem.[28] Nevertheless, it is clear from the above that he did not consider the theory to be a threat to phenomenology; indeed, '[f]ar from it. Husserl speaks as if it marked progress with respect to classical physics, not only from the empirical but from the methodological point of view. The new conception of idealization would be more compatible with the conception of nature and natural science that stems from phenomenology' (Trizio 2021, p. 192).[29]

This is precisely the line that London and Bauer adopt. As we have seen, they argued that quantum theory offers a new epistemology capable of providing an appropriate grounding for our conception of objectivity. It is also worth noting that Husserl emphasized that it is *determinate* nature that can only be calculated at the group level—this allows for Schrödinger's Equation,

[26] As Wiltsche has noted, Husserl actually said very little about the 'new' physics, whether relativistic or quantum (Wiltsche forthcoming).

[27] In addition, Husserl insisted, neither physiology nor biology can be completely reduced to physics (Trizio 2021, p. 192).

[28] This should come as no surprise, given what was noted in Chapter 2.

[29] Nevertheless, as Wiltsche has emphasized, there is still considerable work to be done in further articulating this 'new conception' in the context of quantum physics (forthcoming). As he has argued, although 'Husserl's thesis according to which the mathematical tools underlying physics require simple life-world experiences as their meaning-fundament might be immediately plausible in cases like Galilean proportional geometry' (ibid., p. 29), it is less obviously so when it comes to quantum physics, where we are dealing with 'mathematical concepts that, first, were not introduced with questions of physical applicability in mind, and that, second, do not seem to be connected to pre-theoretical experiences or practices in any obvious way' (ibid., p. 30). Wigner's notion of the 'unreasonable effectiveness' of mathematics in this context might seem to be an obstacle to an appropriate phenomenological understanding of its applicability. However, this effectiveness may not seem so unreasonable once attention is paid to the sequences of moves—both mathematical and physical—that are made in particular cases (Bueno and French 2018). Certainly, however, further consideration is required of the role of the life-world in such moves, via not only the phenomenology of (modern) physics but that of mathematics also (see, for example, Tieszen 2005).

taken as the relevant law, to be applied at the level of the distinct element but what it does not yield at that level is a determinate result.

Having said that, we may also identify a further point of commonality with London and Bauer, in that, as we recall, they described QM as 'truly, a "theory of species"' (1983, p. 257), in the sense that it is 'perfectly adapted' to the task of capturing measurements that do not deal with individual systems but rather with 'species' or kinds of atoms. And they noted that this raises issues as to the extent to which we can maintain the 'everyday concept' of an individual object in this context (ibid.).

This 'everyday concept' plays a crucial role with regard to the notion of 'experience' within phenomenology, insofar as it is characterized as 'the intuition of individual objects, in this case, precisely the spatiotemporal things of our surrounding world' (Trizio 2021, p. 229). Individuation may be obtained via the nexus afforded by spatio-temporal location and causal connections and such individuated objects can then subsumed under 'morphological empirical types' (ibid.). It is the duality between such individual and general types that constitutes the fundamental structure of the world of experience. Now, one might suppose that this stands in opposition to the 'loss' of individuality apparently implied by quantum statistics. However, consider the example of geometry: the elements of the geometrical world, as it were, do not mirror this fundamental structure because they are general ideal types that cannot be perceived.[30] Such geometrical entities are kinds of 'limit-shapes' that result from the idealization of those concepts proceeding from measurements with all reference to individual objects dropped from the very outset. As a result, although this geometrical 'world' is objective, it is not a *world* per se, since it expresses the general spatio-temporal form of an infinite number of possible worlds. Extending, or perhaps, generalizing, this thought from the 'space' of geometry to that of QM, namely Hilbert space, we can see how the apparent opposition can be overcome: the 'inhabitants' of this latter 'world' likewise do not mirror what we try to capture with our 'everyday concept' of individuality, as they too must be regarded as idealities formed, in this case, not by geometrical idealization, but via the new epistemic avenue afforded by quantum statistics.[31]

We might speculate as to the relationship between these passages by Husserl and those in the London and Bauer piece, and whether Husserl consulted

[30] 'The bodies familiar to us in the life-world are actual bodies, but not bodies in the sense of physics' (Husserl 1970b, p. 139; see also Trizio 2021, p. 229).

[31] We can, in fact, maintain individuality in the quantum realm but at a certain (epistemic) cost (we can never tell which individual is which); see French and Krause 2006.

London or the latter recalled this appendix when co-writing their 'little book'. Most likely, however, is that there was a 'common cause' in the form of general considerations of QM as a 'theory of species' that were in the air at the time. More importantly, the further tension generated by the contrast between Husserl's animadversion towards the 'mathematization' of nature and London's own advocacy of the application of group theory to physics can also be ameliorated: London's use of group theory should be understood, from the phenomenological perspective, as a form of idealization in the above sense, yielding certain 'idealities' inhabiting the world (that is objective but not a world per se) of physics.[32] As long as that perspective is adopted, so that the bedrock on which this 'world' is grounded is understood to be the life-world, there is no mis-match.[33]

Thus, the mathematization of modern physics clothes the life-world in a 'garb of symbols' (Husserl 1970b, p. 51) that 'dresses it up as "objectively actual and true nature"' (ibid.).[34] But this 'dressing up' is just that, and does not actually yield another world, in the sense of a 'true being' but rather should be understood as a method or technique whose true meaning has been lost through history through a kind of progressive oblivion (ibid., p. 56; see also Trizio 2021, p. 241).[35] That meaning could only be retained if the scientist were able to enquire back into the 'historical meaning of [the] primal establishment' (Husserl 1970b, p. 56) of this method, together with all the inherited meanings that have accreted to it. Unfortunately, as a 'brilliant technician of the method' she is unable to undertake such reflections since she does not even appreciate the need for such clarification. Indeed, she will dismiss as 'metaphysical' any attempt to encourage these reflections, since she will feel that she, of course, knows what is best when it comes to her work.[36] London, however,

[32] 'The world of physics is not a world, but the hypothetical infinitely determinable character of the world' (Trizio 2021, p. 234). In the case of quantum physics, this infinitely determinable character will embrace the full panoply of possible quantum statistics, including the infinite number of different types of 'parastatistics', all described by group theory, as well as the Bose–Einstein and Fermi–Dirac kinds actually observed.

[33] Recalling the point made previously about the link between what is given to us in perception and the idealized language of physics being provided by the relevant spatial forms, we could accommodate the advance represented by group theory via the extension of what is understood to be the relevant 'space' and take the 'forms' to be group-theoretic.

[34] '[T]he physical entity is at every point transcendent to the material thing, none of whose perceivable determinations enter, as such, into the constructions of physics' (Gurwitsch 1974, p. 179; for a review see Jacobson 1976).

[35] In particular, Husserl insisted, we should not be misled into taking these mathematical formulae and their meaning as the 'true being' of nature itself (1970b, p. 44).

[36] According to Føllesdal the life-world mediates the reference to reality of scientific concepts and acts as the relevant touchstone through scientific revolutions (Follesdal 1999).

was more than just a 'brilliant technician' and did indeed appreciate the need for such a fundamental clarification.

Given the mistaken understanding of the mathematization of nature as revealing its 'true being', rather than as a method of theoretically determining nature, how then should we proceed? By adopting the phenomenological stance, of course:

> It will gradually become clearer, and finally be completely clear, that the proper return to the naiveté of life—but in a reflection which rises above this naiveté—is the only possible way to overcome the philosophical naiveté which lies in the [supposedly] 'scientific' character of traditional objectivistic philosophy. This will open the gates to the new dimension we have repeatedly referred to in advance. (Husserl 1970b, p. 59)

In particular, we must do what Galileo could not do and subsequent generations of physicists have not done (except perhaps Weyl), which is to consider the phenomenological basis of the applicability of mathematics: how is it that geometry can be applied to nature? What is geometry's 'original institution' that gives it its sense? The answers lie in a return to the subjective–relative:

> while the natural scientist is thus interested in the objective and is involved in his activity, the subjective–relative on the other hand is still functioning for him, not as something irrelevant that must be passed through but as that which ultimately grounds the theoretical-logical ontic validity for all objective verification, i.e. as the source of self-evidence, the source of verification. The visible measuring scales, scale-markings etc., are used as actually existing things, not as illusions; thus that which actually exists in the life-world, as something valid, is a premise. (Husserl 1970b, p. 126)[37]

This emphasis on the 'visible' measurement devices and outcomes as elements of the life-world naturally raises the further issue of the relationship between

[37] Here Husserl referred to the famous Michelson–Morley experiment: 'Einstein could make no use whatever of a theoretical psychological-psychophysical construction of the objective being of Mr Michelson; rather he made use of the human being who was accessible to him, as to everyone else in the prescientific world, as an object of straightforward experience, the human being whose existence, with this vitality, in these activities and creations within the common life-world, is always the presupposition for all of Einstein's objective-scientific lines of inquiry, projects, and accomplishments pertaining to Michelson's experiments' (Husserl 1970b, pp. 125–6; the extent to which the Michelson–Morley experiment influenced Einstein in his development of the Special Theory of Relativity is of course disputed.)

the latter and nature, as investigated by the sciences. That, in turn, compels us to reflect on the relationship between the naturalistic and personalistic attitudes (see Trizio 2021, pp. 258ff). These two attitudes should not be regarded as on a par; rather, the former must be regarded as subordinate to the latter, because it is via the operations of the subject that objectivity is achieved, which operations of course cover the ego itself. Furthermore, the personalistic attitude embraces the everyday life of the subject, including their involvement in a community. The naturalistic attitude, on the other hand, results in a view of the world that is shaped by the absolutization of nature which in turn fragments subjectivity and supports a form of objectivity that is ultimately illusory.

The way forward is to adopt the phenomenological stance. Beginning with the individual subject, situated in their environment, and adopting the personalistic attitude, we can construct the spatio-temporal structure that frames our communal life and sets nature as a stratum of this shared environment. From there we reach the objective nature of mathematical physics. From this perspective, we can understand that nature presents itself 'as something constituted in an intersubjective association of persons' (Husserl 1982 II, p. 220). The naturalistic attitude, on the other hand, cannot produce 'higher order personalities' in the form of communities, institutions, nations, and so forth (Trizio 2021, p. 259).

Likewise, the life-world has priority insofar as it is both the horizon of our practical activities, in which we can find the motivational currents that lead to scientific truths and is also historically informed in the sense that everything within it has a meaning that results from the process of 'sedimentation'. Analysis of the relevant invariant structures reveals that all its elements are grounded in the 'stratum' of material nature—not just material objects but also people, localized as they are via their living bodies.[38] This grounding role does not mean that material nature somehow has an independent existence; rather, 'nature', understood by means of an abstract mathematical manifold, *belongs* to the life-world. Indeed, it is by virtue of this that science is even possible, insofar as the life-world both includes scientific truths and is also their 'sense-fundament' (Trizio 2021, p. 267).[39]

[38] This aspect is take up in Merleau-Ponty's work that we will consider in Chapter 8.

[39] As Wiltsche has noted, if the life-world not only provides the justificatory basis for the 'world' of physics (in terms of measurement outcomes, etc.) but also supplies the meaning fundament for the idealities of the latter, then taking these to *represent* the 'innermost structure of material nature' becomes problematic (Wiltsche forthcoming, p. 28).

To comprehend that sense of 'belonging' the phenomenological reduction is required, in the sense of:

> an *epoché* of all participation in the cognitions of the objective sciences, an *epoché* of any critical position-taking which is interested in their truth or falsity, even any position on their guiding idea of an objective knowledge of the world. (Husserl 1970b, p. 135)

It is only through such a transformation, of course, that we can study what natural life and its subjectivity ultimately are and discover the 'universal, absolutely self-enclosed and absolutely self-sufficient correlation between the world itself and world-consciousness' (ibid., p. 151).[40] Following this transformation, the natural world retains its being and the 'objective truths' that it contains, but it now comes 'under our gaze purely as the correlate of the subjectivity which gives it ontic meaning, through whose validities the world "is" at all' (ibid., p. 152).

Thus, to resolve the crisis, science should first appropriately clarify its objective domain. This involves both delimitating the essence of that domain and elucidating its sense of being. As a result the appropriate method will follow. Once that double clarification has been carried out and the appropriate method has been developed, the science in question may be regarded as 'genuine'.

According to Husserl, however, physics as it stood *could not* be regarded as a 'genuine' science because it had lost its grounding in the life-world and transcendental subjectivity.[41] Granted, by means of mathematization, it was able to delineate the essence of material nature in terms of the spatio-temporal characterization outlined above, and develop a corresponding method of measurement, it lacked the resources to appropriately frame the *sense of being* of material nature. As a result, nature and consciousness were split apart, with both interpreted as beings in themselves

[40] By 'world-consciousness' is meant 'the conscious life of the subjectivity which effects the validity of the world, the subjectivity which always has the world in its enduring acquisitions and continues actively to shape it anew' (Husserl 1970b, p. 159).

[41] 'Someone who is raised on natural science takes it for granted that everything merely subjective must be excluded and that the natural-scientific method, exhibiting itself in subjective manners of representation, determines objectively.... But the researcher of nature does not make clear to himself that the constant fundament of his—after all subjective—work of thought is the surrounding life-world; it is always presupposed as the ground, as the field of work upon which alone his questions, his methods of thought, make sense' (Husserl 1970b, p. 295). I am grateful to Philipp Berghofer for pointing me to this passage.

(see Trizio 2021, pp. 280–1).[42] Thus 'material nature' came to be both mathematized and, as conceived in substantival terms, effectively conceptualized as autonomous rather than understood as 'a unit constituted in perceptual intuition for which the possibility of exact mathematical determination is a hypothesis' (ibid., p. 278).

It is only by adopting the phenomenological stance that science in general, and physics in particular, can become 'genuine' in the above sense.[43] Whether this is even possible is not something that Husserl considered but I suggest that we should understand London and Bauer as proposing that *this possibility is realized in QM.*[44] In other words, it is the so-called measurement 'problem' that opens the door to the possibility of physics becoming a 'genuine' science by revealing that the relationship between the observer and observed system must be understood in terms of transcendental subjectivity, with the latter conceived as just such a 'unit constituted in perceptual intuition'.[45] By virtue of being, in itself, a theory of knowledge, QM supplies the resources to frame the sense of being of material nature, allowing us to grasp the sense in which the latter belongs to the life-world.

[42] 'In general we must realize that the conception of the new idea of "nature" as an encapsuled, really and theoretically self-enclosed world of bodies soon brings about a complete transformation of the idea of the world in general. The world splits, so to speak, into two worlds: nature and the psychic world, although the latter, because of the way in which it is related to nature, does not achieve the status of an independent world' (Husserl 1970b, p. 60).

[43] 'Only with the help of such meditations [*sic*] can we disengage the existential sense of the *universe of physics* as objective correlate of the acts in which it is constructed—an objective correlate of a higher degree, because it is conceived and elaborated by starting from that other objective correlate of consciousness which is the *world as we perceive it*. Only thus can the origin and the specific nature of the evidence which is characteristic of the physical sciences be grasped' (Gurwitsch 1974, p. 182).

[44] Alves maintains that 'quantum theory conveys a poor ontology of nature and makes assertions that are not suited for an ontological interpretation at all' (2021, p. 478). Echoing Darwin (see Chapter 2, fn. 39), he argues that what the wave-function posits 'is the *effectiveness* of all *possibilities* of events, but no event at all, as if it was describing a world where nothing happens, a "paralyzed" or "frozen" world, a world that does not have the sense-content for being posited as such' (ibid.). Alves sees this as an example of the 'the ontological indigency of the Copenhagen "spirit" to which London also belongs' (ibid., p. 480). Setting to one side this misalignment of London's position, what I suggest here might go some way towards Alves' concluding suggestion of the need to develop Husserl's final insights.

[45] Heidegger too was concerned with mathematization and took it to be the defining feature of modern science. However, unlike Husserl, he directly engaged with quantum physics through his exchanges with Heisenberg and insisted that 'the counterposition between subject and object in the interpretation of quantum mechanics was specious, for "the divide [between them] is never a divide but precisely a transcendental relation [*Bezug*]"' (Carson 2011, p. 539). For more on Heidegger and QM, see Pris 2014 (who suggested that the 'reduction' of the wave-function has a phenomenological nature in the sense of the Heideggerian *Dasein*) and Sacco 2021 (who argues that a Heideggerian framework is capable of accommodating the 'conceptual novelties' of QM not least in the sense that prior to measurement, what is real should be understood as 'Bestand' or a 'resource' for the appearance of both subject and object); and for more on the influence of Heidegger (and Husserl) on Heisenberg, see Heelan 2013.

And it does this in the context of measurement by demonstrating, as London and Bauer set out, how that material nature comes 'under our gaze purely as the correlate of the subjectivity which gives it ontic meaning'. So understood, QM re-unites nature and consciousness and London and Bauer's 'little book' completes Husserl's project.[46]

[46] '[O]ne must never lose sight... of the fact that it [the *Crisis*] remains unfinished' (Trizio 2021, p. 285).

8
QBism and the Subjective Stance

8.1 Introduction

London himself never elaborated any further on the ideas contained in that 'little book' and appears to have regarded the measurement problem as solved (Gavroglu, private email). Thus, there is no fully-fledged 'London interpretation' of QM to be considered alongside the well-known alternatives. Nevertheless, it is worth considering those which bear certain similarities to the core features of phenomenology: the subjective stance on the one hand and correlationism, on the other. The former is placed front and centre in the 'QBist' approach to QM.[1] Something akin to the latter can be identified in Everett's 'relative state' account (which morphed into the Many Worlds Interpretation) and Rovelli's relational interpretation. Here we shall begin with QBism before turning to the other two in the next chapter.

8.2 QBism: Centring the Agent

The core feature of this position is that it offers an approach to QM that takes the agent and her experience to be fundamental (DeBrota and Stacey 2018; see also Healey 2017; also Stacey 2019). Thus, the wave-function should be understood solely in epistemic terms, as representing not the state of a physical system but rather that of some agent with regard to their possible future experiences.[2] It does this by encoding, as the relevant probabilities, the agent's coherent degree of belief regarding each of certain alternative experiences that result from an act they perform, such as the outcomes of a

[1] Van Fraassen has suggested that one way of evading the critique he has presented of my interpretation of London and Bauer's work (see Chapter 6, fn. 48) would involve 'an entire recasting of QM in terms of information theory of the sort that Chris Fuchs advocates, where the quantum states represent personal states of information. I am not saying that this is impossible, but it is not something signaled in London and Bauer's paper' (private communication; for a phenomenological approach to information-theoretic approaches, see Bilban 2021).

[2] Antecedents can be found in the work of Bitbol, Destouches, and Destouches-Février (de la Tremblaye 2020, p. 248).

A Phenomenological Approach to Quantum Mechanics: Cutting the Chain of Correlations. Steven French, Oxford University Press. © Steven French 2023. DOI: 10.1093/oso/9780198897958.003.0008

measurement procedure, where these beliefs are then updated via some conditionalization rule.[3]

From this perspective the measurement problem simply dissolves: the observation of an outcome becomes nothing more than the acquisition of new information, leading to the reassignment of the 'state'. Since that simply expresses the agent's degrees of belief, any discontinuity between the old 'state' and the new amounts to nothing more than the updating of such credences.[4] Again, the issue of intersubjective agreement looms, as this subjective stance allows that, strictly speaking, two agents observing the same measurement may not have the same experience. However, concordance between credences can be established by appealling to well-known devices that show that updating different prior probabilities in the light of new but common information will lead to convergence (see for example Talbott 2016 for a general overview).

The question now is how to understand, from such a subjectivist perspective, the apparently objective probabilities given by the Born Rule, which, we recall, derives them from the square of the amplitude of the relevant wave-function.

As it turns out, it can be demonstrated that quantum probabilities are just 'objectified' forms of these subjective probabilities (Earman 2018 and 2019).[5] Unfortunately, however, appeal is made here to certain features of the quantum mechanical formalism which, the QBists insist, should not simply be assumed but, on the contrary, recovered *from* the Born Rule, taken as primitive (Fuchs and Stacey 2020, p. 3).[6] It is through the latter that we make contact with the world and it is the subjectivist stance that should come first, 'with the mathematical structure of the theory derivative from it' (Fuchs, in Crease and Sares 2020, p. 558). Hence, QBism should be seen as a form of reconstructive endeavour, rather than an interpretive one (we'll return to this distinction in Chapter 10).

So, let's consider again the example of Schrödinger's Cat: it may appear that the QBist sidesteps the issue of whether the cat should be described as alive or dead before the box is opened because all she is concerned with is the assignment of degrees of belief to the relevant propositions about what will be found (Earman 2019, pp. 415–16). On this view, the 'collapse' of the wave-

[3] The most well known such rule is Bayes Theorem but recently QBists have acknowledged that this is not the only means by which the relevant probabilities can be modified over time (see Stacey 2022, p. 1).

[4] Fuchs has recently indicated that he may be shifting to a more 'voluntaristic' approach whereby statements of subjective probability are not reports on one's psychological state but rather reflect certain epistemic commitments (Fuchs in Crease and Sares 2020, p. 556).

[5] The demonstration hinges on Gleason's Theorem, which shows that for Hilbert spaces with dimensions greater than two, the quantum mechanical probabilities are the only ones that form generalized probability functions (Earman 2018, 2019, pp. 407–8).

[6] We'll return to an explicit consideration of such recovery or reconstructive projects in a phenomenological context in the final chapter; see Berghofer et al. 2021.

function is nothing more than a change in the mathematical representation of an agent's degree of belief upon updating with the new information about the measurement outcome. But then it would seem that she cannot *explain* why the agent experiences a definite outcome, such as observing a live cat, for example. And although this issue can be avoided as long as we think of 'the agent' as nothing more than a disembodied probability calculator fed information by an 'oracle', it comes back to bite us once we think of ourselves as '*physically embodied observers* [my emphasis] . . . whose information acquisition has to be treated quantum mechanically in terms of an interaction with the (measurement apparatus + object system)' (ibid., p. 416; see also McQueen 2017).

However, as far as the QBist is concerned, treating 'information acquisition' in such terms is again to put the correlationist cart before the subjectivist horse. Explanation is certainly not the name of the game, at least not in the sense presupposed here—all they are interested in is their own personal experiences and how they can be related via the probability calculus. As for the allegation that the QBists' subjectivist stance leads to a form of 'solipsistic phenomenalism' insofar as it 'deprives them of the resources to tackle questions about the relation of agents to a non-phenomenalistic world' (Earman 2019, p. 417)—this is precisely where a dose of phenomenology can help. Indeed, as we shall see, it has been argued that the phenomenological understanding of physical embodiment precisely supplies the resources needed.

Before we get there, it is important to be clear that on this view, there is no entanglement of the apparatus with the system as part of the measurement process, with the observer in turn becoming entangled with the joint system that is then formed (Fuchs and Stacey 2020, p. 9). Such a characterization violates one of the core tenets of QBism by ascribing distinct quantum states to the system, the measurement apparatus, and the observer.[7] Instead, the measurement apparatus should be regarded as an *extension* of the observer-as-agent. As a result:

> a sufficiently practiced scientist using an electron microscope to measure atoms might be said to literally 'sense atoms' and not merely be making inferences about them as abstract or hypothetical entities.
>
> (Pienaar 2020, p. 1918)[8]

[7] This also represents a difference with Bohr's position, since by insisting on the necessity of classical language to describe measurement results, the latter introduces a form of mediation between those results and the agent's experience, whereas QBism identifies them.

[8] Again, we recall the thought experiment that involves replacing someone's eyes with electron microscopes.

In this regard we can appropriate the term '*Umsicht*' from Heidegger (ibid., p. 1918, fn. 1), where this means that the agent is only aware of the measurement apparatus circumspectly so that it effectively disappears in its use—until that is, it breaks down, say, when it obtrudes into our awareness and its 'thingness' becomes apparent again.[9]

Obviously this 'extension' of the agent amounts to a shift in the boundary between the agent and the world: 'The World has thus shrunk by losing a System, but the Agent has grown in gaining an Apparatus' (Pienaar 2020, p. 1912). Such an extension is regarded as an act of 'free postulation', in that there is no external criterion in terms of which it can be determined to be 'correct' or not. And the justification for this should by now be familiar, embodied as it is in von Neumann's principle: QM does not prescribe where that boundary should be drawn, only that it must be drawn somewhere, as determined by considerations that lie outwith the theory itself.[10]

Here, then, we see a clear difference with the approach of London and Bauer, arising from this central feature of QBism, namely that entanglement must be derived, rather than presumed:[11]

> QBism indeed regards agents as embodied; how could a disembodied entity take physical actions and experience consequences? The argument that because agents are embodied their interactions with the world must be treated as the generation of entangled states simply presumes its conclusion. (Fuchs and Stacey 2020, p. 9)

8.3 QBism and the World

Two questions now arise: how should we conceive of the relationship between the agent and the world? And how should we understand the notion of embodiment? We'll tackle the second in section 8.6 but with regard to the first, Fuchs maintains that '[w]e believe in a world external to ourselves precisely because we find ourselves getting unpredictable kicks (from the

[9] A tip of the hat to Chris Kenny for helping explain this.

[10] Continuity through such an extension is established by demonstrating that possible measurements post-extension can be obtained from those before the incorporation of the apparatus (Pienaar 2020, p. 1918).

[11] Having said that, a reconciliation might be achieved if the kind of treatment found in London and Bauer's work is understood as simply a presentation of the relevant formalism and not read as conceptually primitive. London and Bauer themselves of course do not appear to display any QBist tendencies!

world) all the time' (Fuchs 2017a, p. 121). However, the QBist cannot adopt a realist stance here, not least because she denies that certain central features of the theory, namely quantum states and their evolution, represent external reality (Fuchs 2017b; Glick 2021, p. 6). Having said that, the Born Rule *is* taken to be objective, in the sense that any agent should use it to find their way in the world, as it were, and in that sense can be understood as corresponding to something we might want to call 'real' (see Fuchs 2017a).[12]

Alternatively, QBism's 'first-person' approach has encouraged the thought that it should be regarded as a kind of 'quasi-idealism' (Glick 2021, p. 8).[13] This is further supported by the emphasis on measurement's creative aspect: 'At the instigation of a quantum measurement, something new comes into the world that was not there before; and that is about as clear an instance of creation as one can imagine' (Fuchs 2010, p. 19). This has been read as suggesting 'a metaphysical picture in which we construct the world via our interactions with it' (Glick 2021, p. 10). Of course, not all such constructions are viable as the history of science demonstrates: the embeddedness of the agent in the world means that, for example, adopting a classical approach to one's expectations regarding measurement outcomes would lead to disaster.[14] It is the combination of features of the world and features of us, as agents, that make it the case that measurements can be regarded as acts of creation from the perspective of the agent (ibid., p. 11).

This is strongly redolent of a phenomenological stance, of course, particularly if we also consider the normative dimension,[15] as expressed relationally via Born's Rule (ibid., pp. 12–13; see also Fuchs in Crease and Sares 2020 pp. 552–6 and Healey 2017). Any metaphysics to be associated with the 'realist' side of things would then only be relevant insofar as it is needed to account for the constraints imposed by that normativity.

[12] Insofar as the rule expresses a relationship between probabilities associated with different sequences of measurements, some form of structural realism might be adopted (Glick 2021, p. 7, fn. 11; see also De Brota, Fuchs, and Schack 2020, p. 1864). Indeed, Fuchs has suggested that QBism could be viewed as a kind of 'normative structural realism', where the structure 'is neither ontic nor epistemic in the sense of representing an objective state of affairs (either reality or knowledge)' (in Crease and Sares 2020, p. 553).

[13] Crease and Sares, on the other hand, argue that by virtue of remaining within the natural attitude, QBism is *too* wedded to realism and that the QBist should follow the phenomenologist in 'bracketing' the reality of the external world (Crease and Sares 2020, p. 542).

[14] Pienaar states that according to QBism 'reality is inherently subjective' (2020, p. 1898) with objectivity secured via the 'holistic structural features of the theory that apply equally to all Agents' (ibid.).

[15] This is something that QBism shares with Healey's pragmatist approach and Fuchs himself has repeatedly drawn attention to the connections with pragmatist philosophers, especially William James, and the idea that 'reality is not ready-made and complete' (in Crease and Sares 2020, p. 548; Healey 2017).

Nevertheless, there is tension here: without an appropriate description of the world, what reason do we have for following that normative constraint? Or, in other words, what grounds the Born Rule? We could just appeal to induction, noting that the rule has been successful in the past (Glick 2021, pp. 13–14). But then it is not clear how to cash out this notion of 'success' in QBist terms since measurement outcomes are not objective features of reality but are particular to the agent's perspective. This is rendered all the more acute by the fact that, understood as relational, the Born Rule as it stands does not make any predictions—it needs to be supplemented, either with a quantum state ascription, which is ruled out in QBist terms, or with the probability of another measurement outcome, which again is understood as subjective. Given that, it seems that QBism doesn't have the resources to ground the rule inductively in a way that would provide a compelling reason for all users of the theory.

Instead, QBists appeal to a form of coherence argument. This is a standard move in the subjectivist camp whereby the axioms of probability theory are justified on the grounds that if they're not accepted, a series of bets could be made for which the agent is guaranteed to lose money, regardless of the outcomes. Thus, QBists claim that not following the rule would lead to similar incoherence. However, given that this holds only in those worlds where QM provides a good guide for agents in them,[16] the issue returns: what is it about *our* world that makes the Born Rule the objectively correct constraint?

We could simply take it to be a 'brute feature of reality' (Glick 2021, p. 15) that represents, as a constraint, the limit of what we can say about the world.[17] Alternatively, in accordance with the acknowledgment that QBism is an ongoing programme, relevant empirical features could be sought that would necessitate the rule. However, given that such features would be manifested via measurement outcomes which, again, are regarded as entirely subjective, it is difficult to see what empirical resources the QBist might draw on.

That would suggest that instead of looking to the *world* for any grounding, the QBist should focus on the nature of the *agent*. As we'll see, the Everettian or 'Many Worlds' interpretation must also deal with this problem of justifying

[16] Pienaar distinguishes it from standard 'Dutch book' coherence and calls it 'World-coherence' (2020, p. 1900).

[17] Fuchs himself suggests as much when he states that the rule plays some ontic role (in Crease and Sares 2020, p. 555). There may also be other 'ontic elements' associated with the Bell and Kochen–Specker theorems, for example, which impose certain structural constraints typically understood in terms of non-locality and contextuality (Crease and Sares 2020, p. 555).

the Born Rule (albeit for different reasons) and recent analyses have likewise proposed a similar shift (Wallace 2012).

Of course, the justification for the rule would then lie with the subjective experiences of the observer and their decision-making, with the attendant issues of explicating a relevant notion of rationality, and so forth, and would be in tension with the suggestion that the rule has an ontic flavour.[18] One way of dissipating the tension would be to step away from such labels as 'realist' and 'idealist' and adopt a phenomenological stance according to which the rule is grounded in our engagement with the 'life-world'. This would then account for its ontic flavour whilst acknowledging its ultimate 'subjective–relative' nature (recall Husserl 1970b, p. 126).[19] Before we get there, however, it would be useful to consider how the QBist treats the 'Wigner's Friend' thought experiment (presented as central to the development of QBism in De Brota, Fuchs, and Schack 2020, p. 1860).

8.4 QBism and Wigner's Friend

We recall the basic set-up: the Friend is in a room with a system and measurement apparatus. Wigner remains outside until the Friend undertakes the measurement and then the two compare notes. We also recall that in the original telling of this little narrative, Wigner asked his Friend what he saw before Wigner entered the room or turned around and insisted that his Friend will reply expressing a definite outcome, since 'the question whether he did or did not see the [definite outcome] was already decided in his mind' (Wheeler and Zurek 1983, p. 176). And it is here that Wigner gave, in support of this latter claim, London and Bauer's assertion of the Friend's 'characteristic and quite familiar faculty' of introspection. Since the issue as to what he saw was already decided in his Friend's mind before he (Wigner) returned to the set-up asked, Wigner concluded that the state immediately after the interaction between his Friend and the system cannot be a superposition and hence consciousness must play a different role in QM than an inanimate measuring device.

[18] According to Bitbol and de la Tremblaye, such tensions within QBism arise because of its 'dual image' of an agent 'really' acting on a 'real' system (forthcoming, p. 10). A thorough 'phenomenologization' of the view is proposed, according to which 'neither the nature of the world nor the nature of its objects is fundamentally different from the nature of experience' (ibid., p. 27).

[19] Destouches-Février (see later) derived the rule from the non-contextuality of the probabilities (that is, their coherence) and the contextuality of the phenomena, understood as 'the fact that the world *cannot* be neatly separated into an observing system and an observed system' (Bitbol and de la Tremblaye forthcoming, p. 19).

From QBism's point of view, however, this is a story about two agents, with the measurement apparatus treated as an extension of the Friend's body (see Fuchs 2010, p. 6). A measurement is then understood in terms of the agent acting on the given system and is represented formally via a set of operators, with the action having a partially predictable consequence for the agent, namely a particular measurement outcome. The agent will then entertain certain degrees of belief about such consequences, where these are represented within the formalism by the wave-function. Furthermore, and importantly (not least for the connection with phenomenology of course) such a consequence is a 'unique creation within the previously existing universe', albeit not subject to the agent's 'whim and fancy' (ibid., p. 6; here we might recall London and Bauer's idea of 'free creation').[20]

Following such a measurement it might be asked, what is the 'correct' quantum state that each agent should have assigned to the system? For the Friend it will be a definite state, either 'spin up' or 'spin down', say. But what about Wigner? Regarding his Friend and the box as just another isolated quantum system, and before interacting with her, Wigner would of course assign an entangled quantum state to this combined system from which the state of the system could then be extracted (using a partial trace operation). However, it would not be that assigned by his Friend. Who, then, is correct?

For the QBist this question simply makes no sense, presuming as it does an agent-independent notion of 'correctness' (Fuchs 2010, p. 7). Since quantum states are entirely subjective and non-representational, the information gained should not be understood to be 'about' some mind-independent reality but rather has to do only with the consequences of the agent's actions upon the system. A slide into idealism is avoided, however, because 'the real world, the one both agents are embedded in—with its objects and events—is taken for granted. What is not taken for granted is each agent's access to the parts of it he has not touched' (ibid., p. 7).[21] As far as Wigner is concerned, when it comes to his interactions with his Friend, or the system, or both, he should

[20] According to Fuchs just as 'a healthy body can be stricken with a fatal disease which to outward appearances is nearly identical to a common yearly annoyance' (2010, p. 1), so quantum theory, perhaps the healthiest 'body' in the history of physics, incorporates what many take to be merely a 'common annoyance' but which may turn out to be a symptom of something fatal, namely the fundamental role of the notions of 'observer' and 'measurement'.

[21] Bitbol has argued that, from the phenomenological perspective, this requires both the reduction to the life-world and the transcendental reduction, since without the latter 'we would have had to find a reason, in the mesoscopic domain of experimental devices and laboratory activities, why only *one* of the two state vectors is valid' (2021, p. 574).

make any decisions regarding the consequences of these interactions according to the prescription given by quantum theory. This prescription will be different as far as the Friend is concerned and as long as we appreciate that difference, the QBist maintains, there is no conflict (see also De Brota, Fuchs, and Schack 2020, pp. 1866–8).[22]

Nevertheless, there remains the worry that the central issue of establishing intersubjectivity here has not been fully addressed. Indeed, the claim that QM 'doesn't give one agent the ability to conceptually pierce the other agent's personal experience' (Fuchs 2010, p. 8), may suggest that QBism still sails a little too close to a form of solipsism (see, for example Crease and Sares 2020, pp. 545–6). In this regard it is worth noting that the QBist acknowledges that the nature of the agent is left out of this picture—indeed, to expect QM to derive the notion of agent is akin to expecting to be able to derive the notion of the user of logic from the formalism itself, or the reader of a probability textbook from its contents—'How could you possibly get flesh and bones out of a calculus for making wise decisions?' (Fuchs 2010, p. 8). As sympathetic as we might be to this resistance to an unwarranted demand, the feeling remains that absent some further consideration of the *nature of the agent*, and their embodiment, the QBist picture is incomplete (and indeed, Fuchs acknowledges in several places that there is more to be said).[23] It is here, of course, that the phenomenologist may step in, particularly given the QBist insistence that it is precisely in this way that QM is different from any theory posed before, namely in being simply an addition to probability theory, understood as normative; that is, as a *theory of knowledge*, just as London and Bauer maintained.[24]

So, let's now consider how the QBist might draw on certain features of the phenomenological stance in order to philosophically underpin their position. As we'll see, the extent to which such a move can be deemed successful depends on how that stance is conceived.

[22] When it comes to Wigner's concern that the ascription of a superposition state to the arrangement that includes his Friend, the apparatus, and the original system implies that his Friend must be regarded as in a 'state of suspended animation', the QBist's response is, bluntly, that Wigner's state ascription simply has no bearing on the state of consciousness of said Friend. Thus, we can both ascribe such a state and grant the Friend the conscious experience of seeing either an alive or dead cat. For further discussion and the QBist response to a recent extension of Wigner's thought experiment (Frauchiger and Renner 2016 and 2018), see (De Brota, Fuchs, and Schack 2020).

[23] 'QBism knows that its story cannot end as a story of gambling agents—that is only where it starts' (Fuchs 2010, p. 27).

[24] And this, Fuchs insists, 'is all the vaccination one needs against the threat that quantum theory carries something viral for theoretical physics as a whole. A healthy body is made healthier still' (2010, p. 9).

8.5 QBism and Phenomenology

Bitbol has highlighted three points of contact between QBism and phenomenology;[25] the first is the most obvious, perhaps, namely that just as the QBist regards QM from a 'first-person' perspective, so phenomenology requires the adoption of the same in order to identify the contribution of consciousness to experience (Bitbol 2020, p. 232).[26]

The second has to do with the shift in attention that we find, in both QBism and phenomenology, away from apparently 'external' objects, whether those of science or the life-world, and towards that contribution of consciousness. It is this shift that marks the phenomenological reduction, of course, and just as the latter is driven, methodologically, by the *epoché*, so in QBism we are urged to suspend our judgement with regard to the referential capacity of the symbols of the formalism of QM (Bitbol 2021, p. 570). Having noted that, we might wonder if the contact is entirely smooth here, given that QBism seems to go further than merely *suspending* judgement by adopting a stance that is closer to instrumentalism in taking these symbols to ultimately represent merely the probabilistic weights that agents assign to the outcomes of experiments (Bitbol 2020, p. 232).

The third similarity proceeds from the second: the QBist insistence that QM only tells us something about the *expectations* we should have concerning the outcomes of experiments is, Bitbol has argued, similar to Husserl's understanding of perception, based as it is on his conception of 'horizontal intentionality' (see also Bitbol 2021, p. 571). We recall that the idea here is that in perception only part of the perceived object is intuitively given to us but we possess an intentional awareness of the other 'profiles' or adumbrations of the object. Our anticipation of our perception of these profiles can be situated in an open manifold of such anticipations that constitutes what Husserl calls the *intentional horizon*.

This third point of contact is then taken up by de la Tremblaye who has used the example of our perception of a cup and suggested that:

> [t]he cup is the analogue of the microsystem, the perceptual horizon parallels the QBist quantum state, the perceptual act corresponds to the physicist's

[25] QBism is the 'most consistent phenomenological approach' towards QM (Bitbol 2021, p. 570).

[26] Having said that, we recall that Pfänder and fellow members of the Munich school urged the inclusion of intersubjective relationships. And as we shall see, including such relationships alters QBism as well.

> measurement and the modification of my possible horizon corresponds to the modification of the state vector after the measurement.
>
> (de la Tremblaye 2020, p. 255)

Thus, just as perception is a matter of updating the horizon of possibilities associated with our present observation of an object, such as a cup, so the quantum state, on a QBist reading, expresses a 'bundle of expectations' (ibid., p. 254). Before a measurement, then, the relevant eigenstates correspond to anticipated possible profiles and '[t]he (probabilistic) estimates of subsequent measurements are… analogous to estimates of future perceptions, namely the internal perceptual horizon of an object' (ibid.). From this horizon only one possible scenario results, of course, and likewise, on a QBist reading, as we've seen, a measurement outcome is considered a personal experience.[27]

Now, the intentional horizon encompasses various anticipated possible profiles; so, for example, in the case of a cup about to fall off a table, we can anticipate either that it will break or will remain undamaged but never both. Likewise, when we perform a spin measurement, via a Stern–Gerlach apparatus, say, we can anticipate either the outcome 'spin up' or 'spin down' but never both (2020, p. 254). In the former case, our anticipations are based on our past experiences with falling cups and on our understanding of the relevant background conditions (whether the floor is carpeted or not, say) and it is on this basis that the horizon of possibilities is determined. In the case of the spin measurement, likewise, the possibilities are determined by our beliefs, at least as far as the QBist is concerned:

> In establishing the state vector, I express in a formal way my beliefs about the future of my measurements; and these beliefs arise by due consideration of my own past experience (including the experience of preparation). The (probabilistic) estimates of subsequent measurements are thus analogous to estimates of future perceptions, namely the internal perceptual horizon of an object. (de la Tremblaye 2020, p. 254)

From this horizon only one of the possible scenarios is perceived, giving priority to the role of that present perception in that, by virtue of the perceptual horizon being an integral part of our experience of the cup, say,

[27] In both cases the processes of knowledge acquisition and decision-making is dynamic, crucially involving an active role on the part of the agent (de la Tremblaye 2020, p. 256).

that present perception has a direct effect on the constitution of that cup, by imposing a determination on that horizon. Likewise, again, in the case of the spin measurement, only one outcome is perceived, with the experience of the flash on the screen, or the click of the counter, taken as analogous to the 'sensory nucleus of perception' in Husserlian terms.

However, there are two worries that arise at this point. The first is whether we can straightforwardly draw parallels between our everyday experiences, embedded as they are in the 'life-world' and those that arise in the form of what the QBist calls 'kicks' from the world as manifested in the spin measurement. The example of the falling cup is, of course, one with which many of us are reasonably familiar, to the extent that we can claim to have fairly well-formed expectations as to the possibilities in play. We don't even have to look to cases of quantum phenomena to note that those expectations are based on certain inductive inferences regarding the phenomena in question. And those inferences may well lead us astray—after all, one of the possibilities compatible with (classical) statistical mechanics is that all the air molecules in the room could suddenly be distributed to gather together beneath the cup as it falls, thereby cushioning it and even lifting it back onto the table!

Of course, that would fall under the 'cup doesn't break' possibility but still, it does give grounds for questioning whether the expectations we form in 'everyday' situations are sufficiently similar to those we could legitimately form in a laboratory, say, so as to justify drawing the above parallel. Granted that the examples that Husserl himself presented to help the reader understand the notion of the intentional horizon were drawn from 'everyday life', the distinction between the life-world and the idealized 'world' of physics, as covered in the previous chapter, raises questions as to whether the notion is sufficiently elastic in this respect.[28]

A further difference is that whereas the cup either has to break or not, such a disjunction does not always hold in the case of spin 'up' and 'down'—indeed, that the range of possibilities should include a superposition of such disjuncts is precisely what lies behind Schrödinger's Cat thought-experiment. Now, the QBist may respond that to assume that the cat could 'really' be in a state of alive-and-dead or the particle in a state of spin-up-and-down is, again, to beg the question: all we have to work with are our personal experiences of a definite outcome, together with the Born Rule, however that is grounded. It

[28] See, however, Alves 2021, pp. 474–5.

is precisely because of this that de la Tremblaye, for example, can draw the comparison that she does.[29]

However, there are further concerns that are generated from within phenomenology itself, particularly with regard to how we should interpret this notion of the intentional horizon. Zahavi, for example, has argued that it actually requires a certain kind of intersubjectivity in that such a profile cannot be seen as future-oriented, nor as a current fiction or product of the imagination but 'must be understood as the noematic correlate of the possible perception of an Other' (1997, p. 3). In other words, it is the perceptions of *another* that underpin the required correlation:

> When I experience someone, I am not only experiencing another living body situated 'there', but also positing the profile which I would have perceived myself if I had been there... Thus, my concrete experience of the Other can furnish my intentional object with an actual co-existing profile. (ibid., p. 3)

Of course, there is an immediate objection: surely my perception of the cup cannot be dependent upon my simultaneous perception of another subject who is also actually perceiving the cup?! Indeed, there would have to be a huge number of such actual subjects, given the number and variety of possible profiles.

Husserl himself was aware of this problem and suggested that this insertion of a form of intersubjectivity leads to a certain 'openness' by virtue of entailing structural references to the perceptions of numerous possible others (Zahavi 1997, p. 4). And furthermore, there is a certain reciprocity involved—in the sense that I must now accept that I am an Other with respect to one of these other perceiving egos—that 'implies a dethronement of my own ego as the sole pole of constitution... and this dethronement has far reaching constitutive implications' (ibid., p. 5). In particular, objectivity, understood as intersubjective validity, can only be established once that reciprocity is acknowledged and the ego perceives itself to be 'one among Others' (and again here we recall the discussion in the previous chapter).[30] Thus:

[29] That our expectations must be governed by the Born Rule might also be alluded to in these considerations but that point doesn't impact on the parallels drawn with regard to the notion of the intentional horizon at least.

[30] This then impacts on the range of the primordial reduction as well: 'If the horizontal co-givenness of the absent profiles refers us to the open intersubjectivity (since these profiles are to be understood as profiles for an open plurality of possible Others), then my horizontal intentionality and, consequently, my awareness of appearing objects imply an a priori reference to the constitutive contribution of

> it is an apodictic transcendental fact that my subjectivity constitutes for itself a world as intersubjective. The other self is therefore a necessary intentional 'object' of the absolutely evident structure of my awareness. Furthermore, this other self is necessarily coequal with my self. My transcendental self, by virtue of its evident structure, perceives itself as without any superiority over the other self. (I am an intentional object for him, as he is for me; he is an absolute constitutive consciousness, as I am.) This is all part of the apodictic facticity of my transcendental subjectivity. It does not depend on the fortuitous constitution of a particular object of valid *Einfühlung* in perception, but is simply an explication of the fact that I do intend a world as necessarily intersubjective. (That is what I mean by calling it a world. If it were not intersubjective, it would not be a world). Strictly it is an (open?) infinity of other subjects which is required by the apodictic factual structure of my transcendental consciousness, not one other subject.
>
> (Husserl in Cairns 1976, pp. 82–3; reproduced in Zahavi 1997, pp. 9–10, fn. 24)

This then raises further concerns about how well grounded the parallels are between phenomenology and QBism. In particular, from this perspective, two interacting agents cannot each consider the other as a 'system' (De Brota, Fuchs, and Schack 2020)—each has to recognize the other as an 'absolute constitutive consciousness', leading to an obvious tension with the 'first-person' perspective. Indeed, Bitbol has acknowledged this when he argues that intersubjective agreement must be based on a shared acknowledgment of the existence of the ordinary objects of the life-world (2021, p. 572). Hence:

> even though QBism is phenomenologically right to claim that the *de jure* basis of scientific knowledge is personal lived experience and verbal communication between subjects of experience, it should also recognize that the de facto basis of quantum physics is Bohr's classical-like domain of ordinary objects and instruments. (ibid.)[31]

foreign subjectivity. Thus, the actual experience of another embodied subject is founded upon an a priori reference to the Other. Prior to my concrete encounter with another subject, intersubjectivity is already present as co-subjectivity. Against this background, it must be concluded that an attempt to implement a transcendental aesthetics primordially . . . is a failure, and, consequently, that the constitution of the Other as an incarnated subject (as a lived body) cannot be undertaken primordially either' (Zahavi 1997, p. 5; see also ibid., p. 7).

[31] As a result, he claims, the types of experience that feature in the probability assignments that QBism takes to its heart are most conveniently expressed in terms of 'classical-like' predicates regarding the relevant instruments, such as the Stern–Gerlach apparatus mentioned both above and by London and Bauer (Bitbol 2021, p. 572).

In effect, then, what is typically portrayed as the basis of QBism—namely the first-person perspective—must be modified to some extent in order to maintain its alignment with phenomenology.

Further impetus for such modifications derives from the need to supply some form of grounding for the Born Rule as already noted. As Bitbol has acknowledged, 'when one is asked to *explain* the structure of the quantum probabilistic predictions, one must go beyond the purely subjectivistic option of QBism' (Bitbol 2021, p. 573). Such an explanation is posited to lie at the 'interface' between 'outer reality' and 'inner subjectivity', involving as it does both the experimental context and the creativity of the agent (ibid.).[32] And this interface, in turn, is manifested in our *embodiment*.

This leads us to the second question posed at the beginning of section 8.3: what is the nature of that embodiment? Here we turn to the work of Merleau-Ponty which has also been drawn upon in establishing a connection between QBism and phenomenology (see Berghofer and Wiltsche 2020).[33] As we'll see, this not only incorporates an explicit consideration of this notion of embodiment, but also offers a detailed analysis of modern physics which makes explicit reference to London and Bauer's little book.

8.6 Quantum 'Flesh'

While a student in Paris, Merleau-Ponty attended Husserl's 1929 Sorbonne lectures (Husserl 1964), as well as Gurwitsch's lectures on Gestalt psychology which he folded into his phenomenological stance,[34] proposing that:

> matter, life, and mind are increasingly integrative levels of Gestalt structure, ontologically continuous but structurally discontinuous, and distinguished by the characteristic properties emergent at each integrative level of complexity. (Toadvine 2019)

[32] Bitbol has drawn on the research of Paulette Destouches-Février (1951), whose background was originally in philosophy and mathematics, but who was awarded the *diplome d'études supérieures* in physics for her thesis on particle indistinguishability (Février 1939). She subsequently published work on the nature of wave mechanics and hidden variables interpretations (see https://fr.wikipedia.org/wiki/Paulette_Destouches-Février). Her husband studied with de Broglie and presented a 'principle of subjectivity' according to which measurement results should not be considered as pertaining to intrinsic properties of the system but as properties of the 'system-apparatus' complex (Bitbol 2001; Pellegrini 2021, pp. 487–96).

[33] This work was influenced by that of Destouches-Février.

[34] He cited approvingly Gurwitsch's claim that Husserl's philosophy 'lead[s] to the threshold of *Gestaltpsychologie*' (Toadvine 2019).

He went on to argue that the most basic form of perceptual experience is the Gestalt but that the increasing determination of the apparently indeterminate and ambiguous elements of perception, as amplified by science, lead eventually to 'the theoretical construction of an objective world of determinate things' (Toadvine 2019). Through the 'transcendental reduction' our naïve belief in an 'objective' world can then be bracketed and the messy, indeterminate, and perspectival 'lived world' brought back to centre stage, to be considered as the origin of meaning.

Central to this process is our perception of our own body, whose orientedness towards the world yields the background against which objective space is constituted, where the body is to be understood as not so much 'in' space as living it (ibid.). This in turn exemplifies the broader relationship between the body and the world, understood as equally active and receptive:

> The properties of things that we take to be 'real' and 'objective' also tacitly assume a reference to the body's norms and its adoption of levels. An object's 'true' qualities depend on the body's privileging of orientations that yield maximum clarity and richness. This is possible because the body serves as a template for the style or logic of the world, the concordant system of relations that links the qualities of an object, the configuration of the perceptual field, and background levels such as lighting or movement.
>
> (Toadvine 2019)[35]

Although the relational aspect was central to Merleau-Ponty's system, he insisted that things cannot be reduced to mere perceptual correlates as they exhibit that resistance that Fuchs, for example, has acknowledged with the idea of the world 'kicking' back. This is coupled with a depth that goes beyond the perspectival limitations of our perception. Thus, for Merleau-Ponty, the body presents a further 'genre' of being that lies between the subject and the object and it is to this that he appealed in tackling the issue of intersubjectivity: 'We perceive others directly as pre-personal and embodied living beings engaged with a world that we share in common' (ibid.). It is through our common corporeality that we obtain a shared social world, thereby to be understood as a permanent feature of our existence.

[35] Von Baeyer has also drawn connections between QBism and phenomenology via the more recent work of Todes who likewise argued that the human body provides the foundation for understanding all human experiences (von Baeyer forthcoming).

This correlative feature is extended to the body itself, which doubles as both sentient and sensible when we touch ourselves, for example (see Merleau-Ponty 1968, p. 133). This demonstrates, on the one hand, an ontological continuity between subject and object in general, but, on the other, reveals that there is always a kind of gap between the sentient and the sensible such that they never coincide (Merleau-Ponty 1968, p. 135; see also Toadvine 2019; Pellegrini 2021, pp. 496–8). It is this bi-directional exchange that underpins Merleau-Ponty's talk of the 'flesh' of things that makes communication between the sensing body and sensed things possible. Such talk should not be taken as anthropocentric: 'the presence of the world is precisely the presence of its flesh to my flesh' (Merleau-Ponty 1968, p. 127).[36]

Neither should 'flesh' here be understood as matter, in the sense of 'corpuscles of being' that make up other beings; rather, it acts as 'a sort of incarnate principle that brings a style of being wherever there is a fragment of being' (ibid., p. 139). Within this framework, we must abandon the old assumptions that place the body in the world and the 'seer', or observer, in the body. As Merleau-Ponty asked—and here we might recall von Neumann's Psychophysical Parallelism—'[w]here are we to put the limit between the body and the world, since the world is flesh?' (ibid., p. 138). Bitbol has described this as 'an ontology of radical *situatedness*: an ontology in which we are not onlookers of a nature given out there, but rather intimately intermingled with nature, somewhere in the midst of it' (2020, p. 236; Pellegrini 2021, pp. 496–9).[37]

Such a 'radical situatedness' encourages comparisons with QM, of course, and Merleau-Ponty himself engaged with the latter in a series of lectures delivered at the Collège de France (Merleau-Ponty 2003; see Barbaras 2001). Here he raised the fundamental question of whether the picture of the world that physics presents could include the physicist qua observer herself (Berghofer and Wiltsche 2020, p. 33). QM, Merleau-Ponty argued, attempts to do precisely this, by placing the relationship between the subject and object in question (ibid.) and can be accommodated by shifting to the phenomenological stance according to which the physicist is 'intermingled' with the world.[38] Thus,

[36] Merleau-Ponty also wrote: 'One can say that we perceive the things themselves, that we are the world that thinks itself—or that the world is at the heart of our flesh' (1968, p. 135, fn. 2), lines that resonate with Wheeler's 'Participatory Anthropic Principle'.

[37] As a result, Bitbol has argued, the role of constituting objectivity is extended to anything that expresses this principle of incarnation—in effect, then, the world as flesh becomes self-objectifying (2020, p. 236). The worry, of course, is that we lose any sense of 'objectivity' in such a move.

[38] Thus, according to Merleau-Ponty, 'no one can truly understand quantum mechanics without accepting a deep transformation of our conception of knowledge' (Bitbol 2020, p. 239). Here again we find an echo of London and Bauer's statement at the beginning of their 'little book'.

a moment comes when the development of physics calls into question the presupposition of an absolute spectator and '"objective" and "subjective" are recognized as two orders hastily constructed within a total experience, whose context must be restored in all clarity' (1968, p. 20). It was just such a moment that arrived with the advent of QM which should be recognized as a physics that situates the physicist physically (!) and 'enjoin[s] a radical examination of our belongingness to the world' (Merleau-Ponty 1968, p. 27; quoted in Berghofer and Wiltsche 2020, p. 33).[39]

In particular, according to this new scientific ontology 'existing things are not individual realities, but generic realities' (ibid., p. 92). And it is here that he cited London and Bauer's comment that QM should be considered a 'theory of species', with the indiscernibility of quantum particles providing grounds for denying their status as individual existents (again, see French and Krause 2006).[40] As a result, the statistics provide the 'maximum image of the object' that we can obtain (Merleau-Ponty 2003, p. 93), with measurement yielding different examples from an ensemble of such entities.

Underpinning this new picture, Merleau-Ponty noted, is the non-classical relation between measurement and the observed thing. Contrary to what the QBists assert, he emphasized that the measurement apparatus cannot be regarded as an extension of our senses, since it does not *present* the object to us, but rather, 'realizes a sample of [the] phenomenon as well as a fixation' (ibid.; see also Berghofer and Wiltsche 2020, p. 34). Here again he drew on London and Bauer, in particular their comparison between the perspectives of the observer and a 'witness' observing the observer and her observation, which of course foreshadowed Wigner's thought experiment.

Significantly, Merleau-Ponty also reproduced the passage containing that famous line about the observer having 'relations of an entirely particular character with himself' which, given the faculty of introspection, yields not a mysterious interaction between the apparatus and the object but rather a separation of the 'I' and the constitution of a new objectivity. Thus, 'the role of the observer is not to make the object pass from the in-itself to the for-itself (as in Descartes)' (Merleau-Ponty 2003, p. 94); rather, as London and Bauer said, it is to break the chain of statistical probabilities and 'make an individual existence emerge in act' (ibid.; here we recall the point about such an

[39] Merleau-Ponty also gave a potted history that drew heavily, but not surprisingly, on de Broglie's work and mentioned von Neumann as trying to 'extract a probabilistic logic within which quantum mechanics would lose its strange character' (Merleau-Ponty 2003, p. 91). Unfortunately little of this is commented on in Barbaras 2001, for example.

[40] 'There is no more individuated being in the system' (Merleau-Ponty 2003, p. 93).

individual existence emerging as a pole in the correlation). And what underlies this existence is 'a thought that annexes itself to the apparatus' (ibid.). Measurement, then, is an *engaged* operation and this is reminiscent of our own situation of embodiment, whereby 'any operation of our own body is an operation within the "flesh of the world"' (Bitbol 2020, p. 239).[41]

As a result, Merleau-Ponty advocated a form of 'participationist' realism drawn from Destouches-Février's work (see Berghofer and Wiltsche 2020, p. 33), that transcends the opposition between object and subject and is broadly structuralist in character in setting the relations presented by the theory at its heart.[42] Having said that, this structuralist understanding must remain grounded in the world of perception, on the basis of which reality is constituted. As Pellegrini has noted:

> For Merleau-Ponty, for the knowledge related to the microscopic world, too, one cannot ignore the world of perception, since it is here that the knowledge of what is real at a microscopic level is understood, through measurements and theoretical work. It is always within the *Lebenswelt* that this work is carried out. (2021, p. 493)[43]

The picture presented by QM only appears contrary to that given by 'natural perception' if we think of the latter as placing before us well-defined beings of 'pure exteriority' (Merleau-Ponty 2003, p. 99). However, the phenomenological analysis of perception demonstrates that this is at best only a half-truth. If we shift attention from the 'isolating attitude', we can discover all sorts of 'ambiguous beings' in the 'natural field'—here Merleau-Ponty gave the

[41] Bitbol has asserted that the situation in QM is an extension of our situation of embodiment, so that '[a]t the end of the day, quantum physics testifies that the world behaves as a big flesh, of which our flesh is a sample' (2020, p. 241). As he then acknowledges, it is thought that cuts the measurement chain, thereby yielding a definite outcome.

[42] These relations can claim a certain objectivity by virtue of being independent of the measurement process but are relative to the 'species' of system being studied and refer, not to objects per se, but to 'certain mathematical forms that are necessary for the description of the relation of the subject to the object' (Merleau-Ponty 2003, p. 98; see again Berghofer and Wiltsche 2020, p. 33). Having said that, the fact that they are determined by the theory confers on them a form of reality going beyond the simply mathematical (see also French 2014, ch. 8). Destouches-Février insisted that such relations 'schematize the general conditions on the observers in their relations with the objects—confer on them a reality not possessed by purely mathematical beings, independent of any sensible meaning' (trans. in Pellegrini 2021, p. 491). The connection with structural realism is explicitly made by Berghofer and Wiltsche 2020, p. 35, who note that although one might be tempted to call Merleau-Ponty's position a form of structural idealism, the fact that he insists that these relations cannot be reduced to the mental suggests that it has a realist flavour.

[43] According to Merleau-Ponty Destouches-Février did not appreciate this and so remained tied to a dualistic view (Pellegrini 2021, p. 494).

example of the wind—as well as non-determinate and even 'negative' beings, 'whose entire essence is to be absence' (ibid., p. 99), as well as those that are neither finite nor infinite.

The point is not so much to say that quantum systems are like the wind (!) but rather to acknowledge that the phenomenological conception of perception can accommodate entities that are not appropriately characterized as unique individuals. Having said that, perception does not contain everything; it is through the internal critique of physics, in terms of the phenomenological reduction, that we become aware that the perceived world is not objectively given. Merleau-Ponty concludes with a sentiment that again recalls London and Bauer, namely that the meaning of physics is to allow us to make 'negative philosophical discoveries', in the sense that we come to appreciate that certain claims that we believed to be well grounded turn out not to be—the idea of things as 'individual realities' being one such. Thus, '[p]hysics destroys certain prejudices of philosophical and non-philosophical thought without, for all that, being a philosophy' (ibid., p. 100; see Bitbol 2020, p. 240 and Pellegrini 2021, p. 486).

As Berghofer and Wiltsche put it:

> there can be no doubt that Merleau-Ponty... accepts the perspectivity of our scientific image of reality. For Merleau-Ponty, however, this claim is not the result of a reflective analysis from outside of physics. Quite the opposite, on Merleau-Ponty's reading, quantum mechanics itself implies the strong ontological claim that the classical picture of a purely objective, observer-independent physical reality is untenable, and that every complete physical description of reality must incorporate the physicist as well as her experience. Seen from this perspective, then, quantum mechanics has the potential to live up to the ideal of a fully rationalized, critical, and ultimately *phenomenological* physics. (2020, p. 37)

It is this *perspectival* feature of Merleau-Ponty's thought that encourages a positive comparison with QBism. However, as we've seen, he also drew on London and Bauer's analysis, with its explicit incorporation of the correlationist aspect, both phenomenologically and physically, as manifested via the notion of entanglement. This is anathema to the QBist, of course, as is Merleau-Ponty's centring of the relations represented by the theory more generally. If, then, the QBist wants to draw on phenomenology to philosophical underpin her position, she is going to have to either modify the latter or exclude the correlationist understanding of the former.

8.7 Conclusion

As we have seen, then, the compatibility of QBism with phenomenology, and hence the extent to which we can obtain a truly 'phenomenological physics', hinges on an understanding of this philosophical stance that brings to the fore its 'first-person' perspective. In that case, not only must intersubjectivity be accommodated, in some manner, but also, crucially, what Fuchs has called the 'kicks' from the world. Within its reconstructive approach to QM as a whole, QBism is forced to conceptualize these in terms of singular experiences with the relationships represented by Schrödinger's notion of entanglement then treated as derivative.[44]

As we've also noted, this sits uncomfortably with phenomenology's correlative aspect, understood in terms of that 'mutually dependent context of being' in which consciousness and the world stand (Beck 1928; see Zahavi 2017).[45] Taking the latter seriously, as I have argued that London and Bauer did, encourages a close examination of those interpretations of QM that emphasize its relational features, such as the Everettian, or Many Worlds account, its 'Many Minds' variant, and so-called Relational QM. As we'll now see, these offer opportunities for the development of an alternative kind of 'phenomenological physics'.

[44] Thus, Bitbol has argued that the claim that QM describes the correlations expressed by the notion of entanglement, for example, 'can only arise from a descriptive, and therefore "realist," construal of quantum states; and therefore, deriving the "reality" of correlations from this argument is a *petitio principii*' (2021, p. 578). Undertaking the transcendental reduction (as incorporated by QBism) then has the consequence that entanglement is nothing but 'a fake descriptive projection of a mathematical property of the predictive symbol of quantum physics' (ibid.), as also is evident from the impossibility of faster than light communication involving such correlations. Bitbol and de la Tremblaye have gone on to draw on the work of Barbaras, who espoused a form of phenomenology that incorporated the primacy of 'belonging' over relation, and present QBism as a kind of 'eco-phenomenology' or radical participatory empiricism (forthcoming).

[45] Following on from the previous footnote, we do not necessarily have to take a realist view of quantum states—at least not in the usual sense—in order to accommodate this correlative aspect, as I hope the discussion of London and Bauer's account has shown.

9

Many Worlds, Many Minds, and (Many) Relations

9.1 Introduction

There is a huge literature on the history of Everett's account, its development into the 'Many Worlds' interpretation, and its assorted philosophical features and implications (see, for example, Barrett 2018; Vaidman 2021; Wallace 2012). What I shall do here is, first of all, briefly go over the origins of the interpretation, rooted as they are in some familiar concerns. Then I shall explore a well-known attempt to accommodate consciousness within it before moving on to consider a further interpretation—so-called 'Relational QM'—which bears certain similarities to Everett's, certainly with regard to the treatment of quantum states. I'll wrap things up with some consideration of the extent to which there might be commonalities between the relevant philosophical features of these accounts and a phenomenological stance that incorporates the correlative aspect emphasized by London and Bauer.

9.2 Everett and Wigner's Friend

Let's return to Schrödinger's Cat once more. We recall that on the 'consciousness causes collapse' view, attributed to von Neumann and, mistakenly, to London and Bauer, prior to the box being opened, the state of the cat is described by a superposition but once the box is opened, and the cat observed, the state collapses to a definite outcome—either alive or dead. We also recall the criticisms of this view, particularly those of Putnam and Shimony, but even earlier, in a 1954 talk at Princeton Einstein had 'colourfully expressed his discomfort with the idea that simple acts of observation can bring about drastic changes in the universe' (Freire Jr 2015, p. 88). This evidently had a considerable impact on Everett (Barrett and

A Phenomenological Approach to Quantum Mechanics: Cutting the Chain of Correlations. Steven French, Oxford University Press. © Steven French 2023. DOI: 10.1093/oso/9780198897958.003.0009

Byrne 2012, p. 15),[1] who, in a later letter, stated that he thought it 'unreal that there should be a "magic" process in which something quite drastic occurred (namely, the "collapse" of the wave function), while in all other times systems were assumed to obey perfectly natural continuous laws' (letter to Jammer, in Jammer 1974, p. 508).

He was also less than satisfied with the Bohrian approach, because of what he took to be its ad hoc stipulation that measurement scenarios had to be understood in classical terms. Having said that, the grounds of Everett's attitude certainly shifted under the influence of his thesis advisor, John Wheeler;[2] in his original, so-called, 'long thesis',[3] he rejected both von Neumann's[4] and Bohr's accounts. However, in the shorter version that was accepted for his PhD[5] and subsequently published in *Reviews of Modern Physics*,[6] their inadequacies were presented more as simply obstacles to the application of QM to field theory and cosmology than as major conceptual deficiencies (see Barrett and Byrne 2012, p. 5).[7]

In their place Everett proposed his 'relative state' account: Instead of taking only one of the various superposition components as selected upon measurement, here *all* are retained, in a sense, each yielding the 'relative state' of the

[1] Freire Jr has also suggested that, with von Neumann and Wigner at Princeton at the time, and Bohm also there a few years previously, a critical attitude towards the orthodox view of QM might occasionally have been expressed (2015, p. 89).

[2] Everett attributed his interest in QM to Charles Misner and Aäge Petersen who, drinking together one evening, said some 'ridiculous things about the implications of quantum mechanics' (Everett 1977). Petersen was one of Bohr's assistants and so not surprisingly, adopted the latter's view of the measurement situation (although he also cited not only von Neumann's book but also papers by Margenau, Shimony, and Wigner; Petersen 1968, p. 174). Charles Misner was also a student of Wheeler's but worked primarily on General Relativity.

[3] Initially called 'Quantum Mechanics by the Method of the Universal Wave Function' and subsequently re-titled 'Wave Mechanics Without Probability'.

[4] Everett does not seem to have considered London and Bauer's piece, whether as a summary of von Neumann's view or an extension of it.

[5] This had the title 'On the Foundations of Quantum Mechanics'. It seems that between 1957 and 1983, only seven people had checked it out from the Princeton University Library, beginning with Shimony in August 1957 (Barrett and Byrne 2012, p. 174). Fifth on the list, from 1966, is the philosopher David Lewis. It may have been around then that Lewis began to develop his realist view of possible worlds (Al Wilson, Helen Beebee, personal emails; see also Wilson 2020). Although he took undergraduate classes in physics while at Swarthmore College (including 'Radiation and Statistical Physics', which covered the 'early' quantum theory and quantum statistics 'with applications'; thanks to Anthony Fisher), the crucial motivating impetus may have been the graduate course he took at MIT in 1965 which was co-taught by none other than Putnam and Shimony. In a 1967 letter that cited both Shimony's 1963 paper, 'Role of the Observer in Quantum Theory', and Everett's 1957 *Review of Modern Physics* piece, Lewis described the latter's view as 'quite convincing' (thanks again to Helen Beebee).

[6] As '"Relative State" Formulation of Quantum Mechanics'.

[7] This shift is usually attributed to Wheeler, anxious not to antagonize Bohr and his followers. Freire Jr has noted that Schrödinger was sent a preprint of this 1957 paper but there is no record of his response (2015, p. 89, fn. 59). Margenau, on the other hand, is reported as having responded quite favourably (ibid., p. 113).

system.[8] So, upon opening the box, relative to one observer, the state of the cat will be observed to be 'alive', but relative to her counterpart, its state will be 'dead'.[9] These components were initially referred to as 'branches' and subsequently, as is well known, came to be interpreted in terms of distinct 'worlds'.[10]

Interestingly, Everett began his 'long' thesis of 1956 with a summary of the von Neumann approach, described as 'the most common form encountered in textbooks and university lectures on this subject' (ibid., p. 73), although not representative of 'the more careful formulations of some writers' (ibid.).[11] He went on to note that, 'the situation becomes quite paradoxical if we allow for the existence of more than one observer' (ibid., p. 73) and then presented a version of the 'Wigner's Friend' scenario as a kind of *reductio* of the 'standard' view (Barrett and Byrne 2012, p. 74). Thus, he took the lesson to be that the Friend cannot be said to have any 'independent objective existence' (ibid., p. 75), prior to Wigner entering the room, but that Wigner himself would have no reason to feel complacent 'since the whole present situation may have no objective existence but may depend upon the future actions of yet another observer' (ibid.; also Barrett and Byrne 2012, pp. 30–2). This showed what was wrong with the 'standard' view as simply stated ('simply', in the sense of being given without explicating the grounds for the 'collapse').[12]

[8] We recall that Becker has argued that von Neumann can be interpreted as adopting a form of relative-state account (Becker 2004).

[9] The idea of an observer's 'counterpart' in another 'world' can be made philosophically robust through Lewis' realist stance towards possible worlds, noted above (Wilson 2020).

[10] At the Xavier conference in 1962, in discussion with Everett, Podolsky suggested that 'Somehow or other we have here the parallel times or parallel worlds that science fiction likes to talk about so much' (Podolsky, Hart, and Werner 2002, Tues: A.M. p. 19). Everett agreed and when, a few lines later, Podolsky remarked that we would have a non-denumerable infinity of worlds, Everett again said 'Yes' and in response to a concern from Shimony regarding the site of the observer's awareness (see Section 9.4), stated that 'Each individual branch looks like a perfectly respectable world where definite things have happened' (ibid., Tues: A.M. p. 22) However, as Barrett and Byrne emphasize, although Everett may have used this term informally, he never did so in either the short or long versions of his thesis (Barrett and Byrne 2012, p. 41).

[11] Von Neumann's book was the main reference of Everett's work (Freire Jr 2015, p. 89 and 93; although in the original manuscript of the long thesis the quotes from von Neumann appear to have been added later; ibid., p. 92). We have previously noted the differences between the original and English (and revised) versions of von Neumann's work and it seems Everett may have read it in the original German (ibid., p. 93, fn. 76).

[12] Although, 'there is little doubt that he [Wigner] had been discussing the problem with his students for many years' (Barrett and Byrne 2012, p. 14), there is no record of his presenting the details of the argument in the class on 'Methods of Mathematical Physics' at Princeton in 1954, which Everett took and where he may have come face-to-face with the measurement problem (Barrett and Byrne 2012, p. 12). Everett actually presented his version some years before Wigner's appeared in print (Barrett and Byrne 2012, p. 29, fn. 2; Freire Jr 2015, p. 95, fn. 83). However, although Everett's work was cited in Margenau (1963b), albeit without comment, it was not mentioned by Wigner in his papers on the measurement problem (Wigner 1962, 1963a; see Barrett 2017, p. 32). In a 1963 letter to Shimony, Wigner was quite dismissive, writing, '[t]he state vector, as he imagines it, does not convey any

Indeed, he insisted that we would only obtain a satisfactory formulation of QM if we could give a consistent account of the Wigner's Friend set-up. Within such a formulation, the probabilistic assertions deduced from the theory would be understood as 'subjective appearances' to observers who should otherwise be treated as perfectly ordinary physical systems always subject to the linear dynamics, thereby placing the theory in correspondence with experience (Barrett 1999, p. 53). What that would yield, then, is a theory that is objectively continuous and causal, while subjectively discontinuous and probabilistic. Everett took this to resolve the issue of nested measurements that feature in the Friend scenario because it would justify our use of the statistical assertions of the orthodox view in a logically consistent manner that allows for the existence of other observers (Everett 1956, 77–8).

9.3 Everett on 'Subjective Appearances'

It is worth noting that Everett characterized theories in terms of mathematical models that could be put in an isomorphic or homomorphic relationship with the 'world of experience',[13] by which he meant 'the sense perceptions of the individual, or the "real world"—depending upon one's choice of epistemology' (Everett, in Barrett and Byrne 2012, p. 169; see Freire Jr 2015, pp. 99–101).[14] The notion of 'experience' here is then cashed out in terms of the observer's memory sequences which are represented by the terms in the expansion of the wave function (Barrett and Byrne 2012, p. 169, fn. cq). However, not all such

information to anyone, and I don't see what its role is in the framework of science as we understand it' (Freire Jr 2015, p. 130, fn. 225). Schrödinger also alluded to the argument in 1958 and apparently adopted an ironical stance towards it (Freire Jr 2015, p. 95, fn. 83).

[13] Although Everett had a good background in mathematics, before switching to physics, there appears to be no evidence that he was aware of this characterization as a general approach to theories and their relationship to experience (Freire Jr personal communication; Barrett personal communication).

[14] In a letter to Philipp Frank, a former physicist who adopted a form of 'empirical pragmatism' (Barrett and Byrne 2012, p. 258; Frank 1957; see Mormann 2017 and Uebel 2011), Everett stated that his formulation of QM had 'the interesting feature... that this correspondence [with experience] can be made only by invoking the theory itself to predict our experience—the world picture presented by the basic mathematical theory being entirely alien to our usual conception of "reality". The treatment of observation itself in the theory is absolutely necessary.' (Barrett and Byrne 2012, p. 258). Frank replied that he had 'always disliked the traditional treatment of "measurement" in Quantum Theory according to which it seems as if "measurement" would be a type of fact which is essentially different from all other physical facts' (ibid., p. 259). Given London and Bauer's concern about the scientific community being seen as 'spiritualistic', it is interesting to note that a section in Frank's own book (Frank 1957, p. 232) is titled 'The "Spiritual Element" in Atomic Physics', where he dismissed the claim made by certain interpretations of QM, 'that a "mental element" is introduced into the physical world and that "materialism" is refuted' (Frank 1957, p. 232).

sequences will be relevant to our particular experience and for Everett it was enough for the theory to be 'empirically faithful' in the sense that such experience is represented by a 'typical' term in that expansion (ibid.).[15] And a measure of 'typicality' here is given by the square of the coefficient of the corresponding term; that is, via the 'Born Rule', thereby ensuring concordance with all the predictions of 'standard' QM.

These subjective perceptions of the observer could be identified with objective properties (Everett 1956, p. 63), and so the observer's mental states should also be described by a wave-function.[16] However, unlike the case of London and Bauer, there is no separation of the ego since there is no room for mental entities in Everett's models (Freire Jr 2015, pp. 100–1)—the difference can be characterized sharply in terms of how we should understand the theory: for Everett this is given by the theory itself, whereas for London and Bauer of course, it is only obtained by adopting the phenomenological stance.

Nevertheless, there is a significant commonality between the two in terms of their *correlational* understanding of QM. For Everett, of course, this is expressed through his relative state conception (Barrett and Byrne 2012, p. 35 and for the relevant passage from Everett's 1957 paper, see Barrett and Byrne ibid., p. 180). The overall 'correlation structure' is characterized by the 'universal' wave-function and all the different ways in which it can be decomposed, given the different choices of basis (Barrett and Byrne ibid., p. 35, fn. 4).[17] This structure will then encompass relative states that describe systems having quasi-classical properties, including some with observers and their determinate measurement records, although these will be very much in the minority.

Thus, Everett wrote:

> after the interaction has taken place there will not, generally, exist a single observer state. There will, however, be a superposition of the composite system states, each element of which contains a definite observer state and a definite relative object-system state. Furthermore ... each of these relative object-system states will be, approximately, the eigenstates of the observation corresponding to the value obtained by the observer which is described by

[15] 'The theory *is* isomorphic with experience when one takes the trouble to see what the theory itself says our experience will be' (Everett to DeWitt, Barrett and Byrne 2012, p. 255).

[16] Everett insisted that when a theory is highly successful, so that we have confidence in it, we tend to identify its 'constructs' with elements of the real world. However, no such construct should be regarded as more real than any other and, in particular, those of classical physics are just as much fictions in our minds as those of any other theory (Freire Jr 2015, p. 99).

[17] He explicitly compared this to the relativity of states found in the theory of Special Relativity.

> the same element of the superposition. Thus, each element of the resulting superposition describes an observer who perceived a definite and generally different result, and to whom it appears that the object-system state has been transformed into the corresponding eigenstate. In this sense the usual assertions [the collapse of the state on measurement] appear to hold on a subjective level to each observer described by an element of the superposition. (Everett 1956, in Barrett and Byrne 2012, p. 36 and p. 78)

Note that final line: as in the London and Bauer case, the 'usual assertions' appear to hold, albeit for different reasons.[18]

9.4 From Many Worlds to Many Minds

Those differences manifest in the explanation of how *determinate* measurement records are obtained, of course (see Barrett and Byrne 2012, p. 37). Here it is the notion of the *relative* state that does the explanatory work: although post-measurement the system does not possess a determinate property and the observer does not record a determinate outcome, each can be 'in' relative states by virtue of the *correlation* between the given property and the outcome. Thus, the observer will obtain a determinate *relative* record, which Everett then identified with the observer's (relative) memory states (Everett 1956, p. 78; Barrett 2018).

This raises a further issue, however—one that was initially posed by Shimony, no less. He asked whether the *awareness* of the observer is associated with just one branch and none of the others, or with each (Podolsky, Hart, and Werner 2002, Tues: A.M. pp. 17–18 and p. 22). Everett replied, somewhat tangentially, that '[e]ach individual branch looks like a perfectly respectable world where definite things have happened' (ibid., p. 22).[19]

This issue also bothered Albert and Loewer (1988), who argued that although an observer's physical state evolves according to the standard unitary dynamics, Everett's 'typicality' measure should be understood as giving the probabilities for the stochastic evolution of the observer's mind. In other

[18] Thus, Everett insisted that what is meant by statements such as 'a hydrogen atom has formed in a box' is just that certain correlations have been established, thereby suggesting a broadly structuralist account of physical objects themselves (Freire Jr 2015, p. 103). Freire Jr has suggested that Everett's mathematical work on correlations 'was probably undertaken independently of his reflection on quantum mechanics' (ibid., p. 102, fn. 113).

[19] Shimony went on to ask whether, within a branch, there is any difference between Everett's view and the standard one, to which Everett responded, '[n]one whatever'.

words, that mind comes to be randomly associated with the measurement records associated with a particular branch, with probabilities given by that measure. This understanding of Everettian branching in mentalistic terms came to be known as the 'Many Minds Interpretation' (see Butterfield 1996).[20]

Unfortunately, however, as we've noted, the linearity of Schrödinger's Equation means that if the spin state of some system, say, is a superposition, then when an observer undertakes a measurement, her brain will also evolve into a superposition. And if the observer's mental states are associated with the corresponding brain state, then she won't be able to report a definite belief, say, when she makes that measurement. We could insist that mental states simply *cannot* be in superpositions.[21] However, that potentially weakens the link between brain states and mental states, thereby opening the door to some form of non-physicalism.

At least there is no physical branching on this view, and of course, that we never 'feel' or otherwise find ourselves in a state of superposition is straightforwardly accounted for (we recall our discussion of Chalmers' approach from Chapter 3). The problem, now, is that with mental states and brain states no longer straightforwardly related, we can no longer tell what an observer believes on the basis of her brain state. Indeed, certain elements of the resulting superposition will correspond to brains without minds (this has become known as the 'mindless hulk' problem; see Lockwood 1996a, p. 175) and which of the elements actually represents a mind will not be determined by the underlying brain state.

[20] An early version of this view has been attributed to Zeh (by Lockwood 1996b; see Zeh 2000). Beginning with the concern about maintaining psychophysical parallelism in the face of quantum superpositions, Zeh insisted that the subjective awareness underlying an observation must be accommodated in some manner, since '(e)pistemologically, any concept of observation must ultimately be based on an observing subject' (ibid., p. 5). Rejecting the 'collapse' account, he then argued that Everett's is the only other option, with the caveat that there is only one *quantum* world, as described by that formalism, but many dynamically coupled components representing classically different worlds. Thus, it is the 'apparent' world that branches and '[o]nce we have accepted the formal part of quantum theory, only our experience teaches us that consciousness is physically determined by (factor) wave functions in certain components of the total wave function' (Zeh 2000, p. 7). The other components, with their separate conscious versions of ourselves, were then dismissed as heuristic fictions.

[21] As Albert and Loewer put it: 'The heart of the problem is that the way we conceive of mental states, beliefs, memories, etc., it simply makes no sense to speak of such states or of a mind as being in a superposition' (Albert and Loewer 1988, p. 203). Furthermore, they insisted, echoing London and Bauer, when we introspect, after a measurement, we never find ourselves in a superposition of thinking that spin is up and thinking that spin is down. Of course, Albert and Loewer noted, this presumes that such introspection is trustworthy but one doesn't have to be a phenomenologist to find that easy to swallow—we might well be misled in various ways (and here we might recall Shimony's misplaced concerns) but none of those ways correspond to the kinds of superpositions that are of concern here.

Of course, physicalism could be preserved by demanding that every mental state of the observer is related to a particular brain state. Furthermore, each brain state to which a mental state is related could then be associated with an infinite number of minds, in the state, together with a measure on the totality of minds associated with all the brain states such that the correct probabilities are obtained (Albert and Loewer 1988, p. 207). Now there would be no brain states without corresponding mental states—in effect enough mental states are 'packed in' to ensure that does not happen.[22] And finally, although individual minds evolve probabilistically, the evolution of the set of minds associated with a particular brain state is deterministic, since the evolution of the measurement process is deterministic and the proportions of minds in various mental states can be read off the final state.[23]

Albert and Loewer contrasted their approach with what they called 'idealist' solutions to the measurement problem:

> which entail that consciousness, by bringing about a collapse or in choosing to measure certain observables, in some mysterious way makes reality (perhaps different realities for different observers).
>
> (Albert and Loewer 1988, p. 209)

Unfortunately, however, according to their account, a given mind's beliefs about the post-measurement state of a system would be typically false, since when an observer measures the spin of a system and one of her minds comes to believe the state of the system is spin-up, say, with some of her other minds believing it is spin-down, *in fact* the state of the system + observer's brain is neither spin-up nor spin-down but a superposition of the two. Consequently, most of our beliefs about the world would be strictly incorrect. Granted that subsequent measurements will not conflict with that initial measurement, at least *not from the perspective of that particular mind* (ibid., p. 209), the mind-independence of the world is here bought at some considerable cost (see also Wallace 2012, p. 106).

[22] An infinite number must be introduced because some measurements may have an infinite number of possible outcomes, such as those of position or momentum, for example.

[23] Lockwood called Albert and Loewer's manoeuvre a 'rather desperate expedient' (1996a, p. 176). His own attempt to restore physicalism involved the introduction of a 'superpositional dimension' along which distinct conscious experiences are distributed, in order to accommodate the superposition of brain states (1996a, p. 179). Whether this constitutes less of a 'desperate expedient' is a matter of taste.

9.5 Relativized Reality

Alternatively, it has been argued that a return to Everett's concept of the *relative state* renders this invocation of consciousness unnecessary (Saunders 1995). If the state of the observer who observes 'spin-up' is understood as relative to the state of the system designated by 'spin-up' and correspondingly for 'spin-down', then to ask which state is *actual* or *real* upon measurement misses the point: 'There is no ordinary notion of reality or actuality, not the position of the moon in the night sky or the thought of supper, which is not a relativized reality' (Saunders 1996, p. 243).

There is still an obvious cost here, but in line with Everett's original vision, what reality is relativized *to* does not have to be consciousness—it could be some suitable 'proxy' for the observer, such as a computer in an appropriately controlled experiment. Of course, 'if we are talking about things that *we* observe or that are connected to *us*, the answer is that they are relativized to ourselves, to our concrete physical forms' (Saunders 1996, p. 243). However, since the reality of matter will be relative, just as the reality of persons is, it is not the case that the matter of the observer's brain is entangled with the apparatus, whereas her consciousness is not. Thus, the motivation for introducing consciousness evaporates.

Nevertheless, we are not quite out of the woods yet: when it comes to examples like Schrödinger's Cat or a 'spin-up'/'spin-down' measurement it may seem plausible to take each term in the superposition as corresponding to a 'branch' which may give rise to a world (see Wallace 2012). The states 'cat alive' and 'cat dead' are here taken as the 'basis vectors', the linear combination of which gives the superposition that describes the state of the cat pre-measurement. However, such a combination can be given in terms of an infinite number of possible basis vectors. Now, these alternatives may seem unnatural in the case of the cat but what appeal to 'naturalness' could we make in the case of spin? Of course, when we measure spin in a certain direction we get the values 'up' or 'down' and not some recondite combination but what determines *that*? There doesn't seem to be anything within the formalism that we can appeal to.

We could insist that what we observe directly is not spin but rather the *position* of particles on a screen, say, in a Stern–Gerlach experiment.[24] However, insofar as such a measurement record describes some form of

[24] This provides the motivation for the Bohmian approach which takes position as the fundamental observable.

spatio-temporal location, as when an elementary particle is recorded as hitting the scintillation screen 'here, now', an explanation needs to be given of why, of all the possible bases into which the wave-function may be decomposed, it is the position basis that is somehow 'preferred'. Once again responses typically go beyond the original terms of Everett's interpretation, with decoherence put forward as the mechanism underpinning this 'preference' and yielding 'quasi-classical' worlds more generally (Wallace 2012, ch. 3).[25] For Everett, however, no such appeal was necessary, since he only required that QM be *empirically faithful* and so, 'all [he] needed to explain a particular actual record was to show that there is some decomposition of the state that represents the modeled observer with the corresponding relative record' (Barrett 2018; see Barrett and Byrne 2012, p. 253).[26]

From a phenomenological perspective, of course, this preference for the position basis can be accounted for in terms of our situation within the life-world, as discussed in Chapter 7. To insist that this is unacceptably subjective is to presume some form of realism that both the phenomenologist and Everett would reject. This is not to say that the former would agree that quantum theory should be regarded as 'empirically faithful' in the latter's sense (not least because of differences regarding the relationship between theory and experience) but there is certainly a degree of commonality here, particularly with regard to the emphasis on the observer's experiences.[27]

Relatedly, Saunders has also suggested that:

> the preferred basis problem only arises in so far as we try to develop a physical theory of epistemology, be it of humans or animals or computers or whatever, a theory of what sorts of things or sequences of relative-states could count as epistemic agents. To specify a preferred basis is, as it were, to specify an epistemic community. (1996, p. 245)

[25] We recall that decoherence occurs when a variable of a system in a superposition state becomes correlated with the environment, thereby suppressing interference effects and allowing the observer to treat the expectation values of that variable as if they were classical.

[26] The difference between his conception and that of Bohr and his followers is explored in Freire Jr 2015, pp. 115–17 (see also p. 123) and the similarities with regard to the relativity of states are considered on pp. 117–18.

[27] Bitbol has tried to bring Everett closer to the phenomenological fold by noting not only the relative nature of the observer's memory states but also their 'situated meaning' that manifests through 'one's own lived experience of some particular measurement outcome' (Bitbol 2021, p. 570). He goes on to note that, '[h]ere again (as in von Neumann) consciousness *does nothing* to the physical world. Instead, the so- called "events of the physical world" are *reinterpreted as* a handy way to express the common focus of the expectations and observations of situated agents endowed with conscious experiences' (ibid.; see also Bitbol 2022, pp. 272–4).

And, as he continued, not only do we *have* to do this but it can then be considered to be part of the theory insofar as the latter is empirical and hence must incorporate observation.

The issue now is how that incorporation should proceed. We could, for example, provide a physical description of typical epistemic agents that is precise enough to specify what would be observed by such agents. Alternatively, a description of the agent's environment could be given, to which she will be adapted and which she can perceive. The first can be associated with Everett's original strategy, articulated in terms of memory records and sequences, whereas the second has a Bohrian flavour and is consonant with decoherence theory (Saunders 1996, p. 247). However, from the phenomenological perspective, the idea that decoherence could furnish the beginnings of a 'theory of experience' (see Dowker and Kent 1996), is to put the cart before the horse: with the life-world taken as primary, the issue is to account for the relationship with the world as presented by physics, as we discussed in Chapter 7.[28] Indeed, from this perspective, we can only talk about those things for which there is the possibility of their being *observed by us*, or being connected to us.[29] If we accept that we cannot claim to have accounted for the 'appearances' in the absence of some consideration of the way that consciousness relates to the world, then we can take QM, as understood phenomenologically through London and Bauer's account, to already be the 'physical theory of epistemology' that Saunders has insisted is required.[30]

9.6 Back to the Born Rule

This still leaves the issue of how to obtain the Born Rule. In the Everettian context this is particularly challenging, since if all possibilities are realized, albeit in distinct branches, what sense does it make to speak of the *probability* of obtaining a given outcome? Appealing to the notion

[28] Interestingly, Dowker and Kent cite Wigner and write, '[i]t is awkward that experience thereby becomes entangled with the quantum formalism at a fundamental level, but of course this could conceivably turn out to be unavoidable—who can tell for sure? Likewise, it is not unknown for interpreters of quantum mechanics to find themselves driven to solipsism, and some physicists find this a respectable scientific position' (1994, p. 1632).

[29] Rocha, Rickles, and Boge place a Kantian spin on Dowker and Kent's discussion by suggesting that in the absence of a (broadly) transcendental explanans there would be no organisms capable of observing what we do in fact observe (Rocha, Rickles, and Boge 2021, p. 23).

[30] Lockwood saw his and Saunders' approaches as complementary, in that the latter begins with the world, as specified in Everettian fashion and works his way 'inwards' to the observer, whereas Lockwood began with consciousness and works his way 'outward' to the world (Lockwood 1996b).

of 'typicality'—whereby, for a 'typical' branch, the frequency of results will be precisely as predicted by 'ordinary' QM[31]—still leaves unexplained what it is about the world that makes it appropriate for us to expect the relative sequence of records of our branch to be 'typical':

> In short, while one can get subjective expectations for future experience by stipulation, the theory itself does not describe a physical world where such expectations might be understood as expectations concerning what will in fact occur. (Barrett 2018)

Now, a hard-line Everettian could simply bite the bullet and insist that the account was never intended to yield *this* sort of description of 'the world'. To demand as much would be to require the theory to be more than just 'empirically faithful'. Nevertheless, those who do require more of the theory have sought to plug this gap.

Thus, perhaps the most widely accepted explanation of the Born Rule has adopted a *subjective* understanding of the relevant probabilities, identifying them with the degrees of belief of (rational) agents, where an agent's degree of belief can be accessed through their betting behaviour (Wallace 2012, ch. 6; here we recall our discussion in the context of QBism):[32] their degree of belief that some event E will occur is set equal to p units if they are willing to pay or receive p for a bet that pays 1 such unit if E occurs and 0 if it doesn't (see Hajek 2019). Incoherence is then avoided by constraining these degrees of belief to obey the usual probability calculus. If we then have an agent, facing the Schrödinger's Cat scenario, say, who is indifferent between receiving €20 on those branches where the box is opened and the cat is alive and €10 on all other branches, then they are deemed to assign probability ½ to those branches containing the record 'cat alive' (Vaidman 2021). It then turns out that the only rationally coherent strategy for such an agent to adopt is for her to assign these probabilities according to the Born Rule (Deutsch 1999; Wallace 2012, ch. 6; see also Bacciagaluppi and Ismael 2015).

We see here a grounding of the Born Rule in the subjective experiences of the observer and their (rational) decision-making (which is not to say that this does not involve consideration of 'kicks from the world', as the QBist would put it). Insofar as decision theory can be understood as embodying the core

[31] As the number of observations goes to infinity, almost all branches will contain frequencies of results in accordance with those predictions.

[32] In effect this marks a shift from the understanding of the Born Rule as a typicality measure to one that takes it as a measure of *credence*.

features of rational behaviour, it is through our understanding of such behaviour that probability makes contact with the world. The phenomenologist can then appropriate that general approach and the relevant formal proofs while imposing her own interpretation on the underlying conception of rationality.

Here *reflection* is a pre-condition for the kind of self-critical deliberation required and, as Zahavi has emphasized:

> [i]f we are to subject our different beliefs and desires to a critical, normative evaluation, it is not sufficient simply to have immediate first-personal access to the states in question. Rather, we need to deprive our ongoing mental activities from their automatic normative force by stepping back from them. (2017, p. 23)

In other words, we need to effect the core phenomenological move by engaging in a reflective self-distancing through which we enter into a critical relationship with our mental states. Indeed, '[t]o live in the phenomenological attitude is... not simply a neutral impersonal occupation, but a praxis of decisive personal and existential significance' (ibid., p. 23).

Of course, to relate this fully to the phenomenological stance requires more consideration of the appropriate account of probability than we have space to go into here. Briefly, however, the phenomenologist takes probabilities to be:

> relational modalizations, comparisons and evaluations of the respective 'weight' of manifold possibilities (and within samplings of such manifolds) emerging from spontaneous 'thematizations' of specific intentional modifications, 'modalizations' of the moment of 'belief'. (Lobo 2019, pp. 515–16)

Here we recall again London and Bauer's emphasis on the *free* creation of objectivity, reminiscent as it is of Husserl's remark from his Paris lectures that new configurations of objects are creations of spontaneous 'I'-activity, as noted in Chapter 6. Insofar as we freely create a new objectivity through the regard that separates the ego-pole from the superposition, it might be suggested that it is the spontaneous 'I'-activity that is ultimately responsible for the relevant quantum probabilities.[33]

In this regard there would need to be due consideration of the 'comparisons and evaluations of the respective "weight"' assigned to these manifold

[33] The notion of freedom being employed here plays a major role with regard to the phenomenological *epoché* in general (see Luft 2004).

possibilities. Now, it might be objected that within the Everettian interpretation these extend far beyond those we might 'thematize' on the basis of the life-world alone. Here, however, we can bring into play the more nuanced understanding of the relationship between 'worlds' that we covered in Chapter 7. Having said that, we could also dispense with the decision-theoretic framework entirely, as the core elements of the above proof don't actually depend on it (Bacciagaluppi and Ismael 2015).[34] Instead we could justify it using Gleason's Theorem, understood as the fundamental representation theorem for quantum probabilities,[35] or simply take the Rule as part of the 'primitive' theoretical structure of QM (Bacciagaluppi and Ismael 2015, p. 142).

Finally, we recall that according to the 'Many Minds' variant, the probabilities here refer to the sequences of states of individual minds. To assert that, for example, the probability of obtaining 'spin-up' upon measurement = ½ is just to say that the probability that a mind associated with the observer will observe 'spin-up' is ½. Again, these probabilities do not emerge from the quantum mechanical formalism[36] but there does not seem to be anything in principle that prevents the phenomenologist from also adapting this or indeed any other justification of the Born Rule.

9.7 From Relative States to Relationalism

As it turns out, Everett was not the first to suggest the relative nature of quantum states. Hermann, for example, wrote that the description of a system offered by QM reveals only one aspect of it, namely, 'the aspect that presents itself to the researcher on the basis of the observation made' (quoted in Bacciagaluppi and Crull forthcoming, p. 123).[37] Indeed, she continued, the theory, 'confirms that physics has access only to structures of connections, and shows *in addition* that these structures of connections *are in each case relative to the experimental situation* by means of which the experimenter gains

[34] Indeed, various critics have suggested that the decision-theoretic axioms fall short of being uniquely rationally compelling. Nevertheless, that does not mean they fail to lend the Born measure a significant degree of naturalness (ibid.). According to Wallace the argument establishes that 'if probability basically makes sense, and has the usual qualitative features, in unitary quantum mechanics, then quantitatively it is given by the Born rule' (2012, p. 155).

[35] As noted in Chapter 8, Earman made the same point in the context of QBism (Earman 2018, 2019, pp. 407–8).

[36] Any concerns that an observer might come to believe that she is in a 'maverick' world where her beliefs do not mesh with the quantum probabilities can be assuaged by showing that the probability of such a belief tends to zero (Albert and Loewer 1988, p. 208).

[37] The similarities between Hermann's view and Everett's has been noted by Lumma in the introduction to Hermann (1999); see also Soler (2009).

knowledge of them' (Soler 2009, p. 343.) Phenomenologically, of course, we can understand these 'structures of connections' in terms of the correlations between consciousness and the world as proposed by London and Bauer. If we then put Husserlian qualms to one side and regard such a correlation as expressing a kind of *relation*, then we can usefully compare this to the so-called 'Relational Interpretation of Quantum Mechanics' (RQM; for an overview, see Laudisa and Rovelli 2019).[38]

In contrast with Everett, who insisted that we should take QM at face value, as it were, by understanding the wave function as directly describing reality and then adjusting our conception of that reality in order to accommodate that understanding, the advocates of 'RQM' have something in common with the QBists when they maintain that:

> quantum mechanics is not a theory of the dynamics of an entity ψ, from which the world of our experience somehow emerges. It is instead a theory about the standard world of our experience, described by values that conventional physical variables take at interactions, and about the transition probabilities that determine which values are likely to be realized, given that others were. (Laudisa and Rovelli 2019)[39]

Taking the theory to be 'about the standard world of our experience' puts this interpretation back on the same playing field as the phenomenological stance, of course.[40]

9.8 Relational Quantum Mechanics

The core idea is that '*all* (contingent) physical variables are relational' (Laudisa and Rovelli 2019), in the sense that '[a]ny value these variables take is always

[38] A further motivation for comparison derives from the suggestion that RQM, in effect, resolves the interpretational puzzles of QM by dismissing certain concepts as misleading and inaccessible to measurement (see Laudisa 2019, p. 225).

[39] Having said that, RQM has been described as 'Everettian' (Laudisa and Rovelli 2019) for the obvious reason that it too takes the quantum state to be relative. However, RQM takes the quantum events that are the actualizations of values of physical quantities as the basic elements of reality, where such events do not refer to a system, but rather to a pair of systems (Laudisa and Rovelli 2019). This brings the view closer to the more traditional framework in which such actualization is the result of 'external observation' (van Fraassen 2010, pp. 391–2; see also Ruyant 2018).

[40] Smerlak and Rovelli urge that the Heisenberg picture be adopted, according to which one takes observables rather than wave-functions to evolve in time (Smerlak and Rovelli 2007). The wave-function, on this view, is understood as encoding past and present events observed by the observer.

(implicitly or explicitly) labelled by a second physical system' (ibid.).[41] So, if two systems S and S' interact and some variable *a* associated with S takes a value as a result, then that value should be understood as obtaining only relative to S'. The latter can be any kind of system and neither measurements nor observers are privileged.[42] Nevertheless, if we take S' to be an 'observer' in a generic sense, then the central claim of this interpretation can be expressed as 'different observers can give different accounts of the same set of events' (Rovelli 1996, p. 1643; see also Rovelli 2011, pp. 1480–1; cf. Mermin 1998).

By indexing outcomes in this way, we avoid the well-known issues to do with attributing definite states to systems and, in effect, resolve the measurement problem. So, if we measure the spin on an electron, say, and obtain a certain value, we cannot conclude that the electron *has* that value of spin; we can only conclude that the electron *as measured by us* has that value of spin. Quantum events only arise from interactions between systems, and the fact that such an event occurs is only true with respect to the systems involved in the interaction.

Furthermore, we recall consistency between different observers is ensured by virtue of the fact that any comparison of measurement outcomes will involve a physical process that will itself be described in quantum mechanical terms. The internal self-consistency of the theory then guarantees that there will be no discrepancy between different observers (see also van Fraassen 2010, pp. 403–4).

Having said that, from the perspective of one observer, there is still a 'break' in the unitary dynamics, such that the development of the system can no longer be described by Schrödinger's Equation, whereas from the point of view of another, this development, of the joint system of course, is entirely unitary. According to the relationist, the difference arises because the first observer is deprived of certain information, *namely that pertaining to the interaction of the system with herself*; any such information must be correlative and 'there is no meaning in being correlated with oneself' (Rovelli 1996, p. 1666). Here we bump up against an obvious contrast with the phenomenological approach. Of course, if by 'correlation' we understand a relationship between distinct entities, then Rovelli is correct and this is also how the term is understood in the phenomenological context, when it comes to the relationship between

[41] As with Everett, Rovelli drew on the analogy with relativity theory (for criticism, see Pienaar 2021a).

[42] Laudisa and Rovelli insist that '[t]here is nothing subjective, idealistic, or mentalistic' in this interpretation and that the suggestion that there is arises from a confusion between 'relative' and 'subjective' (Laudisa and Rovelli 2019).

mind and the world. But that does not preclude the holding of the relation of a 'special character' between the ego and itself such that it can come to know the state it is in. It is by virtue of that special relationship that we can acquire the information that Rovelli denies is accessible. The 'break', then, as far as the phenomenologist is concerned has to do with the *act of judgement* associated with that relation and not with any lack of information.[43]

There is a further worry, however: the above description, of how consistency is established between observers, appears to be from a *non-relative* point of view, thereby threatening incoherence. Here we need to be careful to distinguish the *presentation* of the relational account from a *description* of the same situation by a third observer (van Fraassen 2010, p. 397). What the presentation yields is only the general form that the relevant observers' descriptions can take, given certain measurement interactions. In this respect it is not a description of the world per se, but rather, offers a *transcendental* point of view insofar as it sets out the relevant conditions of possibility (ibid., p. 398). The form of the description by a third observer is different and here we must consider the relations between the descriptions that different observers give when they are observing the same system. It can be shown that when it comes to such relations, coherence is ensured as long as the information registered by the components of a composite system engaged in several measurements are related to each other in such a way that the relevant observables do not have simultaneous sharp values when not measured (van Fraassen 2010, pp. 412–14). This in turn is justified by virtue of there being no measurement procedure that could reveal that possibility.[44]

9.9 Wigner's Friend and Self-Inclusivity

This brings us back to the Wigner's Friend scenario (see again Rovelli 1996 and Di Biagio and Rovelli 2021; also Dieks 2019, p. 56) and here again we can discern a difference from the approach of London and Bauer (see also Laudisa 2019).

[43] And we recall here the comparison with Dalla Chiara's approach to the measurement problem in terms of the realization of the reduction of the wave-function via the action of some 'metatheoretical' object (Dalla Chiara 1977, p. 340).

[44] Van Fraassen understands this to be in conformity with the kind of 'moderate empiricism' that Einstein adopted in his creation of relativity theory and which he takes the Copenhagen physicists to have been inspired by.

In particular, it has been pointed out that RQM does not allow us to include in the description of the observation by Wigner's Friend, the moment at which a definite outcome is obtained. This is because even though it is indeed the case that the Friend and the quantum system need to interact for the former to be able to make the measurement, in the relational framework this interaction will not be directly represented in the Friend's description simply because *she does not include herself in the system to be described* (Weststeijn 2021). And again, that is because on the relationist view, she cannot have information about herself.[45]

We have already noted that this represents a point of disagreement with the phenomenological approach.[46] Indeed, as we've seen, according to London and Bauer, it is by virtue of this 'regard', understood as a relation of 'special character', that the chain of correlations is cut and consequently, we can determine what our state is and hence, meaningfully describe it.[47] Of course, on a narrow construal of what counts as the relevant 'description' in this context, where this is represented mathematically in the standard way, the moment at which the definite outcome occurs is indeed not included. But that's not surprising insofar as, following von Neumann, this moment has to lie beyond the physical. If we adopt a wider construal, however, which includes the observer then, as far as phenomenology is concerned, that must include the 'I' which manifests as the ego-pole of the correlation with the quantum system through that very act of introspection. Of course, this requires acknowledgement of the privileged status of the observer in this respect, which would be in tension with RQM as originally formulated. However, there does not seem to be any obstacle in principle to modifying the interpretation along these lines or to adopting an appropriately phenomenological stance towards the interactions in general.[48]

[45] It is not just that a given system cannot perform a complete self-measurement, where this is understood in physical terms (Breuer 1995), but rather that 'an observer cannot even meaningfully describe her own quantum state' (Weststeijn 2021), which RQM includes as an extra assumption.

[46] In response to the recent extension of Wigner's thought experiment, Narasimhachar has insisted that, for human agents, observational states include 'elements of self-awareness' (Narasimhachar 2020, p. 4), and that '[a]n agent must always reason based on the assumption that their current observational state is definite and not superposed or mixed with other potential states, which the agent does not currently experience' (ibid., p. 3). This principle can be straightforwardly accommodated within the phenomenological stance, of course.

[47] It might be objected that insofar as this is a definite state, it is not a 'quantum' state; that is, it is not a superposition, but then Rovelli's claim is trivial.

[48] Pienaar has also noted RQM's commitment to physicalism in that 'relations in RQM are strictly indexed to *objects*, rather than to (thinking and feeling) *subjects*' (2021a) which, he argues, affords only a weak form of relational objectivity.

9.10 Conclusion: A Comparison

Where do we stand then? As we've seen, QBism, with its emphasis on a first-person perspective, has been explicitly allied with phenomenology. However, it struggles to accommodate the correlationist element of the latter which, I have argued, is central to London and Bauer's approach.[49] Both the Everettian and Relational interpretations do better on that score, although further work is required to bring them fully within the phenomenological framework.

On the one hand, by rejecting von Neumann's 'Processes of the First Kind' the Everettian interpretation appears to leave no room for consciousness but, on the other, in taking the theory just 'as it is'—that is, without some further element that accounts for the observation of definite outcomes—it offers the possibility of a cleaner delineation of the scientific 'world' from the life-world, as discussed in Chapter 7.[50] Subjective elements then do find their way back in, initially in the form of the observer's memory sequences that constitute their experiences (which are represented by the theory homomorphically) and more recently via efforts to account for quantum probabilities in decision-theoretic terms. Furthermore, explicit calls for a 'physical theory of experience' in this context have already been answered by London and Bauer's insistence that QM, understood phenomenologically, is itself a theory of knowledge.

QBism, on the other hand, definitely does not take the theory 'as is' but instead attempts to *reconstruct* it from a purely agent-oriented basis (and we'll come back to that in Chapter 10). Nevertheless, as in the Everettian case, the agent is also characterized decision-theoretically, in that 'they must be capable of reasoning using probabilities, and must strive to make *coherent* decisions' (Pienaar 2021b). If the QBist were to allow ego–world correlations into the

[49] Of course, as a reader of an earlier draft pointed out, there are understandings of phenomenology that downplay that aspect, but I would argue that these do not mesh so smoothly with either London's own philosophical background or the text of his 'little book' with Bauer.

[50] Within that conception of the scientific 'world', the phenomenologist would have to say something about the Everettian 'many worlds', whether those are understood as merely heuristic or as incorporating counterpart Egos. As we recall, on London and Bauer's account, the sense of rejection of the other possibilities encoded in the superposition is not that of somehow consciously preventing them from becoming actual but rather that of determining, post-separation, that terms in the quantum mechanical description that would be applicable 'from outside' as it were, do not correspond to what has been observed. In other words, as London and Bauer themselves go on to say, this right to choose and thereby create their own objectivity can be attributed to the observer in virtue of their immanent knowledge of their own state. Again, it is not a 'right' that the observer possesses whilst 'in' the superposition, as it were, but is one that can only be attributed post-separation, when they have that certain knowledge of their own state.

reconstruction, then this account could also be accommodated within London and Bauer's framework.

Relational QM can be situated between QBism and the Everettian account insofar as it incorporates a relativist conception of the state of the system, but also understands that state to be perspectival, in a sense. If this is cashed out in such a way as to introduce conscious observers, then again there is the possibility of an interesting accommodation with phenomenology.[51] However, as we've noted, both QBism and RQM would have to drop their insistence that an observer cannot assign a quantum state to themselves. Nevertheless, although both positions incorporate an observer–system distinction, there is flexibility as to where that distinction is drawn:

> In RQM, the physical boundary that encloses the matter which makes up an observer can be redefined; in QBism, the sensory boundary that defines the limits of the agent's perceptions (i.e., their body) can similarly be redefined. (Pienaar 2021b)

Here a new avenue opens up for phenomenological exploration by developing an account of these boundaries in terms of Merleau-Ponty's 'ontology of radical situatedness' (Bitbol 2020; Bitbol and de la Tremblaye forthcoming). If we understand our 'intermingling' with the world that this involves in terms of the superposition of both the ego and the physical system, which separate as 'poles' through the reflective 'regard', then we can perhaps see how we might build on, and go beyond, London and Bauer's 'little book'.

[51] Rovelli has recently suggested that relationalism supports a 'very mild' version of panpsychism, which he likewise regards as perspectival (Rovelli 2021, p. 34). This then helps alleviate the tension between the mental and the physical by shifting away from the traditional substance + properties metaphysics on both sides. In this respect, then, the phenomenologically inclined may well agree with Rovelli's rejection of the claim that 'the subject of experience is an irreducible entity' (ibid.).

10
Interpretation or Reconstruction?

10.1 Introduction

Let's return to where we started, with the measurement problem. As we've seen, the usual approach to coming up with a solution is to begin with the formalism of QM, whether as presented by von Neumann or in some textbook, and then introduce some extra feature in order to account for the determinate outcomes that are observed. For Bohr and his followers, this had to do with the macroscopic measurement context which, it was argued, could only be described in classical terms. For the likes of Wigner, it was consciousness, taken, contentiously, to somehow 'collapse' the wave-function. Everett, by contrast, eschewed any such feature and tried to crack the problem by taking the theory at face value, but then found that he too had to go beyond the formalism in order to recover the relevant probabilities.[1]

London and Bauer also made an appeal to consciousness but, I have argued, avoided any 'collapse' by situating the theory within a phenomenological framework. The extent to which this can then be regarded as an 'extra feature', is debatable, since from this perspective, 'the world is a world of knowledge, a world of consciousness' (Husserl 1970b, p. 265) and hence such a framework is in fact *required.* Once the phenomenological stance is adopted, the very idea of leaving consciousness out of the picture is simply absurd (ibid.) and so, to think of it as something over and above the formalism is to slip back into the natural attitude that we had stepped away from via the *epoché.* As a result, we must examine our attitude to the measurement problem, and quantum physics more generally, in a whole new light, one that aligns with London and Bauer's understanding of QM as a 'theory of knowledge'.

[1] For Everett this was through the notion of 'typicality', as we have seen, whereas today's Everettians appeal to decision theory.

A Phenomenological Approach to Quantum Mechanics: Cutting the Chain of Correlations. Steven French, Oxford University Press. © Steven French 2023. DOI: 10.1093/oso/9780198897958.003.0010

10.2 Don't Interpret, Reconstruct

So, the sort of approach to the problem that has been adopted by Bohr, Wigner, and almost everyone else—namely, start with the theory as given and add whatever is necessary to account for determinate observations—chimes with the *interpretive* approach towards scientific theories more generally. *Reconstructive* attempts, on the other hand, turn this approach on its head: they begin with certain basic postulates and then try to derive the formalism from them. The most well-known example of this is QBism, which, as we discussed in Chapter 8, begins with the agent and her experience and attempts to obtain the quantum formalism from that basis. As we've also seen, some account must then be given of the Born Rule, reflecting, as it does, the 'kicks from the world'.

Such attempts are fraught with difficulties. And in a sense, that should come as no surprise, given not only the specific details of the formalism but also the complex history of the development of QM. A comparison is sometimes made with Special Relativity, as characterized in terms of Einstein's derivation of the Lorentz equations from the principle of relativity and the postulate of the constancy of the speed of light (Chiribella and Spekkens 2016, p. 3). Likewise, it is suggested, all we need do is identify similar fundamental postulates from which the rest of quantum theory can be derived. But the historical development of the latter is more akin to that of General Relativity, with its false starts and application of novel mathematical devices, and to expect to be able to derive the formalism of QM from a few plausible physical postulates might well be regarded as entertaining high hopes.[2]

Nevertheless, the idea of reconstructing a less mathematically freighted form of the theory has been taken to resonate with the sentiments expressed by Husserl in the *Crisis* (Berghofer, Goyal, and Wiltsche 2021, p. 423). Thus, the aim of this reconstructive programme 'is to formulate *physical principles*—ideally, principles with an intuitive comprehensibility comparable to those underlying classical physics—from which the quantum formalism can demonstrably be systematically derived' (ibid., p. 424). Two questions immediately arise: what is to be taken as 'the quantum formalism'? And from what basis is such a systematic derivation to be attempted?

The answer to the first is that typically it is von Neumann's formalism that is the focus of such attempts, although as Mitsch has noted, this should be

[2] We recall that both Everett and Rovelli also drew on an analogy with Special Relativity and that this has been criticized by Pienaar (2021a).

understood as an attempt to formally express a *specific* account of only certain quantum phenomena (Mitsch 2022, p. 84). As for the second, reconstruction usually begins from an operational base, as we've noted in the QBist case, in terms of which a 'system' undergoes 'measurement' by some 'agent', yielding a particular 'outcome', with 'system', 'measurement', 'agent', and 'outcome' all taken as primitive (Berghofer, Goyal, and Wiltsche 2021, p. 424). Now, such terms can be taken to represent elements of the life-world as it pertains to the practice of physics—albeit with further phenomenological analysis required, of course—but obviously to proceed from that basis to a set of logico-mathematical axioms, something further needs to be introduced and this is where mathematics is applied like a suit of clothes, draped over these terms. Different mathematizations can also be adopted but it is notable that recent reconstructive efforts have co-opted the framework of information theory, for example, with the measurement taken to provide information about the world and quantum theory itself understood as 'a compact codification of the regularities that we discover in that information' (ibid., p. 425).[3] Using this framework, various features of the formalism of QM can then be recovered, or so it is claimed.

10.3 Reconstruction and Perspectivalism

Grounding such a reconstruction on this kind of operational basis generates a *perspectival* account of measurement, in the sense that:

> [f]irst, it only provides information about *certain* degrees of freedom of the state of the system. Second, one only receives limited information about *these* degrees of freedom. (Berghofer, Goyal, and Wiltsche 2021, p. 428)

As a result, the agent's choice of what initial measurement to make determines what she learns, or not, about the system and also how the state of the latter is affected by that measurement.

This perspectival understanding also resonates with a phenomenological stance according to which the agent does not have some privileged, 'objective' standpoint and is not separate from the world on which she acts (ibid., p. 431; see also Massimi 2021). This is directly related to the phenomenological disclosure of the horizontal structure of experience, which, we recall, has to

[3] Such a characterization of the theory is itself contentious of course.

do with the way that perceptual experiences go beyond what is immediately given. We apprehend the core of what is given to us as surrounded by a horizon of 'co-givenness' that is shaped by our previous experiences, background beliefs, and so on (Berghofer, Goyal, and Wiltsche 2021, pp. 432–3). As a result, the sense of the objects that we engage with is constituted by us, *in a correlational manner.*

More broadly, such reconstructive approaches can be interpreted as uncovering and building on the 'sense giving foundations' of the quantum formalism (Berghofer, Goyal, and Wiltsche 2021, pp. 429–30). We recall that Husserl's criticism of Galileo, standing in for physicists in general, was that he confused his method for representing reality via mathematics with reality itself and failed to appreciate the distorting impact of the idealization involved. Of course, Husserl urged a return to the life-world as providing such foundations, whereas advocates of reconstruction seek them in *certain physical principles* (ibid., p. 430). As a result, this approach operates closer to the practice of physics, not least by virtue of abandoning the effort to incorporate properties of classical mechanics within this set of principles. This could be used to widen the difference between such approaches and phenomenology, insofar as such properties could arguably be seen as more directly relatable to features of the life-world. However, any gap could also be bridged by paying due attention to the relationship between the life-world and the scientific 'world' that was considered in Chapter 7.

Likewise, granted the above point regarding the primacy of the agent, we again recall that for Husserl, 'an experience, like any intentional act, is an "intentional relation of consciousness to object" where we have "the ego as one pole of the relation in question, while the other pole is the object" such that "[j]ust like any object-pole, the Ego-pole is a pole of identity"' (Berghofer, Goyal, and Wiltsche 2021, p. 430). If this intentional relation is understood as a form of correlation, then a phenomenological reconstruction needs to take that into account as well.

10.4 Blurring the Line

Following Zahavi I have emphasized the importance of this correlational feature within phenomenology and placed it at the heart of my understanding of London and Bauer's approach to the measurement problem. This suggests that instead of taking the subject/agent as primitive, we should adopt such an attitude towards the *subject–object relationship*, as represented, within the formalism of QM, by the relevant superposition.[4] Undertaking this shift then removes the motivation for co-opting devices such as information theory in order to transform the set of primitive notions expressed in natural language into the basis for a derivation of the formalism. In its place we have the transformation effected by entanglement (understood here as also applying to the conscious subject) which, if we were to draw on the analogy with Special Relativity, is akin to the core principles of the latter and, as we have noted, was regarded by Schrödinger as *the* characteristic trait of QM.

What we then obtain is something that is closer to RQM than QBism. And as result, the line between *reconstruction* and *interpretation* becomes blurred. If we have a relevant physical principle, such as entanglement, already to hand, why seek something else that is deemed in some sense to be more basic? Again, as we have noted, the QBist has to supplement their agential bedrock with the Born Rule. Instead, we can take entanglement, understood as applying to consciousness and thereby capturing Husserl's relationship between the 'ego-pole' and the 'object-pole' and, for example, co-opt the Everettian's use of decision theory to get the quantum probabilities.

Having said that, we should refrain from regarding such principles as providing the 'sense-giving foundations' of the quantum formalism. Bluntly, there is no need for that. Husserl, of course, urged a return to the life-world as providing such foundations and indeed, would have insisted that any appeal to physical principles merely puts off this return—their sense too must be grounded in the life-world. The point for Husserl was not to be *sceptical* about the mathematization of nature, much less to suggest that the mathematics of modern physics should be abandoned or replaced, but rather that we should become aware of its *idealizing* effect and as a result, adopt the phenomenological stance which restores the true basis of meaning that has been obscured. From this perspective *there is no need to reconstruct the quantum formalism* but, rather, we should fully appreciate that its sense is given in the life-world, as he maintained in the *Crisis*, the central theme of which, as we have seen, can be extended into the quantum domain through the work of London and Bauer.

However, neither does this amount to a mere *interpretation* of QM, along the lines of the Everettian or other approaches because all such interpretations fail to incorporate that which, according to London and Bauer, QM itself reveals to us: namely the centrality of the relationship between system and

[4] Furthermore, according to that understanding, the apparent irreducibility of the subject is manifested only *after* the ego-pole has separated from the correlation with the object.

observer which is manifested in the so-called measurement 'problem'. That is, such approaches fail to recognize what London and Bauer emphasized at the very beginning of their 'little book': QM itself is a *theory of knowledge*.

10.5 Conclusion: Between 'ψ-epistemic' and 'ψ-ontic'

This might suggest that their account should be aligned with the so-called 'ψ-epistemic' approaches which we touched on in considering Langevin's introduction. However, the correlational aspect means that the wave-function can be understood as describing the state of the object as well; that is, London and Bauer's account could also be considered to be 'ψ-ontic' in a sense. This might seem impossible given the claim that the wave-function has to be regarded as one or the other and cannot be both (Harrigan and Spekkens 2010). The proof of this depends on the point that if the state of a system could be represented by either of two different wave-functions, depending on what is known about the system, then those wave-functions must be regarded as epistemic; and if not, then the relevant wave-function is ontic.[5]

However, this proof assumes that the wave-function either represents the state of the world in the realist sense (that is, via a 1-1 mapping) and is ontic, or it represents the state of the observer, and is epistemic (Hance, Rarity, and Ladyman 2022).[6] But, as we've already seen, the phenomenological stance sits askew to the realism–antirealism debate and rejects the 'philosophical absolutizing' of the world that sits at its heart. Instead, the world and consciousness must be seen as inseparable correlates of one another, constitutively bound together. And as we have noted repeatedly, it is only through the introspective regard that these correlates can be said to emerge as 'poles' and *only then* can we speak of the 'ontic' state of the system and the 'epistemic' state of the observer. But such talk is really just a *façon de parler* because phenomenologically the wave-function is neither ontic nor epistemic insofar as it describes that fundamental correlative relationship which is encapsulated in the slogan, '[n]o object without a subject and no subject without an object' (Zahavi 2017, p. 102).

Thus, the London and Bauer approach, understood phenomenologically as it should be, opens up a third-way between the 'ψ-ontic' and 'ψ-epistemic'

[5] It has also been argued that 'ψ-epistemic' approaches must, in some respect, contradict QM (Pusey, Barrett, and Rudolph 2012).

[6] Thanks again to Philipp Berghofer for bringing this paper to my attention.

accounts and, more generally, between interpretation and reconstruction. It can do this because, as we have observed, it takes QM to be, not just a piece of physics, but a *theory of knowledge itself*. This idea was lost in the multi-dimensional shift in physics that was both geographical and attitudinal, with the new 'pole' represented by Everett, who eschewed such reflections. In a sense, however, it was presaged by Weyl's shift in feelings towards his unified theory of gravitation and electromagnetism, mentioned in Chapter 4. Ryckman has argued that this reflected the theory's 'ambiguous character as lying in the intersection of physics and philosophy' (Ryckman 2005, p. 159). The London and Bauer 'solution' to the measurement problem also has an ambiguous character, albeit of a different order, insofar as whether or not it can be considered to lie on this intersection is itself unclear and, at best, determined by historical contingency. At the time it was published, many physicists, including of course London and Bauer themselves, took the issue of accommodating observation within QM to be a matter for *physics* to resolve. Having said that, as far as London was concerned, all such issues had to be approached from the phenomenological perspective and for him, the very notion of such an 'intersection' would have seemed problematic. At some point in the post-war period, this issue became 'the measurement problem' and was filed under 'philosophy' by most physicists (Margenau and Wigner being notable exceptions). In a sense, then, this entire exercise, exploratory as it has been, represents an attempt to 're-file' the problem and suggest that London and Bauer's idea of QM as both a theory of physics and a theory of knowledge deserves further consideration. Whether you are sympathetic to the phenomenological stance or not, I would hope that you would agree with that much. Time, then, to elevate their 'little book' from its place in the footnotes of history and treat it as offering a radical new advance in the understanding of our relationship to the world.

References

Albert, D. (1992), *Quantum Mechanics and Experience.* Cambridge, MA: Harvard University Press.

Albert, D. and Loewer, B. (1988), 'Interpreting the Many Worlds Interpretation', *Synthese* 77, pp. 195–213.

Albert, D. and Putnam, H. (1995), 'Further Adventures of Wigner's Friend', *Topoi* 14, pp. 17–22.

Ales Bello, A. (2018), 'The Sense of Mystical Experience According to Gerda Walther', in A. Calcagno (ed.), *Gerda Walther's Phenomenology of Sociality, Psychology and Religion*, Women in the History of Philosophy and Sciences 2; Cham: Springer, pp. 135–48.

Allen, C. and Trestman, M. (2020), 'Animal Consciousness', in E. N. Zalta (ed.), *The Stanford Encyclopedia of Philosophy*, https://plato.stanford.edu/archives/win2020/entries/consciousness-animal/

Alves, P. (2021), 'Fritz London and the Measurement Problem: A Phenomenological Approach', *Continental Philosophy Review* 54, pp. 453–81.

Anderson, P. (2005), 'Thinking Big', *Nature* 437, p. 29.

Anta, J. (2022), 'A Philosopher against the Bandwagon: Carnap and the Informationalization of Thermal Physics', *HOPOS* 12, pp. 43–67.

Ash, M.G. (1998), *Gestalt Psychology in German Culture, 1890–1967: Holism and the Quest for Objectivity.* Cambridge: Cambridge University Press.

Atmanspacher, H. (2004), 'Quantum Theory and Consciousness: An Overview with Selected Examples', *Discrete Dynamics in Nature and Society* 1, pp. 51–73.

Atmanspacher, H. (2015), 'Quantum Approaches to Consciousness', in E. N. Zalta (ed.), *The Stanford Encyclopedia of Philosophy*, https://plato.stanford.edu/archives/sum2015/entries/qt-consciousness/

Bacciagaluppi, G. (2016), 'The Role of Decoherence in Quantum Mechanics', in E. N. Zalta (ed.), *The Stanford Encyclopedia of Philosophy*, https://plato.stanford.edu/archives/fall2016/entries/Quantum Mechanics-decoherence/

Bacciagaluppi, G. and Crull, E. (forthcoming), *The Einstein Paradox: The Debate on Non-Locality and Incompleteness in 1935.* Cambridge: Cambridge University Press.

Bacciagaluppi, G. and Ismael, J. (2015), 'Review of *The Emergent Multiverse*', *Philosophy of Science* 82, pp. 129–48.

Bächtold, M. (2008), 'Five Formulations of the Quantum Measurement Problem in the Frame of the Standard Interpretation', *Journal for General Philosophy of Science* 39, pp. 17–33.

Baracco, F. (2019), 'Weyl's Phenomenological Constructivism', *Meta: Research In Hermeneutics, Phenomenology, and Practical Philosophy* XI, pp. 589–617.

Barbaras, R. (2001), 'Merleau-Ponty and Nature', *Research in Phenomenology* 31, pp. 22–38.

Barber, M. (2022), 'Alfred Schutz', in E. N. Zalta (ed.), *The Stanford Encyclopedia of Philosophy*, https://plato.stanford.edu/archives/spr2022/entries/schutz/

Barrett, J.A. (1999), *The Quantum Mechanics of Minds and Worlds*, Oxford: Oxford University Press.

Barrett, J.A. (2017), 'Typical Worlds', *Studies in the History and Philosophy of Modern Physics*, 58, pp. 31–40.

Barrett, J.A. (2018), 'Everett's Relative-State Formulation of Quantum Mechanics', in E. N. Zalta (ed.), *The Stanford Encyclopedia of Philosophy*, https://plato.stanford.edu/archives/win2018/entries/qm-everett/

Barrett, J.A. and Byrne, P. (2012), *The Everett Interpretation of Quantum Mechanics*, Princeton University Press.

Barua, A. (2017), 'Investigating the "Science" in "Eastern Religions": A Methodological Enquiry', *Zygon* 52 pp. 124–45.

Bauer, E. (1913), 'Les Quantités Élémentaires D'Énergie et D'Action', in E. Bauer et. al. (eds) *Les Idées Modernes sur la Constitution de la Matiére*, Paris: Gauthier-Villars, pp. 115–47.

Bauer, E. (1963a), Interview of Edmond Bauer by Thomas S. Kuhn and Theo Kahan on 8 January 1963, Niels Bohr Library & Archives, American Institute of Physics, College Park, MD USA, www.aip.org/history-programs/niels-bohr-library/oral-histories/4498-1

Bauer, E. (1963b), Interview of Edmond Bauer by Thomas S. Kuhn and Theo Kahan on 14 January 1963, Niels Bohr Library & Archives, American Institute of Physics, College Park, MD USA, www.aip.org/history-programs/niels-bohr-library/oral-histories/4498-2

Bauer, E. and Meijer, P. (1962/2004), *Group Theory; The Application to QUANTUM MECHANICS*, Amsterdam: North-Holland 1962; Garden City, NY: Dover 2004.

Beck, M. (1928), 'Die Neue Problemlage der Erkenntnistheorie', *Deutsche Vierteljahrsschrift für Literaturwissenschaft und Geistesgeschichte* 6, pp. 611–39.

Becker, A. (2018), *What Is Real? The Unfinished Quest for the Meaning of Quantum Physics.* London: John Murray.

Becker, L. (2004), 'That von Neumann Did Not Believe in a Physical Collapse', *British Journal for the Philosophy of* Science 55, pp. 121–35.

Bell, D. (1990), *Husserl*, London and New York: Routledge.

Bell, J.L. and Korté, H. (Winter 2016), 'Hermann Weyl', in E. N. Zalta (ed.), *The Stanford Encyclopedia of Philosophy* https://plato.stanford.edu/archives/win2016/entries/weyl/

Beller, M. (1999), *Quantum Dialogues: The Making of a Revolution*. University of Chicago Press.

Bennett, J.B. (1978), 'The Tacit in Experience: Polanyi and Whitehead', *The Thomist: A Speculative Quarterly Review* 42, pp. 28–49.

Berghofer, P., Goyal, P., and Wiltsche, H. (2021), 'Husserl, the Mathematization of Nature, and the Informational Reconstruction of Quantum Theory', *Continental Philosophy Review* 54, pp. 413–36.

Berghofer, P. and Wiltsche, H. (2020), 'Phenomenological Approaches to Physics: Mapping the Field', in H. Wiltsche and P. Berghofer (eds.), *Phenomenological Approaches to Physics*, Cham: Springer, pp. 1–47.

Bergson, H. (1944), *Creative Evolution*. New York: Random House.

Beyer, C. (2018), 'Edmund Husserl', in E. N. Zalta (ed.), *The Stanford Encyclopedia of Philosophy*, <https://plato.stanford.edu/archives/sum2018/entries/husserl/>.

Beyler, R. (1996), 'Targeting the Organism: The Scientific and Cultural Context of Pascual Jordan's Quantum Biology 1932–1947', *Isis* 87, pp. 248–73.

Bhattacharya, A. (2021), *The Man from the Future*. London: Allen Lane.

Bierman, D. (2003), 'Does Consciousness Collapse the Wave-Packet?' *Mind and Matter* 1, pp. 45–57.

Bierman, D. (2006), 'Empirical Research on the Radical Subjective Solution of the Measurement Problem: Does Time Get its Direction through Conscious Observation?' *AIP Conference Proceedings* 863, pp. 238–59.

Bierman D.J. and Whitmarsh S. (2006), 'Consciousness and Quantum Physics: Empirical Research on the Subjective Reduction of the Statevector', in J.A. Tuszynski (ed.), *The Emerging Physics of Consciousness: The Frontiers Collection*. Berlin, Heidelberg: Springer, pp. 27–48.

Bilban, T. (2013), 'Husserl's Reconsideration of the Observation Process and Its Possible Connections with Quantum Mechanics', *Prolegomena* 12, pp. 459–86.

Bilban, T. (2020), 'The phenomenological approach to quantum mechanics : a better understanding of contemporary philosophy of quantum mechanics by revisiting Bohr and Husserl', *Horizon: fenomenologičeskie issledovaniâ* 9, pp. 216–234.

Bilban, T. (2021), 'Informational Foundations of Quantum Theory: Critical Reconsideration from the Point of View of a Phenomenologist', *Continental Philosophy Review* 54, pp. 581–94.

Birch, J., Schnell, A., and Clayton, N. (2020), 'Dimensions of Animal Consciousness', *Trends in Cognitive Sciences* 24, pp. 789–801.

Birch, J. et. al. (2022), 'How Should We Study Animal Consciousness Scientifically?', *Journal of Consciousness Studies* 29, pp. 8–28.

Birtwhistle, G. (1929), *The New Quantum Mechanics*. Cambridge: Cambridge University Press.

Bitbol, M. (2000), *Physique et philosophie de l'esprit*. Paris: Flammarion.

Bitbol, M. (2001), 'Jean-Louis Destouches: théories de la prévision et individualité' *Philosophia Scientiae* 5, pp. 1–30.

Bitbol, M. (2020), 'A Phenomenological Ontology for Physics: Merleau-Ponty and QBism,' in P. Berghofer and H. Wiltsche (eds), *Phenomenological Approaches to Physics*, Cham: Synthese Library, Springer, pp. 227–42.

Bitbol, M. (2021), 'Is the Life-World Reduction Sufficient in Quantum Physics?', *Continental Philosophy Review* 54, pp. 563–80.

Bitbol, M. (2022), 'The Roles Ascribed to Consciousness in Quantum Physics: A Revelator of Dualist (or Quasi-Dualist) Prejudice', in S. Gao (ed.), *Consciousness and Quantum Mechanics*, Oxford: Oxford University Press, pp. 260–81.

Bitbol, M. and de la Tremblaye, L. (forthcoming), 'QBism: An Eco-Phenomenology of Quantum Physics' in H. Wiltsche and P. Berghofer (eds), *Phenomenological Approaches to QBism*.

Bloch, F. (1964), Interview of Felix Bloch by Thomas S. Kuhn on 14 May 1964, Niels Bohr Library & Archives, American Institute of Physics, College Park: MD, www.aip.org/history-programs/niels-bohr-library/oral-histories/4509

Bloch, F. (1981), Interview of Felix Bloch by Lillian Hoddeson on 15 December 1981, Niels Bohr Library & Archives, American Institute of Physics, College Park: MD, www.aip.org/history-programs/niels-bohr-library/oral-histories/5004

Blundell, S. (2011), 'The Forgotten Brothers', *Physics World*, April, pp. 26–9.

Bohr, N. (1934), *Atomic Theory and the Description of Nature*. Cambridge: Cambridge University Press.

Bohr, N. and Rosenfeld, L. (1933), 'Zur Frage der Messbarkeit der elektromagnetischen Feldgrössern', *Det Kgl. Danske Videnskabernes Selskab. Mathematisk-fysiske Medelelser*. XII. pp. 1–65 (repr. in R.S. Cohen and J. Stachel (eds), *Selected Papers of Léon Rosenfeld*. Boston Studies in the Philosophy of Science, Vol. 21, Dordrecht: Springer.

Born, M. (1962), Interview of Max Born by Thomas S. Kuhn and Friedrich Hund on 17 October 1962, Niels Bohr Library & Archives, American Institute of Physics, College Park: MD, www.aip.org/history-programs/niels-bohr-library/oral-histories/4522-3

Born, M. (1971), *The Born–Einstein Letters*, New York: Macmillan.

Born, M. (1978), *My Life: Recollections of a Nobel Laureate*, London: Taylor & Francis.

Breazeale, D. (2018), 'Johann Gottlieb Fichte', in E. N. Zalta (ed.), *The Stanford Encyclopedia of Philosophy*, https://plato.stanford.edu/archives/sum2018/entries/johann-fichte/

Breuer, T. (1995), 'The Impossibility of Accurate State Self-Measurements', *Philosophy of Science* 62, pp. 197–214.

Brooks, M. (2012), 'Reality: How Does Consciousness Fit In?', *New Scientist* 45, pp. 42–3.

Brunschvicg, L. (2006), 'Physics and Metaphysics' (trans. J-L Lafrance), *Philosophical Forum* 37, pp. 65–74.

Bueno, O. (2016), 'Von Neumann, Empiricism, and the Foundations of Quantum Mechanics', in D. Aerts, C. de Ronde, H. Freytes, and R. Giuntini (eds), *Probing the Meaning and Structure of Quantum Mechanics: Superpositions, Semantics, Dynamics and Identity*, Singapore: World Scientific, pp. 192–230.

Bueno, O. (2019), 'Is There a Place for Consciousness in Quantum Mechanics?', in A. Acácio de Barros and C. Montemayor (eds), *Quanta and Mind: Essays on the Connection Between Quantum Mechanics and the Consciousness*, Cham: Springer, pp. 129–50.

Bueno, O. and French, S. (2018), *Applying Mathematics: Immersion, Inference, Interpretation*. Oxford: Oxford University Press.

Bueno, O., French, S., and Ladyman, J. (2002), 'On Representing the Relationship between the Mathematical and the Empirical', *Philosophy of Science* 69, pp. 452–73.

Bueno, O., French, S., and Ladyman, J. (2012), 'Empirical Factors and Structural Transference: Returning to the London Account', *Studies in History and Philosophy of Modern Physics* 43, pp. 95–103.

Butterfield, J. (1996), 'Whither the Minds', *British Journal for the Philosophy of Science* 47, pp. 200–21.

Cairns, D. (1976), *Conversations with Husserl and Fink*. The Hague: Martinus Nijhoff.

Camilleri, K. (2009), 'Constructing the Myth of the Copenhagen Interpretation', *Perspectives on Science* 17, pp. 26–57.

Carr, D. (1999), *The Paradox of Subjectivity: The Self in the Transcendental Tradition*. Oxford: Oxford University Press.

Carson, C. (2010), *Heisenberg in the Atomic Age: Science and the Public Sphere*. Cambridge: Cambridge University Press.

Carson, C. (2011), 'Modern or Antimodern Science? Weimar Culture, Natural Science, and the Heidegger–Heisenberg Exchange', in C. Carson, A. Kojevnikov, and H. Trischler (eds), *Weimar Culture and Quantum Mechanics: Selected Papers by Paul Forman and Contemporary Perspectives on the Forman Thesis*. London: Imperial College Press, pp. 523–42.

Cartwright, N. (1985), 'Review of *Quantum Theory and Measurement*', *Philosophy of Science* 52, pp. 480–1.

Cassirer, E. (1944), 'The Concept of Group and the Theory of Perception', *Philosophy and Phenomenological Research* 5, pp. 1–36.

Cassirer, E. (1956), *Determinism and Indeterminism in Modern Physics*. Yale University Press (1936).

Cauchois, Y. (1964), 'Hommage à Edmond Bauer', *Journal de Chimie Physique et de Physico-Chimie Biologique* 61, pp. 963–8.

Chalmers, D. and McQueen, K. (2022), 'Consciousness and the Collapse of the Wave Function', in. S. Gao (ed.), *Consciousness and Quantum Mechanics*, Oxford: Oxford University Press, pp. 11–63.

Chayut, M. (2001), 'From the Periphery: The Genesis of Eugene P. Wigner's Application of Group Theory to Quantum Mechanics', *Foundations of Chemistry* 3, pp. 55–78.

Chevalley, C. (1999), 'Why Do We Find Bohr Obscure?', in D. Greenberger, W.L. Reiter, and A. Zeilinger (eds), *Epistemological and Experimental Perspectives on Quantum Physics*, Dordrecht: Kluwer Academic, pp. 59–73.

Chiribella, G. and Spekkens, R. (2016), 'Introduction', in G. Chiribella and R. Spekkens (eds), *Quantum Theory: Informational Foundations and Foils*, Dordrecht: Springer, pp. 1–18.

Ciocan, C. (2017), 'Husserl's Phenomenology of Animality and the Paradoxes of Normality', *Human Studies* 40, pp. 175–90.

Clauser, J.F., Horne, M.A., Shimony, A., and Holt, R.A. (1969), 'Proposed Experiment to Test Local Hidden-Variable Theories, *Physics Review. Letters*, 23, pp. 880–4.

Costa de Beauregard, O. (1976), 'Time Symmetry and Interpretation of Quantum Mechanics', *Foundations of Physics* 22, pp. 121–35.

Costa de Beauregard, O. (1992), 'Is There a Reality Out There?' *Foundations of Physics* 6, pp. 539–59.

Crane, T. and French, C. (2017), 'The Problem of Perception', in E. N. Zalta (ed.), *The Stanford Encyclopedia of* Philosophy, https://plato.stanford.edu/archives/spr2017/entries/perception-problem/

Crease, R.P., Kamins, D.A., and Rubery, P. (2021), 'Introduction: Phenomenology of Quantum Mechanics', *Continental Philosophy Review* 54, pp. 405–12.

Crease, R.P. and Sares, J. (2020), 'Interview with Physicist Christopher Fuchs', *Continental Philosophy Review* 54, pp. 541–61.

Crowell, S. (2001), *Husserl, Heidegger and the Space of Meaning*. Evanston: Northwestern University Press.

Crull, E. (2022), 'Grete Hermann's Interpretation of Quantum Mechanics', in O. Freire Jr et. al. (eds), *Oxford Handbook of the History of Interpretations of Quantum Physics*. Oxford: Oxford University Press, pp. 567–586.

Cushing, J. (1994), *Quantum Mechanics: Historical Contingency and the Copenhagen Hegemony*. Chicago: Chicago University Press.

Dalla Chiara, M. (1977), 'Logical Self Reference, Set Theoretical Paradoxes and the Measurement Problem in Quantum Mechanics', *Journal of Philosophical Logic* 6, pp. 331–47.

Daneri, A., Loinger, A., and Prosperi, G.W. (1962), 'Quantum Theory of Measurement and Ergodicity Conditions', *Nuclear Physics* 33, pp. 297–319.

Darrow, K.K. (1964), 'Obituary: Edmond Bauer', *Physics Today* 17, pp. 86–7.

Darwin, C. (1929), 'A Collision Problem in the Wave Mechanics', *Proceedings of the Royal Society London A* 124, pp. 375–94.

Darwin, C.G. (1931). *The New Conceptions of Matter*. London: Bell & Sons, Ltd.

De Broglie, L. (1957), *The Theory of Measurement in Quantum Mechanics (Usual Interpretation and Causal Interpretation)*, Trans. D.H. Delphenich. Paris: Gauthier-Villars.

de la Tremblaye, L. (2020), 'QBism from a Phenomenological Point of View: Husserl and QBism', in P. Berghofer and H. Wiltsche (eds), *Phenomenological Approaches to Physics*. Cham: Synthese Library, Springer, pp. 243–60.

DeBrota, J.B., Fuchs, C.A., and Schack, R. (2020), 'Respecting One's Fellow: QBism's Analysis of Wigner's Friend', *Foundations of Physics* 50, pp. 1859–74.

DeBrota, J.B. and Stacey, B.C. (2018), 'FAQBism', arXiv:1810.13401.

Destouches-Février, P. (1951), *La structure des théories physiques*. Paris: Presses Universitaires de France.

Deutsch, D. (1999), 'Quantum Theory of Probability and Decisions', *Proceedings of the Royal Society of London A* 455, pp. 3129–37.

Di Biagio, A. and Rovelli, C. (2021), 'Stable Facts, Relative Facts', *Foundations of Physics* 51: 30.

Dieks, D. (2019), 'Quantum Mechanics and Perspectivalism', in O. Lombardi, S. Fortin, C. López, and F. Holik (eds), *Quantum Worlds: Perspectives on the Ontology of Quantum Mechanics*, Cambridge University Press (online), pp. 51–70.

Dirac, P.A.M. (1930), *The Principles of Quantum Mechanics*, Oxford: Oxford University Press.

Dowe, P. (1993), 'The Anti-Realism of Costa De Beauregard', *Foundations of Physics Letters* 6, pp. 469–75.

Dowker, A.F. and Kent, A. (1996), 'On the Consistent Histories Approach to Quantum Mechanics', *Journal of Statistical Physics* 82, pp. 1575–646.

Duck, I. and Sudarshan, E.C.G. (1997), *Pauli and the Spin-Statistics Theorem*, Singapore: World Scientific.

Duncan, A. and Janssen, M. (2013), '(Never) Mind Your P's and Q's: Von Neumann Versus Jordan on the Foundations of Quantum Theory', *European Physics Journal H* 38, pp. 175–259.

Dyson F. (2013), 'A Walk Through Johnny von Neumann's Garden', *Notices of the American Mathematical Society* 60, pp. 154–61.

Earman, J. (2008), 'Superselection Rules for Philosophers', *Erkenntnis* 69, pp. 377–414.

Earman, J. (2018), 'The Relation between Credence and Chance: Lewis' 'Principal Principle' is a Theorem of Quantum Theory', http://philsci-archive.pitt.edu.

Earman, J. (2019), 'Quantum Bayesianism Assessed', *Monist* 102, pp. 403–23.

Ehlers, J. (1971), 'Comments and Questions Concerning Shimony's Lecture', in B. d'Espagnat (ed.), *Foundations of Quantum Measurement*, Cambridge, MA: Academic Press, p. 478.

Einstein, A. (1924), 'Quantentheorie des einatomigen idealen Gases', Sitzungsberichte *der Preussischen Akademie der Wissenschaften* 22, pp. 261–7.

Einstein, A. (1925), 'Quantentheorie des einatomigen idealen Gases', Sitzungsberichte *der Preussischen Akademie der Wissenschaften* 3, pp. 3–16.

Einstein, A., Podolsky, B. and Rosen, N. (1935), 'Can Quantum-Mechanical Description of Reality Be Considered Complete?', *Physical Review* 47, pp. 777–80.

Esfeld, M. (1999a), 'Wigner's View of Physical Reality', https://www.unil.ch/files/live/sites/philo/files/shared/DocsPerso/EsfeldMichael/1999/SHPMP99.pdf

Esfeld, M. (1999b), 'Wigner's View of Physical Reality', *Studies in History and Philosophy of Modern Physics* 30, pp. 145–54.

Everett, H., (1956), 'The Theory of the Universal Wave Function', first printed in DeWitt and Graham (1973), pp. 3–140. Reprinted as cited here in Barrett and Byrne (2012), pp. 72–172.

Everett, H. (1957), '"Relative State" Formulation of Quantum Mechanics', *Reviews of Modern Physics* 29, pp. 454–62.

Everett, H. (1977), Interview of Hugh Everett by Charles Misner on May 1977, Niels Bohr Library & Archives, American Institute of Physics, College Park: MD, www.aip.org/history-programs/niels-bohr-library/oral-histories/31230

Everitt, C.F.W. (1996), 'A Physicist's Journey: Review of *Fritz London. A Scientific Biography* by Kostas Gavroglu', *Science* 272, pp. 1273–4.

Faye, J. (2019), 'Copenhagen Interpretation of Quantum Mechanics', in E. N. Zalta (ed.), *The Stanford Encyclopedia of Philosophy*, https://plato.stanford.edu/archives/spr2019/entries/qm-copenhagen/.

Faye, J. and Jaksland, R. (2021), 'What Bohr Wanted Carnap to Learn from Quantum Mechanics', *Studies in History and Philosophy of Science* 88, pp. 110–19.

Fevrier, P. (1939), 'Sur l'indiscernabilité des corpuscules', *Journal de Physique et Le Radium*, 10, pp. 307–12.

Feyerabend, P. (1957), 'On the Quantum Theory of Measurement', in S. Körner and M.H.L. Pryce (eds.), *Observation and Interpretation:A Symposium of Philosophers and Physicists.* London and New York: Butterworths Scientific Publishing, pp. 121–130.

Feyerabend, P. (1958), 'Review of John von Neumann, *Mathematical Foundations of Quantum Mechanics...*' *British Journal for the Philosophy of Science* 8, pp. 343–7; repr. In S. Gattei and J. Agassi (eds), *Physics and Philosophy: Philosophical Papers Vol. 4*, Cambridge: Cambridge University Press, pp. 276–94.

Feyerabend, P. (1962), 'Problems of Microphysics', in R.G. Colodny (ed.), *Frontiers of Science and Philosophy*, Pittsburgh: University of Pittsburgh Press, pp. 189–283.

Feyerabend, P. (1966), 'Dialectical Materialism and the Quantum Theory', *Slavic Review* 25, pp. 414–17.

Feyerabend, P. (1968), 'On a Recent Critique of Complementarity: Part I', *Philosophy of Science* 35, pp. 309–31.

Feyerabend, P. (1995), *Killing Time: The Autobiography of Paul Feyerabend.* Chicago: University of Chicago Press.

Fine, A. (1986), The Shaky Game: Einstein, Realism and the Quantum Theory. Chicago: University of Chicago Press.

Fine, A. and Ryckman, T. (2020), 'The Einstein–Podolsky–Rosen Argument in Quantum Theory', in E. N. Zalta (ed.), *The Stanford Encyclopedia of Philosophy*, https://plato.stanford.edu/archives/sum2020/entries/qt-epr/

Fisette, D. (Spring 2019), 'Carl Stumpf', in E. N. Zalta (ed.), *The Stanford Encyclopedia of Philosophy*, https://plato.stanford.edu/archives/spr2019/entries/stumpf/

Føllesdal, D. (1999), 'The *Lebenswelt* in Husserl', in L. Haaparanta, M. Kusch and I. Niiniluoto (eds), *Language, Knowledge and Intentionality*, Helsinki: Acta Philosophica Fennica 49.

Forman, P. (2011), 'Scientific Internationalism and the Weimar Physicists: The Ideology and its Manipulation in Germany after World War I', in C. Carson, A. Kojevnikov, and H. Trischler (eds), *Weimar Culture and Quantum Mechanics: Selected Papers by Paul Forman and Contemporary Perspectives on the Forman Thesis.* London: Imperial College Press and World Scientific, pp. 27–56.

Frank, P. (1957), *Philosophy of Science: The Link between Science and Philosophy.* Englewood Cliffs: Prentice-Hall.

Franklin, A. and Perovic, S. (2021), 'Experiment in Physics', in E. N. Zalta (ed.), *The Stanford Encyclopedia of Philosophy*, https://plato.stanford.edu/archives/sum2021/entries/physics-experiment/

Frauchiger, D. and Renner, R. (2016), 'Single-World Interpretations of Quantum Theory Cannot be Self-Consistent', arXiv:1604.07422.

Frauchiger, D. and Renner, R. (2018), 'Quantum Theory Cannot Consistently Describe the Use of Itself', *Nature Communications* 9, 3711, https://doi.org/10.1038/s41467-018-05739-8

Freire Jr, O. (2015), *The Quantum Dissidents: Rebuilding the Foundations of Quantum Mechanics (1950–1990).* Berlin: Springer-Verlag.

French, S. (2002), 'A Phenomenological Approach to the Measurement Problem: Husserl and the Foundations of Quantum Mechanics', *Studies in History and Philosophy of Modern Physics* 33, pp. 467–91.

French, S. (2014), *The Structure of the World: Metaphysics and Representation*. Oxford: Oxford University Press.

French, S. (2020), 'From a Lost History to a New Future: Is a Phenomenological Approach to Quantum Physics Viable?', in H. Wiltsche and P. Berghofer (eds), *Phenomenological Approaches to Physics*. Cham: Springer, pp. 205–26.

French, S. and Krause, D. (2006), *Identity in Physics: A Formal, Historical and Philosophical Approach*. Oxford: Oxford University Press.

French, S. and Ladyman, J. (1997), 'Superconductivity and Structures: Revisiting the London Account', *Studies in History and Philosophy of Modern Physics* 28, pp. 363–93.

French, S. and Saatsi, J. (eds) (2020), *Scientific Realism and the Quantum*. Oxford: Oxford University Press.

Friedman, M. (2013), 'Philosophy of Natural Science in Idealism and Neo-Kantianism', in K. Ameriks (ed.), *The Impact of Idealism: The Legacy of Post-Kantian German Thought, Vol. 1, Philosophy and Natural Science*, Cambridge: Cambridge University Press, pp. 72–104.

Fuchs, C.A. (2010), 'QBism, the Perimeter of Quantum Bayesianism', arXiv:1003.5209.

Fuchs, C.A. (2017a), 'On Participatory Realism', in I.T. Durham and D. Rickles (eds), *Information and Interaction: Eddington, Wheeler and the Limits of Knowledge*. Cham: Springer, pp. 113–34.

Fuchs, C.A. (2017b), 'Notwithstanding Bohr, the Reasons for QBism', *Mind and Matter*, 15, pp. 245–300.

Fuchs, C.A. and Stacey, B. (2020), 'QBians Do Not Exist', arXiv: 2012.14375.

Gamow, G. (1970), *My World Line: An Informal Autobiography*. New York: Viking Press.

Gao, S. (2022), *Consciousness and Quantum Mechanics*. Oxford: Oxford University Press.

Gauthier, Y. (1969), 'La notion théorétique de structure', *Dialectica* 23, pp. 217–27.

Gauthier, Y. (1971), 'The Use of the Axiomatic Method in Quantum Physics', *Philosophy of Science* 38, pp. 429–37.

Gauthier, Y. (2019), 'De l'observateur local à l'observateur transcendental: De Kant et Husserl aux fondements de la physique contemporaine', *Philosophiques* 46, pp. 155–77.

Gavroglu, K. (1995), *Fritz London: A Scientific Biography*. Cambridge: Cambridge University Press.

Gieser, S. (2005), *The Innermost Kernel: Depth Psychology and Quantum Physics. Wolfgang Pauli's Dialogue with C.G. Jung*. Berlin: Springer-Verlag.

Glick, D. (2021), 'QBism and the Limits of Scientific Realism', *European Journal for the Philosophy of Science* 11, https://doi.org/10.1007/s13194-021-00366-5

Glymour, C. and Eberhardt, F. (2016), 'Hans Reichenbach', in E. N. Zalta (ed.), *The Stanford Encyclopedia of Philosophy*, https://plato.stanford.edu/archives/win2016/entries/reichenbach/

Goff, P. (2017), *Consciousness and Fundamental Reality*. Oxford: Oxford University Press.

Goff, P., Seager, W., and Allen-Hermanson, S. (2020), 'Panpsychism', in E. N. Zalta (ed.), *The Stanford Encyclopedia of Philosophy*, https://plato.stanford.edu/archives/sum2020/entries/panpsychism/

Goldstein S. (2009), 'Projection Postulate', in D. Greenberger, K. Hentschel, and F. Weinert (eds), *Compendium of Quantum Physics*. Berlin, Heidelberg: Springer, pp. 499–501.

Goldstein, S. (2021), 'Bohmian Mechanics', in E. N. Zalta (ed.), *The Stanford Encyclopedia of Philosophy*, https://plato.stanford.edu/archives/fall2021/entries/qm-bohm/

Gooday, G. and Mitchell, D. (2013), 'Rethinking "Classical Physics"', in J. Z. Buchwald and R. Fox (eds), *The Oxford Handbook of the History of Physics*, Oxford: Oxford University Press, pp. 721–64.

Grathoff, R. (ed.) (1989), *Philosophers in Exile: The Correspondence of Alfred Schutz and Aron Gurwitsch 1939–1959*. Bloomington: Indian University Press.

Greenberger, D. (1985), 'Review of *Quantum Theory and Measurement*', *American Scientist* 73, pp. 193–4.

Grünbaum, A. (1950), 'Realism and Neo-Kantianism in Professor Margenau's Philosophy of Quantum Mechanics', *Philosophy of Science* 17, pp. 26–34.

Gurwitsch, A. (1941), 'A Non-Egological Conception of Consciousness', *Philosophy and Phenomenological Research* 1, pp. 325–38.

Gurwitsch, A. (1946), 'Review of *Prévoir et Savoir. Etudes sur l'Idée de Nécessité dans la Pensée Scientifique et en Philosophie*, by Yves Simon', *Philosophy and Phenomenological Research*, 7, pp. 339–42.

Gurwitsch, A. (1964), *The Field of Consciousness*. Pittsburgh: Duquesne University Press.

Gurwitsch, A. (1974), *Phenomenology and the Theory of Science* (ed. L. Embree). Evanston: Northwestern University Press.

Gutland, C. (2018), 'Husserlian Phenomenology as a Kind of Introspection', *Frontiers of Psychology*, 6 June 2018, https://doi.org/10.3389/fpsyg.2018.00896

Hájek, A. (2019), 'Interpretations of Probability', in E. N. Zalta (ed.), *The Stanford Encyclopedia of Philosophy*, https://plato.stanford.edu/archives/fall2019/entries/probability-interpret/

Hall, J., Kim, C., McElroy, B., and Shimony, A. (1977), 'Wave-Packet Reduction as a Medium of Communication', *Foundations of Physics* 7, pp. 759–67.

Halpern, P. (2015), https://www.pbs.org/wgbh/nova/article/schrodingers-cat-lives-on-or-not-at-the-age-of-80/

Halvorson, H. (2010), 'The Measure of all Things: Quantum Mechanics and the Soul', in M. Baker and S. Goetz (eds), *The Soul Hypothesis*. London: Continuum, pp. 138–63.

Halvorson, H. and Butterfield, J. (2023), 'John Bell on Subject and Object: An Exchange', *Journal for General Philosophy of Science* 54, pp. 305–324.

Hance, J.T., Rarity, J., and Ladyman J. (2022), 'Could Wave Functions Simultaneously Represent Knowledge and Reality?', *Quantum Studies: Mathematical Foundations*, https://doi.org/10.1007/s40509-022-00271-3

Hardy, L. (2013), *Nature's Suit: Husserl's Phenomenology of the Physical Sciences*. Athens: Ohio University Press.

Hardy, L. (2021), 'One Table or Two? Scientific Anti-realism and Husserl's Phenomenology', *Continental Philosophy Review* 54, pp. 437–52.

Harrigan, N. and Spekkens, R.W. (2010), 'Einstein, Incompleteness, and the Epistemic View of Quantum states', *Foundations of Physics* 40, pp. 125–57.

Healey, R. (2017), 'Quantum-Bayesian and Pragmatist Views of Quantum Theory', in E. N. Zalta (ed.), *The Stanford Encyclopedia of Philosophy*, https://plato.stanford.edu/archives/spr2017/entries/quantum-bayesian/

Heelan, P. (2013), 'Phenomenology, Ontology, and Quantum Physics', *Foundations of Science* 18, pp. 379–85.

te Heesen, A. (2020), 'Thomas S. Kuhn, Earwitness: Interviewing and the Making of a New History of Science' *Isis* 111, pp. 86–97.

te Heesen, A. (2022), *Revolutionäre im Interview: Thomas Kuhn, Quantenphysik und Oral History*, Berlin: Wagenbach.

Heidegger, M. (1977), 'Science and Reflection', in *The Question Concerning Technology and Other Essays*, trans. William Lovitt. New York: Harper & Row.

Heims, S. (1991), 'Fritz London and the Community of Quantum Physicists', in W.R. Woodward and R. S. Cohen (eds), *World Views and Scientific Discipline Formation*, Dodrecht: Kluwer Academic, pp. 177–90.

Heisenberg, W. (1927), 'Über den anschaulichen Inhalt der quantentheoretischen Kinematik und Mechanik', *Zeitschrift für Physik* 43, pp. 172–98.

Heisenberg, W. (1952), *Philosophical Problems of Nuclear Science.* New York: Pantheon.

Heisenberg, W. (1955), 'The Development of the Interpretation of Quantum Theory', in W. Pauli (ed.), *Niels Bohr and the Development of Physics*, New York: McGraw-Hill.

Heisenberg, W. (1958), *Physics and Philosophy.* New York: Harper Row.

Heitler, W. (1963), Interview of Walter Heitler by John L. Heilbron on 18 March 1963, Niels Bohr Library & Archives, American Institute of Physics, College Park: MD, www.aip.org/history-programs/niels-bohr-library/oral-histories/4662-1

Herbert, N. (1994), *Elemental Mind: Human Consciousness and the New Physics.* London: Penguin.

Hermann, G. (1999), 'The Foundations of Quantum Mechanics in the Philosophy of Nature', *Harvard Review of Philosophy* VII, pp. 35–44; trans of G. Hermann, 'Die Naturphilosophischen Grundlagen der Quantenmechanik', *Die Naturwisserischaftrl* 4*12* (1935), p. 721.

Hesse, M. (1961), *Forces and Fields: The Concept of Action at a Distance in the History of Physics.* London: Nelson.

Hooker C.A. (1972), 'The Nature of Quantum Mechanical Reality: Einstein Versus Bohr', *Paradigms and Paradoxes: The Philosophical Challenge of the Quantum Domain*, University of Pittsburgh Press, pp. 67–208.

Hopp, W. (2020), *Phenomenology: A Contemporary Introduction.* New York and London: Routledge.

Hossenfelder, S. (2021), 'What Did Einstein Mean by "Spooky Action at a Distance"?' http://backreaction.blogspot.com/2021/05/what-did-einstein-mean-by-spooky-action.html

Howard, D. (1990), 'Nicht Sein Kann Was Nicht Sein Darf' Or The Prehistory of EPR, 1909–1935: Einstein's Early Worries About the Quantum Mechanics of Composite Systems', in A. Miller (ed.), *Sixty Two Years of Uncertainty: Historical, Philosophical and Physical Enquiries into the Foundations of Quantum Mechanics.* London: Plenum, pp. 61–111.

Howard, D. (2004), 'Who Invented the "Copenhagen Interpretation"? A Study in Mythology', *Philosophy of Science* 71, pp. 669–82.

Howard, D. (2013), 'Quantum Mechanics in Context: Pascual Jordan's 1936 *Anschauliche Quantentheorie*', in M. Badino and J. Navarro (eds), *Research and Pedagogy: A History of Quantum Physics through Its Textbooks* Online version at http://edition-open-access.de/studies/2/

Hoyningen-Huene, P. (1995), 'Two Letters of Paul Feyerabend to Thomas S. Kuhn on a Draft of The Structure of Scientific Revolutions', *Studies in History and Philosophy of Science* 26, pp. 353–87.

Hu, H. and Wu, M. (2010), 'Current Landscape and Future Direction of Theoretical & Experimental Quantum Brain/Mind/Consciousness Research', *Journal of Consciousness Exploration and Research* 1, pp. 888–97.

Hudson, R. (1951), 'Review of F. London, *Superfluids: Macroscopic Theory of Superconductivity*, Vol. I', *Science* 113, p. 447.

Husserl, E. (1959), *Erste Philosophie (1923/24). Zweiter Teil: Theorie der phänomenologischen Reduktion*, ed. R. Boehm. Den Haag: Martinus Nijhoff.

Husserl, E. (1964), *The Paris Lectures*, trans. P. Koestenbaum (1929). Leiden: Martinus Nijhoff.

Husserl, E. (1970a), *Logical Investigations I-II*, trans. J.N. Findlay (1900–1901). London: Humanities Press.

Husserl, E. (1970b), *The Crisis of European Sciences and Transcendental Phenomenology*, trans. D. Carr (1954), Evanston, IL: Northwestern University Press.

Husserl, E. (1973a), *Zur Phänomenologie der Intersubjektivität. Texte aus dem Nachlass. Erster Teil 1905–1920*, ed. I. Kern. Den Haag: M. Nijhoff.

Husserl, E. (1973b), *Zur Phänomenologie der Intersubjektivität. Texte aus dem Nachlass. Dritter Teil 1929–1935*, ed. I. Kern. Den Haag: M. Nijhoff.

Husserl, E. (1977), *Phänomenologische Psychologie: Vorlesungen Sommersemester 1925*, ed. W. Beimel. The Hague: Martinus Nijhoff, 1962, pp. 2–234; trans J. Scanlon, *Phenomenological Pyschology: Lectures, Summer Semester 1925*; The Hague: Martinus Nijhoff 1977, pp. 237–349.

Husserl, E. (1978), *Formal and Transcendental Logic*, trans. D. Cairns. The Hague: Martinus Nijhoff.

Husserl, E. (1982), *Ideas: General Introduction to Pure Phenomenology*, trans. F. Kersten (1913). Dordrecht: Kluwer Academic.

Husserl, E. (1991), *On the Phenomenology of the Consciousness of Internal Time (1893–1917)*, trans. J. B. Brough. Dordrecht: Kluwer Academic.

Husserl, E. (2002), *Einleitung in die Philosophie: Vorlesungen 1922/23*, ed. B. Goossens. Dordrecht: Kluwer Academic.

Husserl, E. (2003), *Transzendentaler Idealismus: Texte aus dem Nachluss (1908–1921)*, ed. R. Rollinger. Dodrecht: Kluwer Academic.

Jacobsen, A.S. (2011), 'Crisis, Measurement Problems and Controversy in Early Quantum Electrodynamics: The Failed Appropriation of Epistemology in the Second Quantum Generation', in A. Kojevnikov, C. Carson, and H. Trischler (eds), *Quantum Mechanics and Weimar Culture: Revisiting the Forman Thesis, with Selected Papers by Paul Forman*. London: Imperial College Press, pp. 375–96.

Jacobsen, A.S. (2012), *Leon Rosenfeld: Physics, Philosophy, and Politics in the Twentieth Century*. Singapore: World Scientific.

Jacobson, P. (1976), '"Dirty Work": Gurwitsch on the Phenomenological Theory of Science and Constitutive Phenomenology', *Research in Phenomenology* 6, pp. 191–7.

Jähnert, M. and Lehner, C. (2022), 'The Early Debates about the Interpretation of Quantum Mechanics', in O. Freire Jr et. al. (eds), *The Oxford Handbook of the History of Quantum Interpretations*, Oxford: Oxford University Press, pp. 135–72.

Jammer, M. (1974), *The Philosophy of Quantum Mechanics*, New York: Wiley.

Jauch, J.M. (1971), 'Foundations of Quantum Measurement', in B. d'Espagnat (ed.), *Foundations of Quantum Measurement*, Cambridge, MA: Academic Press, pp. 20–55.

Jauch, J.M. and Baron, J.G. (1972), 'Entropy, Information and Szilard's Paradox', *Helvetica Physica Acta* 45, pp. 220–32.

Jauch, J.M., Wigner, E.P., and Yanase, M.M. (1967), 'Some Comments Concerning Measurements in Quantum Mechanics', *Il Nuovo Cimento* 48B, pp. 144–51; reprinted in *The Collected Works of Eugene Paul Wigner. Part B. Historical, Philosophical, and Socio-Political Papers. Vol. VII. Historical and Biographical Reflections and Syntheses*, ed. Jagdish Mehra, Berlin, Heidelberg, New York: Springer-Verlag, 2001, pp. 181–8.

Jeffers, S. (2003), 'Physics and Claims for Anomalous Effects Related to Consciousness', *Journal of Consciousness Studies* 10, pp. 135–52.

Jha, S. (2011), 'Wigner's "Polanyian" Epistemology and the Measurement Problem: The Wigner–Polanyi Dialog on Tacit Knowledge', *Physics in Perspective* 13, pp. 329–58.

Joaquim, L., Freire Jr, O., and El-Hani, C.N. (2015), 'Quantum Explorers: Bohr, Jordan, and Delbrück Venturing into Biology', *Physics in Perspective* 17, pp. 236–50.

Jordan, P. (1932), 'Die Quantenmechanik und die Grundprobleme der Biologie und Psychologie', *Naturwissenschaften* 20, 815–21.

Jordan, P. (1936), *Anschauliche Quantentheorie. Eine Einführung in die moderne Auffassung der Quantenerscheinungen*. Berlin: Julius Springer.

Jordan, P. (1949), 'On the Process of Measurement in Quantum Mechanics', *Philosophy of Science* 16, pp. 269–78.

Jordan, P. (1951), 'Reflections on Parapsychology, Psychoanalysis, and Atomic Physics', *Journal of Parapsychology* 15, pp. 278–81.

Jordan, P. (1963), Interview of Ernst Pascual Jordan by Thomas S. Kuhn on 17, 18, 19, and 20 June 1963, Niels Bohr Library & Archives, American Institute of Physics, College Park, MD USA, http://repository.aip.org/islandora/object/nbla:269316

Kaiser, D. (2011), *How the Hippies Saved Physics: Science, Counter-Culture, and the Quantum Revival.* New York and London: Norton.

Kaufmann, F. (1949), 'Cassirer, Neo-Kantianism, and Phenomenology', in P.A. Schilpp (ed.), *The Philosophy of Ernst Cassirer*, Carbondale, IL: Library of Living Philosophers, pp. 799–854.

Kirkwood, J. (1955), 'Review of F. London, *Superfluids: Macroscopic Theory of Superconductivity*, Vol. II', *Journal of the American Chemical Society* 77, p. 3679.

Kisiel, T. (1967), 'Review of *Phenomenology and Physical Science: An Introduction to the Philosophy of Physical Science* by Joseph J. Kockelmans', *Philosophy and Phenomenological Research* 28, pp. 138–9.

Kockelmans, J. (1966), *Phenomenology and Physical Science.* Pittsburgh, PA: Duquesne Press.

Kockelmans, J. (1975), 'Gurwitsch's Phenomenological Theory of Natural Science', *Research in Phenomenology* 5, pp. 29–35.

Koehler, E. (2013), 'Why von Neumann Rejected Carnap's Dualism of Information Concepts', in M. Rédei and M. Stöltzner (eds), *John von Neumann and the Foundations of Quantum Physics*, Dordrecht: Kluwer Academic, pp. 97–134.

Köhler, W. (1927), *The Mentality of Apes*, trans. E. Winter (2nd ed.), London: Kegan Paul, Trench, Trubner and Co. Ltd.

Kojevnikov, A. (2020), *The Copenhagen Network: The Birth of Quantum Mechanics from a Postdoctoral Perspective.* Cham: Springer Nature.

Körner, S. and Pryce, M. H. L. (eds.) (1957), *Observation and Interpretation:A Symposium of Philosophers and Physicists*, London and New York: Butterworths Scientific Publishing.

Kožnjak, Boris (2019), 'The Missing History of Bohm's Hidden Variables Theory: The Ninth Symposium of the Colston Research Society, Bristol, 1957', *Studies in History and Philosophy of Modern Physics* 62, p. 85–97.

Kragh, H. (2013), 'Paul Dirac and *The Principles of Quantum Mechanics*', in M. Badino and J. Navarro (eds), *Research and Pedagogy: A History of Quantum Physics through Its Textbooks*, http://edition-open-access.de/studies/2/

Kronz, F. (1991), 'Quantum Entanglement and Nonideal Measurements: A Critique of Margenau's Objections to the Projection Postulate', *Synthese* 89, pp. 229–51.

Kuby, D. (2021), 'Feyerabend's Reevaluation of Scientific Practice: Quantum Mechanics, Realism and Niels Bohr', in Karim Bschir and Jamie Shaw (eds), *Interpreting Feyerabend: Critical Essays*, Cambridge: Cambridge University Press, pp. 132–56.

Kuhn, T.S. (1978), *Black-Body Theory and the Quantum Discontinuity: 1984-1912*, Oxford: Oxford University Press.

Kurti, N. (1968), Interview of Nicholas Kurti by Charles Weiner on 11 September 1968, Niels Bohr Library & Archives, American Institute of Physics, College Park, MD USA, www.aip.org/history-programs/niels-bohr-library/oral-histories/4725-1

Kuzemsky, A.L. (2008), 'Works by D.I. Blokhintsev and the Development of Quantum Physics', *Physics of Particles and Nuclei* 39, pp. 137–72.

Ladyman, J. (1998), 'What Is Structural Realism?', *Studies in History and Philosophy of Science Part A* 29, pp. 409–424.

Ladyman, J. (2018), 'Intension in the Physics of Computation: Lessons from the Debate about Landauer's Principle', *Physical Perspectives on Computation, Computational Perspectives on Physics*, Vol. 1, Cambridge: Cambridge University Press, pp. 219–39.

Landauer, R. (1961), 'Irreversibility and Heat Generation in the Computing Process', *IBM Journal of Research and Development* 5, pp. 183–91.

Lande, A. (1965/2015), *New Foundations of Quantum Mechanics*, Cambridge: Cambridge University Press.

Laudisa, F. (2019), 'Open Problems in Relational Quantum Mechanics', *Journal for General Philosophy of Science* 50, pp. 215–30.

Laudisa, F. and Rovelli, C. (2019), 'Relational Quantum Mechanics', in E. N. Zalta (ed.), *The Stanford Encyclopedia of Philosophy*, https://plato.stanford.edu/archives/win2019/entries/qm-relational/

Lenzen, V.F. (1933), 'Review of *Science and First Principles*', *The Philosophical Review* 42, pp. 320–2.

Leshan, L. (1974/2003), *The Medium, the Mystic, and the Physicist: Toward a General Theory of the Paranormal.* New York: Allworth Press.

Leshan, L. and Margenau, H. (1982), *Einstein's Space and Van Gogh's Sky.* New York: Scribner.

Leydesdorff, L. (2016), 'The Code of Mathematics', https://www.leydesdorff.net/vonneumann/

Licauco, J. (2014), 'A 'larger truth' about quantum physics', *Lifestyle.Inq*, https://lifestyle.inquirer.net/167704/a-larger-truth-about-quantum-physics/

Linschitz, H. (1988), Interview of Henry Linschitz by Steven Heims on 23 February 1988, Niels Bohr Library & Archives, American Institute of Physics, College Park: MD, www.aip.org/history-programs/niels-bohr-library/oral-histories/5039

Lobo, C. (2019), 'Husserl's Logic of Probability: An Attempt to Introduce in Philosophy the Concept of "Intensive" Possibility', *Meta: Research in Hermeneutics, Phenomenology, and Practical Philosophy* 11, pp. 501–46.

Lockwood, M. (1996a), '"Many Minds" Interpretation of Quantum Mechanics', *British Journal for the Philosophy of Science* 47, pp. 159–88.

Lockwood, M. (1996b), '"Many Minds" Interpretations of Quantum Mechanics: Replies to Replies', *British Journal for the Philosophy of Science* 47, pp. 445–61

London, F. (1935), 'Macroscopical Interpretation of Supraconductivity', *Proceedings of the Royal Society (London)* A152, pp. 24–34.

London, F. (1937), 'A New Conception of Supraconductivity', *Nature* 140, pp. 793–6 and 834–6.

London, F. (1950), *Superfluids*, Vol. I. New York: John Wiley and Sons.

London, F. (1954), *Superfluids*, Vol. II. New York: John Wiley and Sons.

London, F. (1938a), 'The λ-Phenomenon of Liquid Helium and the Bose–Einstein Degeneracy', *Nature* 141, pp. 643–4.

London, F. (1938b), 'On the Bose–Einstein Condensation', *Physical Review* 54, pp. 947–54.
London, F. and Bauer, E. (1939), *La Théorie de L'Observation en Mécanique Quantique*. Paris: Hermann.
London, F. and Bauer, E. (1983), 'The Theory of Observation in Quantum Mechanics', in J.A. Wheeler and W.H. Zurek (eds), *Quantum Theory and Measurement*. Princeton: Princeton University Press, pp. 217–59.
London, F. and London, H. (1935), 'The Electromagnetic Equations of the Supraconductor', *Proceedings of the Royal Society (London)* A149, pp. 71–88.
Lopez McAlister, L. (1995), 'Gerda Walther', in M. E. Waithe (ed.) *A History of Women Philosophers Volume* 4. Dordrecht: Kluwer Academic, pp. 189–206.
Luft, S. (2004), 'Husserl's Theory of the Phenomenological Reduction: Between Life-World and Cartesianism', *Research in Phenomenology* 34, pp. 198–234.
Magat, M. (1964), 'Oeuvres Scientifiques de Edmond Bauer', *Journal de Chimie Physique et de Physico-Chimie Biologique* 61, pp. 970–5.
Marcelle, D. (2019), 'The Freiburg Encounter: Aaron Gurwitsch and Edmund Husserl on Transformations of Consciousness', in M.B. Ferri (ed.), *The Reception of Phenomenology in North America*. Cham: Springer, pp. 47–70.
Margenau, H. (1935), 'Methodology of Modern Physics', *Philosophy of Science* 2, pp. 164–87.
Margenau, H. (1937), 'Critical Points in Modern Physical Theory', *Philosophy of Science* 4, pp. 337–70.
Margenau, H. (1949), 'Reality in Quantum Mechanics', *Philosophy of Science* 16, pp. 287–302.
Margenau, H. (1950), *The Nature of Physical Reality*. New York: McGraw-Hill.
Margenau, H. (1952), 'Physics and Ontology', *Philosophy of Science* 19, pp. 342–5.
Margenau, H. (1956), 'Physics and Psychic Research', *Newsletter of the Parapsychology Foundation* 3, pp. 14–15.
Margenau, H. (1957), 'A Principle of Resonance, *Newsletter of the Parapsychology Foundation* 4, pp. 3–6.
Margenau, H. (1963a), 'Measurements in Quantum Mechanics', *Annals of Physics* 23, pp. 469–85.
Margenau, H. (1963b), 'Measurements and Quantum States: Part I', *Philosophy of Science* 30, pp. 1–16.
Margenau, H. (1964), Interview of Henry Margenau by Bruce Lindsay and W. James King on 6 May 1964, Niels Bohr Library & Archives, American Institute of Physics, College Park: MD, www.aip.org/history-programs/niels-bohr-library/oral-histories/4757
Margenau, H. (1966), 'ESP in the Framework of Modern Science', *Journal of the American Society for Psychical Research* 60, pp. 214–28, https://www.survivalafterdeath.info/articles/margenau/framework.htm
Margenau, H. (1978a), 'Phenomenology and Physics', in H. Margenau (ed.), *Physics and Philosophy: Selected Essays*, Dordrecht: D. Reidel, pp. 317–28.
Margenau, H. (1978b), 'Note on Quantum Mechanics and Consciousness', in H. Margenau (ed.), *Physics and Philosophy: Selected Essays*, Dordrecht: D. Reidel, pp. 373–4.
Margenau, H. (1978c), *Physics and Philosophy: Selected Essays*, Dordrecht: D. Reidel.
Margenau, H. and Wigner, E. (1962), 'Discussion: Comments on Professor Putnam's Comments', *Philosophy of Science* 29, pp. 292–3.
Margenau, H. and Wigner, E. (1964), 'Discussion: Reply to Professor Putnam', *Philosophy of Science* 31, pp. 7–9.

Marin, J. (2009), 'Mysticism in Quantum Mechanics: The Forgotten Controversy', *European Journal of Physics* 30, pp. 807–22.

Massignon, D. (1970), 'Bauer, Edmond', in C.C. Gillespie (ed.), *The Dictionary of Scientific Biography*, Vol. I, New York: C. Scribner & Sons, pp. 519–20.

Massimi, M. (2021), *Perspectival Realism*. Oxford: Oxford University Press.

McCoy, C. (2022), 'The Constitution of Weyl's Pure Infinitesimal World Geometry', *HOPOS: Journal of the International Society for the History of Philosophy of Science* 12, pp. 189–208.

McQueen, K. (2017), 'Is QBism the Future of Quantum Physics?' arXiv:1707.02030v1 [quant-ph].

Mehra, J. (2012), *The Quantum Principle: Its Interpretation and Epistemology*, Dordrecht: D. Reidel

Merleau-Ponty, M. (1968), *The Visible and the Invisible*. Evanston, IL: Northwestern University Press.

Merleau-Ponty, M. (2003). *Nature: Course Notes from the Collège de France*. Evanston, IL: Northwestern University Press.

Mermin, D. (1998), 'What is Quantum Mechanics Trying to Tell Us?', *American Journal of Physics* 66, pp. 753–767.

Mitsch, C. (2022), 'Hilbert-Style Axiomatic Completion: On von Neumann and Hidden Variables in Quantum Mechanics', *Studies in History and Philosophy of Science* 95, pp. 84–95.

Mohanty, J. (1994), 'The Unity of Aron Gurwitsch's Philosophy', *Social Research* 61, pp. 937–54.

Mohanty, J. (1995), 'The Development of Husserl's Thought', in B. Smith and D. Woodruff Smith (eds), *The Cambridge Companion to Husserl*. Cambridge: Cambridge University Press, pp. 45–77.

Moldauer, P.A. (1964), 'Problem of Measurement', *American Journal of Physics* 32, p. 172.

Moore, W. (1989), *Schrödinger: Life and Thought*. Cambridge: Cambridge University Press.

Moran, D. (2012), *Husserl's Crisis of the European Sciences and Transcendental Phenomenology: An Introduction*. Cambridge: Cambridge University Press.

Moreira dos Santos, F. and Pessoa Jr, O. (2011), 'Delineando o problema da medição na mecânica quântica: o debate de Margenau e Wigner versus Putnam', *Scientiæ Studia* 9, pp. 625–44.

Mormann, T. (1991), 'Husserl's Philosophy of Science and the Semantic Approach', *Philosophy of Science* 58, pp. 61–83.

Mormann, T. (2017), 'Philipp Frank's Austro-American Logical Empiricism', *HOPOS: Journal of the International Society for the History of Philosophy of Science* 7, pp. 56–87.

Morris, D. (2005), 'Animals and Humans, Thinking and Nature', *Phenomenology and the Cognitive Sciences* 4, pp. 49–72.

Morrison, M. (2007), 'Spin: All is Not What it Seems', *Studies in History and Philosophy of Modern Physics* 38, pp. 529–57.

Muller, F.A. (1997a), 'The Equivalence Myth of Quantum Mechanics I', *Studies in History and Philosophy of Modern Physics* 28, pp. 35–61.

Muller, F.A. (1997b), 'The Equivalence Myth of Quantum Mechanics II', *Studies in History and Philosophy of Modern Physics* 28, pp. 219–47.

Murgueitio Ramírez, S. (2022), 'On How *Epistemological Letters* Changed the Foundations of Quantum Mechanics', in Olival Freire Jr et al. (eds), *The Oxford Handbook of the History of Quantum Interpretations,* Oxford: Oxford University Press, pp. 755–75.

Narasimhachar, V. (2020), 'Agents Governed by Quantum Mechanics Can Use it Intersubjectively and Consistently', arXiv:2010.01167v1 [quant-ph].

Navarro, J. (2009), '"A Dedicated Missionary": Charles Galton Darwin and the new Quantum Mechanics in Britain', *Studies in History and Philosophy of Modern Physics* 40, pp. 316–26.

Nordheim, L. (1962), Interview of Lothar Nordheim by John Heilbron on 30 July 1962, Niels Bohr Library & Archives, American Institute of Physics, College Park: MD, www.aip.org/history-programs/niels-bohr-library/oral-histories/4799

Northrop, F.S.C. (1931), *Science and First Principles*. New York: Macmillan, repr. 1979, Ox Bow Press.

Norton Wise, M. (1994), 'Pascual Jordan: Quantum Mechanics, Psychology, National Socialism', in M. Renneburg and M. Walker (eds), *Science, Technology and National Socialism*, Cambridge: Cambridge University Press, pp. 224–54.

Okon, E. and Sebastián, M.A. (2016), 'How to Back up or Refute Quantum Theories of Consciousness', *Mind and Matter* 14, pp. 25–49.

Okon, E. and Sebastián, M.A. (2020), 'A Consciousness-Based Quantum Objective Collapse Model', *Synthese* 197, pp. 3947–67.

Oppenheimer, J.R. (1965), 'Interview', *Voices of the Manhattan Project*, Atomic Heritage Foundation, https://www.manhattanprojectvoices.org/oral-histories/j-robert-oppenheimers-interview

Pais, A. (1982), *Subtle is the Lord*. Oxford: Oxford University Press.

Parker, R. (2017), 'Gerda Walther and the Phenomenological Community', *Acta Mexicana De Fenomenología Revista De Investigación FilosóFicA Y Científica* 2, pp. 43–64.

Parker, R. (2021), 'The Idealism–Realism Debate and the Great Phenomenological Schism', in R. Parker (ed.), *The Idealism–Realism Debate among Edmumd Husserl's Early Followers and Critics*, Cham: Springer, pp. 1–26.

Pashby, T. (2020), 'Sensible Quantum Experiences: Encounters with Stein's Philosophy of Quantum Mechanics', *Studies in History and Philosophy of Modern Physics* 69, pp. 128–41.

Paty, M. (2003), 'The Concept of Quantum State: New Views on Old Phenomena', in J. Renn, L. Divarci, and P. Schröter (eds), *Revisiting the Foundations of Relativistic Physics: Festschrift in Honor of John Stachel*, Dordrecht: Kluwer, pp. 451–78.

Pauli, W. (2013), *Writings on Physics and Philosophy*, ed. C.P. Enz and K.v. Meyenn (trans. R. Schlapp). Berlin: Springer–Verlag.

Peierls, R. (1985), *Bird of Passage: Recollections of a Physicist*. Princeton University Press.

Pellegrini, P. (2021), 'Merleau-Ponty's Phenomenological Perspective on Quantum Mechanics', *Continental Philosophy Review* 54, pp. 483–502.

Pellegrino, M.P. (2018), 'Gerda Walther: Searching for the Sense of Things, Following the Traces of Lived Experiences', in A. Calcagno (ed.), *Gerda Walther's Phenomenology of Sociality, Psychology, and Religion*, Women in the History of Philosophy and Sciences 2, Cham: Springer. pp. 11–24.

Pendle, G. (2005), *Strange Angel: The Otherworldly Life of Rocket Scientist John Whiteside Parsons*. New York: Harcourt.

Peres, A. (2002), 'Karl Popper and the Copenhagen Interpretation', *Studies in History and Philosophy of Modern Physics* 33, pp. 23–34.

Petersen, A. (1968), *Quantum Physics and the Philosophical Tradition*. Cambridge, MA: MIT.

Pienaar, J. (2020), 'Extending the Agent in QBism', *Foundations of Physics* 50, pp. 1894–920.

Pienaar, J. (2021a), 'A Quintet of Quandaries: Five No-Go Theorems for Relational Quantum Mechanics', *Foundations of Physics* 51: 97.
Pienaar, J. (2021b), 'QBism and Relational Quantum Mechanics Compared', *Foundations of Physics* 51: 96.
Pietschmann, H. (2020), 'Hans Thirring: A Personal Recollection', https://www.oeaw.ac.at/ikt/das-institut/news-detail/hans-thirring-a-personal-recollection
Podolsky, B., Hart, J. B., and Werner, F. G. (2002), 'Conference Manuscript', *Conference on the Foundations of Quantum Mechanics (1962).* Book 1, http://www.exhibit.xavier.edu/conf_qm_1962/1
Polanyi, M. (1962), 'The Unaccountable Element in Science', *Philosophy* 37, pp. 1–14.
Primas, H. and Esfeld, M. (unpublished), 'A Critical Review of Wigner's Work on the Conceptual Foundations of Quantum Theory'.
Pris, F-I. (2014), 'Heidegger's Phenomenology and Quantum Physics', *Philosophical Investigations* (Russian e-journal) 4, pp. 46–67.
Przibram, K. (ed.) (1967), *Letters on Wave Mechanics: Schrödinger, Planck, Einstein, Lorentz*, trans. from the German with an introduction by Martin J. Klein. New York: Philosophical Library.
Pusey, M.F., Barrett, J., and Rudolph, T. (2012), 'On the Reality of the Quantum State', *Nature Physics* 8, pp. 475–8.
Putnam, H. (1961), 'Comments on the Paper of David Sharp', *Philosophy of Science* 28, pp. 234–7.
Putnam, H. (1964), 'Discussion: Comments on Comments on Comments, a Reply to Margenau and Wigner', *Philosophy of Science* 31, pp. 1–6.
Putnam, H. (1965), 'A Philosopher Looks at Quantum Mechanics', in R. G. Colodny (ed.), *Beyond the Edge of Certainty: Essays in Contemporary Science and Philosophy*, Englewood Cliffs, NJ: Prentice-Hall, pp. 75–101. Repr. in *Mathematics, Matter and Method* (1975), pp. 130–58.
Putnam, H. (1965/1975), 'Philosophy of Physics', in F. H. Donnell, Jr (ed.), *Aspects of Contemporary American Philosophy*, Würzburg: Physica-Verlag, Rudolf Liebing K.G., 1965, pp. 27–40. Repr. in *Mathematics, Matter and Method*, 1975, pp. 79–92.
Putnam, H. (1981), 'Quantum Mechanics and the Observer', *Erkenntnis* 16, pp. 193–219.
Putnam, H. (1991), 'Il Principio di Indeterminazione e il Progresso Scientifico', *Iride* 7, pp. 9–27.
Putnam, H. (2005), 'A Philosopher Looks at Quantum Mechanics (Again)', *British Journal for the Philosophy of Science* 56, pp. 615–34.
Radin, D., Michel, Leena, and Galdamez, Karla et al. (2012), 'Consciousness and the Double-Slit Interference Pattern: Six Experiments', *Physics Essays* 25, pp. 157–71.
Rédei, M. (1996), 'Why John von Neumann did Not Like the Hilbert Space Formalism of Quantum Mechanics (and What He Liked Instead)', *Studies in History and Philosophy of Modern Physics* 27, pp. 493–510.
Rédei, M. (2005), *John von Neumann: Selected Letters*. London Mathematical Society.
Reichenbach, H. (1927), 'A New Model of the Atom', in Reichenbach 1978, pp. 219–25.
Reichenbach, H. (1930), 'The Philosophical Significance of Modern Physics', in Reichenbach 1978, pp. 304–23.
Reichenbach, H. (1944), *Philosophic Foundations of Quantum Mechanics*. Berkeley: University of California Press.
Reichenbach, H. (1978), *Selected Writings 1909–1953*, Vol. 1 (ed. M. Reichenbach and R.S. Cohen). Dordrecht: Reidel.

Rocha, G.R., Rickles, D., and Boge, F.J. (2021), 'A Brief Historical Perspective on the Consistent Histories Interpretation of Quantum Mechanics', arXiv:2103.05280v1 [physics.hist-ph].

Rosenfeld, L. (1963), Interview of Leon Rosenfeld by Thomas S. Kuhn and John L. Heilbron on 19 July 1963, Niels Bohr Library & Archives, American Institute of Physics, College Park, MD USA, www.aip.org/history-programs/niels-bohr-library/oral-histories/4847-2

Rouse, J. (1987), 'Husserlian Phenomenology and Scientific Realism', *Philosophy of Science* 54, pp. 222–32.

Rovelli, C. (1996), 'Relational Quantum Mechanics', *International Journal of Theoretical Physics* 35, pp. 1637–78.

Rovelli, C. (2011), '"Forget time" Essay written for the FQXi contest on the Nature of Time', *Foundations of Physics* 41, pp. 1475–1490.

Rovelli, C. (2021), 'Relations and Panpsychism', *Journal of Consciousness Studies* 28, pp. 32–5.

Ruyant, Q. (2018), 'Can We Make Sense of Relational Quantum Mechanics?', *Foundations of Physics* 48, pp. 440–55.

Ryckman, T. (2005), *The Reign of Relativity: Philosophy in Physics 1915–1925*. Oxford: Oxford University Press.

Ryckman, T. (2022), 'Quantum Interpretations and 20th Century Philosophy of Science', in O. Freire Jr et. al. (ed.), *The Oxford Handbook of the History of Quantum Interpretations*, Oxford: Oxford University Press, pp. 777–95.

Sacco, D. (2021), 'The Phenomenality of the Phenomenon: Heidegger on Physics', *Continental Philosophy Review* 54, pp. 503–19.

Salice, A. (2016), 'The Phenomenology of the Munich and Göttingen Circles', in E. N. Zalta (ed.), *The Stanford Encyclopedia of Philosophy*, https://plato.stanford.edu/archives/win2016/entries/phenomenology-mg/

Sartre, J-P. (1936–7), 'La Transcendance de l'Ego', *Recherches Philosophiques* 6, pp. 85–123.

Sartre, J-P. (1943), *Being and Nothingness*. Paris: Gallimard (English translation 1956).

Saunders, S. (1995), 'Relativism', in R. Clifton (ed), *Perspectives on Quantum Realit.*, Dordrecht: Kluwer Academic, pp. 125–142.

Saunders, S. (1996), 'Comments on Lockwood', *British Journal for the Philosophy of Science* 47, pp. 241–8.

Schrödinger, E. (1935a), 'Die gegenwärtige Situation in der Quantenmechanik, I–III', *Die Naturwissenschaften* 23: 48–50: pp. 807–12, 823–8, 844–9.

Schrödinger, E. (1935b), 'Discussion of Probability Relations between Separated Systems', *Proceedings of the Cambridge Philosophical Society, Mathematical and Physical Sciences* 31: 4, pp. 555–63.

Schrödinger, E. (1952a), 'Are There Quantum Jumps: Part I', *British Journal for the Philosophy of Science* 3, pp. 109–23.

Schrödinger, E. (1952b), 'Are There Quantum Jumps: Part II', *British Journal for the Philosophy of Science* 3, pp. 233–242.

Schrödinger, E. (1958), *Mind and Matter*. Cambridge: Cambridge University Press.

Schroer, B. (2003), 'Pascual Jordan: His Contributions to Quantum Mechanics and his Legacy in Contemporary Local Quantum Physics', https://arxiv.org/pdf/hep-th/0303241.pdf

Schuhmann, K. (1992), 'Husserl and Indian Thought', in D.P. Chattopadhyaya, L. E. Embree, and J. Mohanty (eds), *Phenomenology and Indian Philosophy*. Albany: SUNY, pp. 20–43.

Schwitzgebel, E. (Winter 2016), 'Introspection', in E.N. Zalta (ed.), *The Stanford Encyclopedia of Philosophy*, Stanford: Metaphysics Research Lab, https://plato.stanford.edu/archives/win2016/entries/introspection/

Selleri, F. and Tarozzi, G. (1981), 'Quantum Mechanics Reality and Separability', *Rivista del Nuovo Cimento* 4, pp. 1–53.

Seth, S. (2010), *Crafting the Quantum: Arnold Sommerfeld and the Practice of Theory 1890–1926*. Cambridge, MA: MIT.

Sharp, D.H. (1961), 'The Einstein-Podolsky-Rosen Paradox Re-Examined', *Philosophy of Science* 28, pp. 225–33.

Shimony, A. (1963), 'Role of Observer in Quantum Theory', *American Journal of Physics* 31, pp. 755–77.

Shimony, A. (1971), 'Replies to Comments', in B. d'Espagnat (ed.), *Foundations of Quantum Measurement*, Cambridge, MA: Academic Press, pp. 478–80.

Shimony, A. (2002), Interview of Abner Shimony by Joan Bromberg on 9 and 10 September 2002, Niels Bohr Library & Archives, American Institute of Physics, College Park: MD, www.aip.org/history-programs/niels-bohr-library/oral-histories/25643

Shimony, A. (2004), 'Wigner's Contributions to the Quantum Theory of Measurement', *Acta Physica Hungarica* B, 20: 1–2 pp. 59–72.

Shimony, A. (undated), 'The Character of Howard Stein's Work in Philosophy and History of Physics', http://strangebeautiful.com/other-texts/shimony-intro-stein-festschrift.pdf

Shoenberg, D. (1971), 'Heinz London', *Biographical Memoirs of Fellows of the Royal Society* 17, pp. 440–61.

Simon, J. (2022), 'Form and Meaning: Textbooks, Pedagogy, and the Canonical Genres of Quantum Mechanics', in Olival Freire Jr et al. (eds), *The Oxford Handbook of the History of Quantum Interpretations*, Oxford: Oxford University Press, pp. 709–33.

Skagerstam, B.S.K. (1975).'On the Notions of Entropy and Information', *Journal of Statistical Physics*, 12, pp. 449–62.

Smerlak, M. and Rovelli, C. (2007), 'Relational EPR', *Foundations of Physics* 37, pp. 427–45.

Smith, B. and Woodruff Smith, D. (1995), 'Introduction', in B. Smith and D. Woodruff Smith (eds), *The Cambridge Companion to Husserl*, New York: Cambridge University Press, pp. 1–44.

Smith, J. (2020), 'Self-Consciousness', in E. N. Zalta (ed.), *The Stanford Encyclopedia of Philosophy*,<https://plato.stanford.edu/archives/sum2020/entries/self-consciousness/>

Soler, L. (2009), *The Convergence of Transcendental Philosophy and Quantum Physics: Grete Henry-Hermann's 1935 Pioneering Proposal, Constituting Objectivity*, The Western Ontario Series in Philosophy of Science, Vol. 74, II, Part 5, pp. 329–44.

Spiegelberg, H. (1982), *The Phenomenological Movement: A Historical Introduction*, 3rd edn. The Hague: Martinus Nijhoff.

Stacey B.C. (2016), 'Von Neumann Was Not a Quantum Bayesian', *Philosophical. Transactions of the Royal Society A* 374, 2068: 2015.0235, http://dx.doi.org/10.1098/rsta.2015.0235

Stacey B.C. (2019), 'Ideas Abandoned en Route to QBism', arXiv:1911.07386.

Stacey, B.C. (2022), 'The Status of the Bayes Rule in QBism', arXiv preprint arXiv:2210.10757

Stapp, H. (2009), *Mind, Matter and Quantum Mechanics*, 3rd edn, Berlin: Springer-Verlag.

Stapp. H. (1982), 'Mind, Matter and Quantum Mechanics', *Foundations of Physics* 12, pp. 363–99.

Stein, H. (1982), 'On the Present State of the Philosophy of Quantum Mathematics', *PSA: Proceedings of the Biennial Meeting of the Philosophy of Science Association*, Vol. 2, Lansing, MI: Philosophy of Science Association, pp. 563–81.

Stöltzner, M. (1999), 'What John von Neumann Thought of the Bohm Interpretation', in D. Greenberger, W. L. Reiter, and A. Zeilinger (eds), *Epistemological and Experimental Perspectives on Quantum Physics*, Dordrecht: Kluwer, pp. 257–62.

Stöltzner, M. (2006), 'John von Neumann', in J. Pfeifer and S. Sarkar (eds), *Routledge Encyclopaedia of Philosophy of Science*, Abingdon: Routledge, pp. 503–10.

Stöltzner, M. (2013), 'Opportunistic Mathematics: von Neumann on the Methodology of Mathematical Physics', in M. Redéi and M. Stöltzner (eds), *John von Neumann and the Foundations of Quantum Physics*, Dordrecht: Springer.

Szanton, A. (1992), *The Recollections of Eugene P. Wigner*. New York: Basic Books.

Szilard, Leo (1929), 'Über die Entropieverminderung in einem thermodynamischen System bei Eingriffen intelligenter Wesen', *Zeitschrift für Physik* 53, pp. 840–56.

Takaki, K. (2011), 'Enactive Realism', *Tradition and Discovery* 38, pp. 43–59.

Talbott, W. (2016), 'Bayesian Epistemology', in E. N. Zalta (ed.), *The Stanford Encyclopedia of Philosophy*, <https://plato.stanford.edu/archives/win2016/entries/epistemology-bayesian/>

Taylor, M. (1998), *Husserl and the Cartesian Epistemological Problematic*. PhD thesis, University of Leeds.

Thirring, H. (1963), Interview of Hans Thirring by Thomas S. Kuhn on 4 April 1963, Niels Bohr Library & Archives, American Institute of Physics, College Park: MD, www.aip.org/history-programs/niels-bohr-library/oral-histories/4912

Thomasson, A.L. (2007), 'In What Sense is Phenomenology Transcendental?', *Southern Journal of Philosophy* 45, pp. 41–54.

Tieszen, R. (2005), *Phenomenology, Logic, and the Philosophy of Mathematics*. Cambridge: Cambridge University Press.

Timm, L. (2012), 'Thermodynamics ≠ Information Theory: Science's Greatest Sokal Affair', *Journal of Human Thermodynamics* 8, pp. 1–120.

Tisza, L. (1988), Interview of Laszlo Tisza by Kostas Gavroglou on 12 January 1988, Niels Bohr Library & Archives, American Institute of Physics, College Park: MD, www.aip.org/history-programs/niels-bohr-library/oral-histories/4915-2

Toadvine, Ted (2019), 'Maurice Merleau-Ponty', in E. N. Zalta (ed.), *The Stanford Encyclopedia of Philosophy*, <https://plato.stanford.edu/archives/spr2019/entries/merleau-ponty/>

Tonietti, T. (1988), 'Four Letters of E. Husserl to Hermann Weyl and Their Context', in W. Deppert (ed.), *Exact Sciences and Their Philosophical Foundations*, Frankfurt am Main: Lang, pp. 343–84.

Tononi, G. (2008), 'Consciousness as Integrated Information: A Provisional Manifesto', *Biological Bulletin* 215, pp. 216–42.

Trizio, E. (2021), *Philosophy's Nature: Husserl's Phenomenology, Natural Science, and Metaphysics*. Abingdon: Routledge.

Uebel, T. (2011), 'Beyond the Formalist Criterion of Cognitive Significance: Philipp Frank's Later Antimetaphysics', *HOPOS: Journal of the International Society for the History of Philosophy of Science* 1, pp. 47–72.

Uffink, J. (2020), 'Schrödinger's Reaction to the EPR Paper', in M. Hemmo and O. Shenker (eds), *Quantum, Probability, Logic: Itamar Pitowsky's Work and Influence*, Berlin: Springer, pp. 545–566.

Ullmo, J. (1963), Interview of Jean Ullmo by Thomas S. Kuhn and Theo Kahan on 7 January 1963, Niels Bohr Library & Archives, American Institute of Physics, College Park: MD, www.aip.org/history-programs/niels-bohr-library/oral-histories/4923

Vaidman, L. (2021), 'Many-Worlds Interpretation of Quantum Mechanics', in E. N. Zalta (ed.), *The Stanford Encyclopedia of Philosophy*, https://plato.stanford.edu/archives/fall2021/entries/qm-manyworlds/

van Fraassen, B. (1985), 'Empiricism in the Philosophy of Science', in P.M. Churchland and C.A. Hooker (eds), *Images of Science: Essays on Realism and Empiricism, with a Reply by Bas C. van Fraassen*, Chicago: University of Chicago Press, pp. 245–308.

van Fraassen, B. (2010), 'Rovelli's World', *Foundations of Physics* 40, pp. 390–417.

Vergani, M. (2021), 'Husserl's Hesitant Attempts to Extend Personhood to Animals', *Husserl Studies* 37, pp. 67–83.

Vickers, P. (2013), *Understanding Inconsistent Science*. Oxford: Oxford University Press.

von Baeyer, H.C. (forthcoming), 'On the Consilience between QBism and Phenomenology', in H. Wiltsche and P. Berghofer (eds), *Phenomenological Approaches to QBism*.

von Kármán, T. (1962), Interview of John Heilbron by Theodore Von Karman on 29 June 1962, Niels Bohr Library & Archives, American Institute of Physics, College Park: MD, www.aip.org/history-programs/niels-bohr-library/oral-histories/4935

von Neumann, J. (2018), *Mathematical Foundations of Quantum Mechanics*. Princeton University Press (originally published in German in 1932; translated by R.T. Beyer in 1955 and edited by N. A. Wheeler in 2018).

von Neumann, J. (2000), *The Computer and the Brain*. New Haven: Yale University Press (originally published in 1958).

von Shrenk-Notzing, A. (1920), *Phenomena of Materialisation: A Contribution to the Investigation of Mediumistic Teleplastics*. London: Kegan Paul, Trench, Trubner.

Wallace, D. (2012), *The Emergent Multiverse*. Oxford: Oxford: University Press.

Walther, G. (1923), *Zur Phänomenologie der Mystik*. Tübingen: M. Niemeyer.

Wasserman, G.D. (1955), 'Some Comments on Methods and Statements in Parapsychology and Other Sciences', *British Journal for the Philosophy of Science* 6, pp. 122–40.

Werkmeister, W.H. (1951), 'Professor Margenau and the Problem of Physical Reality', *Philosophy of Science* 18, pp. 183–92.

Wessels, L. (1980), 'Review of *Physics and Philosophy: Selected Essays* by Henry Margenau', *Philosophical Review* 89, pp. 300–5.

Weststeijn, N. (2021), 'Wigner's Friend and Relational Quantum Mechanics: A Reply to Laudisa', *Foundations of Physics* 51: 86.

Weyl, H. (1918), *Raum, Zeit, Materie*. Berlin: J. Springer, 3rd edn, essentially revised; Berlin: J. Springer, 1919, 4th edn, essentially revised; Berlin: J. Springer, 1921, 5th edn, revised; Berlin: J. Springer, 1923, 7th edn, edited (with notes) by J. Ehlers; Berlin: Springer, 1988, *Temps, espace, matière* (from the 4th German edn); A. Blanchard, Paris, 1922, *Space, Time, Matter* (from the 4th German edn), Methuen, London, 1922.

Weyl, H. (1934), *Mind and Nature*. Philadelphia: University of Pennsylvania Press.

Weyl, H. (1949; 2009), *Philosophy of Mathematics and Natural Science*. Princeton: Princeton University Press.

Wheeler, J.A. and Zurek, W.H. (eds) (1983), *Quantum Theory and Measurement*. Princeton: Princeton University Press.

Wightman, A.S. (1995), 'Superselection Rules: Old and New', *Il Nuovo Cimento* 110, pp. 751–69.

Wigner, E. (1962), 'Remarks on the Mind–Body Question', in I.J. Good (ed.), *The Scientist Speculates*, Portsmouth, NH: Heinemann, pp. 284–302.

Wigner, E. (1963a), 'The Problem of Measurement', *American Journal of Physics* 31, pp. 6–15 (see Wheeler and Zurek 1983, pp. 324–41).

Wigner, E. (1963b), Interview of Eugene Wigner by Thomas S. Kuhn on 21 November 1963, Niels Bohr Library & Archives, American Institute of Physics, College Park, MD USA, www.aip.org/history-programs/niels-bohr-library/oral-histories/4963-1

Wigner, E. (1963c), Interview of Eugene Wigner by Thomas S. Kuhn on 3 December 1963, Niels Bohr Library & Archives, American Institute of Physics, College Park, MD USA, www.aip.org/history-programs/niels-bohr-library/oral-histories/4963-2

Wigner, E. (1963d), Interview of Eugene Wigner by Thomas S. Kuhn on 14 December 1963, Niels Bohr Library & Archives, American Institute of Physics, College Park, MD USA, www.aip.org/history-programs/niels-bohr-library/oral-histories/4963-3

Wigner, E. (1964), 'Two Kinds of Reality', *Monist* 48, pp. 248–64.

Wigner, E. (1968), 'A Physicist Looks at the Soul', repr. in Jagdish Mehra (ed.), *The Collected Works of Eugene Paul Wigner, Part B: Historical, Philosophical, and Socio-Political Papers. Vol. VII. Historical and Biographical Reflections and Syntheses*, Berlin: Springer-Verlag, 2001, pp. 41–3.

Wigner, E. (1971), 'The Subject of Our Discussions', in B. d'Espagnat (ed.), *Foundations of Quantum Measurement*, Cambridge, MA: Academic Press, pp. 1–19, repr. in Jagdish Mehra (ed.), The Collected Works of Eugene Paul Wigner, Part B: Historical, Philosophical, and Socio-Political Papers. Vol. VII. Historical and Biographical Reflections and Syntheses, Berlin: Springer-Verlag, 2001, pp. 199–217.

Wigner, E. (1972), 'On Some of Physics' Problems', *Main Currents in Modern Thought*, 28, pp. 75–8; repr. in Jagdish Mehra (ed.), *The Collected Works of Eugene Paul Wigner, Part B: Historical, Philosophical, and Socio-Political Papers. Vol. VII. Historical and Biographical Reflections and Syntheses*, Berlin: Springer-Verlag, 2001, pp. 578–83.

Wigner, E. (1983), 'The Limitations of Determinism'; repr. in Jagdish Mehra (ed.), *The Collected Works of Eugene Paul Wigner, Part B: Historical, Philosophical, and Socio-Political Papers. Vol. VII. Historical and Biographical Reflections and Syntheses*, Berlin: Springer-Verlag, 2001, pp. 133–8.

Wigner, E. (1984), 'Review of the Quantum Mechanical Measurement Problem', in D.M. Kerr et al. (eds), *Science, Computers and the Information Onslaught*, New York: Academic Press, pp. 63–82; repr. in Jagdish Mehra (ed.), *The Collected Works of Eugene Paul Wigner, Part B: Historical, Philosophical, and Socio-Political Papers. Vol. VII. Historical and Biographical Reflections and Syntheses*, Berlin: Springer-Verlag, 2001, pp. 225–42.

Wilson, A. (2020), *The Nature of Contingency: Quantum Realism and Modal Realism*. Oxford: Oxford University Press.

Wiltsche, H. (2012), 'What is Wrong With Husserl's Scientific Anti-Realism', *Inquiry* 55, pp. 105–30.

Wiltsche, H. (2015), 'Review of Lee Hardy: *Nature's Suit*', *Husserl Studies* 31, pp. 175–82.

Wiltsche, H. (2021), 'Physics with a Human Face: Husserl and Weyl on Realism, Idealism and the Nature of the Coordinate System', in H. Jacobs (ed.), *The Husserlian Mind*, London and New York: Routledge, pp. 468–77.

Wiltsche, H. (forthcoming), 'The Coordination Problem: A Challenge for Transcendental Phenomenology of Science', *The New Yearbook for Phenomenology and Phenomenological Philosophy*, 22.

Woodruff Smith, D. (2021), 'Review of "Phenomenology: A Contemporary Introduction"', *Notre Dame Philosophical Reviews*, https://ndpr.nd.edu/reviews/phenomenology-a--contemporary-introduction/

Wüthrich, C. (forthcoming), 'Putnam Looks at Quantum Mechanics (Again and Again)', in M. Frauchiger (ed.), *Themes from Putnam*, Lauener Library of Analytical Philosophy, Vol. 5, W.K. Essler and M. Frauchiger (eds), Berlin: De Gruyter (arXiv preprint arXiv:1406.5737).

Yoshimi, J. (2021), 'Two Conceptions of Husserlian Phenomenology: A Review of Walter Hopp's *Phenomenology: A Contemporary Introduction*', *Husserl Studies*, https://doi.org/10.1007/s10743-021-09290-1

Zahavi, D. (1997), 'Horizontal Intentionality and Transcendental Intersubjectivity', *Tijdschrift voor Filosofie* 59, pp. 304–21.

Zahavi, D. (2017), *Husserl's Legacy*. Oxford: Oxford University Press.

Zahavi, D. (2018), 'Intersubjectivity, Sociality, Community: The Contribution of the Early Phenomenologists', in D. Zahavi (ed.), *The Oxford Handbook of Phenomenology*, Oxford: Oxford University Press. pp. 734–52.

Zahavi, D. (2021), 'Applied Phenomenology: Why It Is Safe to Ignore the Epoché', *Continental Philosophy Review* 54, pp. 59–273.

Zaner, R.M. (2010), 'Editorial Introduction', in R.M. Zaner and L. Embree (eds), *The Collected Works of Aron Gurwitsch, Vol. III*. Dordrecht: Springer, pp. xv–xxxv.

Zeh, H.D. (2000), 'The Problem of Conscious Observation in Quantum Mechanical Description', *Foundations of Physics Letters* 13, p. 221, quant-ph/9908084 (originally circulated in 1981 via *Epistemological Letters of the Ferdidand-Gonseth Association*).

Zinkernagel, H. (2015), 'Are We Living in a Quantum World? Bohr and Quantum Fundamentalism', in F. Aaserud and H. Kragh (eds), *One Hundred Years of the Bohr Atom: Proceedings from a Conference*. Scientia Danica. Series M: Mathematica et physica, Vol. 1. Copenhagen: Royal Danish Academy of Sciences and Letters, 2015, pp. 419–34.

Zyga, L. (2009), 'Quantum Mysticism: Gone But Not Forgotten', *Phys.org* 8 June, https://phys.org/news/2009-06-quantum-mysticism-forgotten.html.

Name Index

For the benefit of digital users, indexed terms that span two pages (e.g., 52–53) may, on occasion, appear on only one of those pages.

Subject Index

For the benefit of digital users, indexed terms that span two pages (e.g., 52–53) may, on occasion, appear on only one of those pages.